Practical MATLAB Modeling with Simulink

Programming and Simulating Ordinary and Partial Differential Equations

Sulaymon L. Eshkabilov

Apress®

Practical MATLAB Modeling with Simulink: Programming and Simulating Ordinary and Partial Differential Equations

Sulaymon L. Eshkabilov
Ag & Biosystems Engineering Department, North Dakota State University, Fargo, USA

ISBN-13 (pbk): 978-1-4842-5798-2   ISBN-13 (electronic): 978-1-4842-5799-9
https://doi.org/10.1007/978-1-4842-5799-9

Managing Director, Apress Media LLC: Welmoed Spahr
Acquisitions Editor: Steve Anglin
Development Editor: Matthew Moodie
Coordinating Editor: Mark Powers

Cover designed by eStudioCalamar

Cover image from rawpixel.com

Distributed to the book trade worldwide by Springer Science+Business Media, 1 New York Plaza, New York, NY 10004, U.S.A. Phone 1-800-SPRINGER, fax (201) 348-4505, e-mail orders-ny@springer-sbm.com, or visit www.springeronline.com. Apress Media, LLC is a California LLC and the sole member (owner) is Springer Science + Business Media Finance Inc (SSBM Finance Inc). SSBM Finance Inc is a **Delaware** corporation.

For information on translations, please e-mail editorial@apress.com; for reprint, paperback, or audio rights, please e-mail bookpermissions@springernature.com.

Apress titles may be purchased in bulk for academic, corporate, or promotional use. eBook versions and licenses are also available for most titles. For more information, reference our Print and eBook Bulk Sales web page at www.apress.com/bulk-sales.

Any source code or other supplementary material referenced by the author in this book is available to readers on GitHub via the book's product page, located at www.apress.com/9781484257982. For more detailed information, please visit www.apress.com/source-code.

Printed on acid-free paper

To the memory of my father.

To my mother.

To my wife, Nigora, after 25 wonderful years together.

Table of Contents

About the Author

Dr. Sulaymon L. Eshkabilov is currently a visiting professor in the Department of Agriculture and Biosystems Engineering at North Dakota State University. He obtained his ME diploma from Tashkent Automobile Road Institute in 1994, his MSc from Rochester Institute of Technology, USA in 2004, and his PhD from Academy Sciences of Uzbekistan in 2005. He was an associate professor at Tashkent Automobile Road Institute from December 2006 to January 2017. He also held visiting professor and researcher positions at Ohio University from 2010 to 2011 and Johannes Kepler University, Austria, from January to September 2017. He has taught the following courses: "MATLAB/Simulink Applications for Mechanical Engineering and Numerical Analysis" and "Modeling of Engineering Systems" for undergraduate students, an "Advanced MATLAB/Mechatronics" seminar/class, and "Control Applications," "System Identification," "Experimentation and Testing with Analog and Digital Devices" for graduate students.

His research areas are mechanical vibrations, control, mechatronics, system dynamics, image processing, and microstructure analysis of materials. He is the author of more than 30 research papers and 5 books. Three of the five books are devoted to MATLAB/Simulink applications for mechanical engineering students and numerical analysis. From 2009 to 2020, he was an external academic expert for the European Commission, assessing academic projects.

About the Technical Reviewer

 **Karpur Shukla** is a research fellow at the Centre for Mathematical Modelling at FLAME University in Pune, India. His current research interests focus on topological quantum computation, nonequilibrium and finite-temperature aspects of topological quantum field theories, and applications of quantum materials effects for reversible computing. He received an MSc in physics from Carnegie Mellon University, with a background in theoretical analysis of materials for spintronics applications as well as Monte Carlo simulations for the renormalization group of finite-temperature spin lattice systems.

Acknowledgments

I express my special gratitude to the technical reviewers, proofreaders, and editors of Apress for their very thorough work while reviewing the content and code in this book. Without their critical insights and corrections at many points of the book, I would not have been able to complete it with this quality. In addition, I would like to express my special gratitude to Mark Powers for his on-time and well-planned correspondence throughout this book project.

My cordial gratitude goes to my mother for her limitless support and love. Up until the very last point of this book, she was always checking in about my progress.

I would like to thank my wife, Nigora, because without her great support, I would not have been able to take up the challenging task of writing this book. I have spent many weekends in my office writing and editing the book content and the MATLAB/Simulink scripts and models. In addition, I would like to thank our children, Anbara, Durdona, and Dovud, for being such delightful people and being the inspiration for my efforts while writing this book.

Introduction

This book covers the most essential and hands-on tools and functions of the MATLAB and Simulink packages and the Symbolic Math Toolbox (MuPAD notes) to solve, model, and simulate ordinary differential equations (ODEs) and partial differential equations (PDEs). It explains how to solve ODEs and PDEs symbolically and numerically via interactive examples and case studies. The main principle of the book is "learn by doing," moving from the simple to the complex. This book contains dozens of solved problems and simulation models embedded in MATLAB scripts and Simulink models, which will help you to master programming and modeling essentials, as well as learn how to program and model more difficult and complex problems that involve ODEs and PDEs.

Practical MATLAB Modeling with Simulink explains various practical issues of programming and modeling in parallel by comparing the programming tools of MATLAB to the modeling tools of Simulink. By studying this book, you'll be proficient at using the MATLAB/Simulink packages and at using the source codes and models from the book's examples as templates for your own projects to solve modeling and simulation, or engineering problems with ODEs and PDEs.

What You Will Learn

By the end of the book, you'll have learned how to do the following:

- Model complex problems using MATLAB and Simulink

- Use MATLAB and Simulink to solve ODEs and PDEs

- Use numerical methods to solve first-, second-, and higher-order and coupled ODEs

- Solve stiff and implicit ODEs

- Employ numerical methods to solve first- and second-order linear PDEs

- Solve ODEs symbolically with the Symbolic Math Toolbox of MATLAB

- Understand the applications and modeling aspects of differential equations in solving various simulation problems

This book is aimed at engineers, programmers, data scientists, and students majoring in engineering, applied/industrial math, data science, and scientific computing. This book continues where Apress' *Beginning MATLAB and Simulink* leaves off.

The book is composed of three parts. Part 1 consists of the following chapters:

- Chapter 1 is dedicated to general formulations and solving ODEs symbolically using The Symbolic Math Toolbox functions and MuPAD note commands.

- Chapter 2 covers the most essential programming aspects to solve first-order ODEs numerically by writing scripts based on the Euler, Runge-Kutta, Milne, Adams-Bashforth, Ralston, and Adams-Moulton methods. You'll write the scripts using MATLAB's built-in ODE solvers, such as ode23, ode23t, ode45, ode15s, ode113, odeset, etc., and Simulink modeling.

- Chapter 3 is dedicated to solving second-order ODEs symbolically and numerically using the Euler, Runge-Kutta, Milne, Adams-Bashforth, Ralston, and Adams-Moulton methods. You'll do this using MATLAB's built-in ODE solvers, such as ode23, ode23t, ode45, ode15s, ode113, odeset, etc., and Simulink modeling.

- Chapter 4 addresses the issues of how to solve stiff ODEs by adjusting a step size, specifying a solver setting, selecting an appropriate solver type, and so forth.

- Chapter 5 is dedicated to solving higher-order and coupled ODEs.

- Chapter 6 is devoted to solving implicitly defined IVPs and differential algebraic equations using MATLAB's ode15i and Simulink modeling.

- Chapter 7 is dedicated to a comparative analysis of solving ODEs with MATLAB's ODE solvers, Simulink modeling, script writing, and computing analytical solutions with the Symbolic MATH Toolbox (MuPAD notes).

Part 2 consists of the following chapter:

- Chapter 8 covers most essential tools of solving boundary value problems of ODEs numerically by using MATLAB solvers such as bvp4c, bvp5c, bvpinit, deval, etc.

Part 3 consists of the following chapters:

- Chapter 9 is devoted to applications of ODEs, specifically, modeling and simulations of spring-mass-damper systems.

- Chapter 10 is devoted to applications of ODEs, specifically, modeling and simulations of mechanical and electromechanical systems.

- Chapter 11 is devoted to applications of ODEs, specifically, modeling and simulations of trajectory problems.

- Chapter 12 is devoted to applications of ODEs, specifically, modeling and simulations of several simulation problems, such as the Lotka-Voltera and Lorenz systems.

Part 4 consists of the following chapter:

- Chapter 13 is devoted to solving partial differential equations using PDE toolbox functions, such as pdepe() and the Laplacian operator del2(), and writing scripts based on Gauss-Seidel and finite difference methods.

How to Access the Source Code

All of the source code (such as M/MN files, Simulink models, and SLX/MDL files) discussed in the book is available to readers via the Download Source Code button at www.apress.com/9781484257982.

INTRODUCTION

A Note to Users

The given scripts in the context of the book may not be the best solutions to the given problems, which was done intentionally in some cases to emphasize methods used to improve them; in other cases, the given scripts are the most appropriate solutions to the best knowledge of the author. Should I spot any better alternative solutions to exercises given in the context of the book, I intend to publish them via the MathWorks MATLAB Central User Community's file exchange via my file exchange link there—under my name.

No matter how hard we have worked to proofread the content of the book, it is inevitable that there might be some typographical errors that have slipped through and will appear in print. My apologies.

Sulaymon L. Eshkabilov
January 2020

PART I

Ordinary Differential Equations

CHAPTER 1

Analytical Solutions for ODEs

Many modeling problems in engineering applications can be formulated using ordinary differential equations. There are a few different definitions of *differential equation*; one of the simplest ones is "A differential equation is any equation which contains derivatives, either ordinary derivatives or partial derivatives" given in [1]. From this definition, we can derive that there are two types of differential equations: ordinary differential equations (ODEs) and partial differential equations (PDEs). ODEs contain one type of derivative or one independent variable, while PDEs contain two or more derivatives or independent variables. For example, a general form for first-order ODEs can be expressed by:

$$\frac{dy}{dx} = f(y, x) \qquad (1\text{-}1)$$

where $y(x)$ is a dependent variable whose values depend on values of the independent variable of x. Another good example of an ODE is Newton's second law of motion formulated by:

$$ma = \frac{dp}{dt} = \frac{mdv}{dt} = F(t, v) \qquad (1\text{-}2)$$

where $F(t, v)$ is the force, which is a function of time (t) and velocity (v); $\frac{dv}{dt}$ is a velocity change rate (acceleration) of a moving object; m is the mass of a moving object; a is an acceleration of a moving object; p is momentum; and $\frac{dp}{dt}$ is its derivative.

© Sulaymon L. Eshkabilov 2020
S. L. Eshkabilov, *Practical MATLAB Modeling with Simulink*, https://doi.org/10.1007/978-1-4842-5799-9_1

The previous formulation of Newton's second law can be also rewritten in the following way:

$$\frac{md}{dt}\left(\frac{dx}{dt}\right) = \frac{md^2x}{dt^2} = F\left(t, x, \frac{dx}{dt}\right) \tag{1-3}$$

where the derivative $\left(\dfrac{dx}{dt}\right)$ of the displacement (x) of a moving object is the velocity (v). In other words, the velocity is a change rate of the displacement $x(t)$ of a moving object in time. This can be visualized with the flowchart displayed in Figure 1-1.

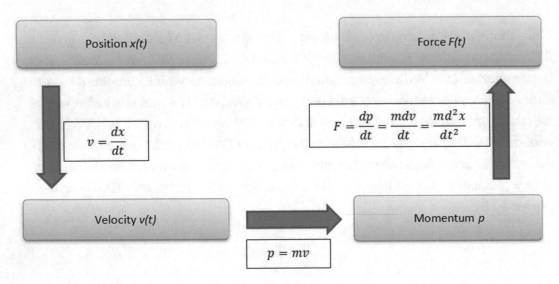

Figure 1-1. *Flowchart expressing motion and exerted force of a moving object*

Classifying ODEs

There are two classifications of ODE-related problems.

- **Initial value problems (IVPs):** Here's an example: $\ddot{x} = xt - 3\dot{x}$ with these initial conditions: $x(0) = 3$, $\dot{x}(0) = 1$.

- **Boundary value problems (BVPs):** Here's an example: $\ddot{x} = xt - 3\dot{x}$ with these boundary conditions: $x(0) = 3$, $x(2) = 1.50$.

IVPs are defined with ODEs together with a specified value, called the *initial condition*, of the unknown function at a given point in the solution domain. In the IVP of ODEs, there might be a unique solution, no solution, or many solutions. By definition, the IVP of ODEs can be explicitly defined or implicitly defined. Most IVPs are explicitly defined.

We will start with explicitly defined IVPs and then move on to implicitly defined ones. Besides being categorized by solution type (how the solution values change over the solution search space), IVPs are divided into stiff and nonstiff problems. Moreover, ODEs are either linear or nonlinear and are either homogeneous or nonhomogeneous.

Here are some specific examples of different ODE types, categories, and groups:

$$\text{Stiff ODEs: } \begin{cases} \dot{x} = -0.04x + 10^4\,yz \\ \dot{y} = 0.04x - 10^4\,yz - 3*10^7\,y^2 \qquad t \in [0, 40] \\ \dot{z} = 3*10^7\,y^2 \end{cases}$$

Nonstiff ODEs: $\dot{y} + 2y = 2t,\ \ddot{w} + \dot{w} = 5$

Linear ODEs: $\dfrac{dv}{dt} = 9.81 - 0.198v,\ \ddot{x} + 3\dot{x} + 5x = 0$

Nonlinear ODEs: $\dfrac{dv}{dt} = 9.81 - 0.198v^2,\ \ddot{x} + 3x\dot{x} + 5x^2 = 0$

Homogeneous ODEs: $\dot{y} + 2y = 0,\ \ddot{x} + 3\dot{x} + 5x = 0$

Nonhomogeneous ODEs: $\dot{y} + 2y = \sin(x),\ \ddot{x} + 3\dot{x} + 5x = e^{2t}$

This chapter contains several examples for ODEs and their application areas.

Example 1

This example shows an exponential growth problem. This equation could describe unconstrained growth of biological organisms (bacteria), values of real estate or investments, membership of a popular networking site, growth in retail business, positive feedback of electrical systems, or generated chemical reactions. It is formulated by the following first-order ODE:

$$\frac{dy}{dt} = \mu y \ \text{ has a solution: } y(t) = y_0 e^{\mu t}$$

Example 2

This example shows exponential decay. This equation could describe many phenomena in nature and engineering, such as radioactive decay, washout of chemicals in a reactor, discharge of a capacitor, and decomposition of material in a river. Exponential decay is expressed with the following first-order ODE:

$$\frac{dy}{dt} = -\mu y \text{ has a solution: } y(t) = y_0 e^{-\mu t}$$

Examples 1 and 2 are two simple examples of first-order ODEs.

Example 3

The motion of a falling object is expressed in the following way using Newton's second law:

$$\frac{md^2 y}{dt^2} = mg - \frac{\gamma dy}{dt}$$

This is a second-order ODE that has a general solution in the following form:

$$y(t) = C_1 e^{-\left(\frac{\gamma t}{m}\right)} - \frac{m(mg - g\gamma t)}{\gamma^2} + C_2$$

where m is the mass of a falling object; g is gravitational acceleration; and γ is an air-drag coefficient of a falling object. There are three parameters. Specifically, m, g, and γ are constant, and two parameters change over time: $\frac{d^2 y}{dt^2}$ (acceleration) and $\frac{dy}{dt}$ (velocity). In the general solution of a falling object using the equation of motion, C_1 and C_2 are arbitrary numbers that are dependent on the initial conditions or, in other words, can be computed considering the initial conditions of a falling object.

There are a few methods to evaluate the analytical solutions of ODEs, including the separation of variables, introduction of new variables, and others. We show via specific examples of these types of ODEs and explain how to evaluate their analytical solutions and compute numerical solutions by employing different techniques in the MATLAB/Simulink environment through scripts and building models. We evaluate analytical solutions of ODEs via specific examples and demonstrate how to use the built-in functions of the Symbolic Math Toolbox[1] and MuPAD[2] notebooks. We put more emphasis on the Symbolic Math Toolbox's command syntaxes rather than MuPAD notebooks. The reason for this is that in future releases of MATLAB, a technical support for MuPAD notebooks will removed.

There are a number of numerical methods, including Euler (forward, backward, modified), Heun, the midpoint rule, Runge-Kutta, Runge-Kutta-Gill, Adams-Bashforth, Milne, Adams-Moulton, Taylor series, and trapezoidal rule methods. Some of these methods are explicit, and others are implicit. To demonstrate how to employ these methods, we will first describe their formulations concisely, and then we will work on their implementation algorithms for writing scripts (programs) explicitly. We do not attempt to derive any of the formulations used in these numerical methods, and there are many literature sources [2, 3, 4, 5] explaining the theoretical aspects of these methods.

In solving an IVP with numerical methods, we first start from an initial point (initial conditions) and then take a step (equal step size or varying step size) forward in time to compute numerical solutions. Some of the previously named numerical methods (e.g., Euler's methods) are single-step methods, and others (Runge-Kutta, Adams-Bashforth, Milne, Adams-Moulton, Taylor series) are multistep methods. Single-step methods refer to only one previous point and its derivative to determine the current value. Other methods, such as Runge-Kutta methods, take some intermediate steps to obtain a higher-order step and then drop off values before taking the next step. Unlike single-step methods, multistep methods keep and use values from the previous steps instead of discarding them. In this way, multistep methods link a few previously obtained values (solutions) and derivative values. All of these methods, such as the single-step and multistep methods, are assessed based on their accuracy and efficiency in terms of computation time and resources (e.g., machine time) spent to compute numerical solutions for specific types of IVPs of ODEs.

---

[1]Symbolic Math Toolbox is a registered trademark of The MathWorks Inc.
[2]MuPAD is a registered trademark of The MathWorks Inc.

Analytical Solutions of ODEs

The Symbolic MATH Toolbox (or MuPAD notebooks) has several functions capable of evaluating analytical solutions of many analytically solvable ODEs. There are two commands (built-in functions)—dsolve() and ilaplace/laplace—with which analytical solutions of some ODEs can be evaluated.

Note that in this section we demonstrate—via a few examples of first- and second-order ODEs and systems of coupled differential equations—how to compute analytical solutions of ODEs.

dsolve()

dsolve() is an ODE solver tool to compute an analytical (or general) solution of any given ODE in MATLAB. dsolve() can be used with the following general syntaxes:

```
Solution = dsolve(equation)
Solution = dsolve(equation, conditions)
Solution = dsolve(equation, conditions, Name, Value)
[y1,...,yN] = dsolve(equations)
[y1,...,yN] = dsolve(equations, conditions)
[y1,...,yN] = dsolve(equations, conditions, Name, Value)
```

Example 4

Given an ODE, $\dot{y} + 2ty^2 = 0$. Note that initial or boundary conditions are not specified. Here is the command by which we can compute a general analytical solution of the given example.

```
>> y_solution=dsolve('Dy=-2*y^2*t')
```

```
Y_solution=-1/(C3-t^2)
```

Note that C3 is defined from the initial or boundary conditions of the given ODE.

There is an alternative command. The given problem with newer versions of MATLAB (starting with MATLAB 2012) can be solved by using the following command syntax:

```
>>syms y(t); y_sol=dsolve(diff(y) == -2*y^2*t)
 y_sol =
           0
 -1/(-t^2 + C3)
```

Example 5

Given an ODE, $\dot{y} + 2ty^2 = 0$, with the initial condition $y(0) = 0.5$.

```
>> Solution=dsolve('Dy=-2*y^2*t', 'y(0)=0.5')
Solution =
 1/(t^2 + 2)
```

Here is the alternative command syntax for newer versions of MATLAB:

```
>> syms y(t); Solution=dsolve(diff(y) == -2*y^2*t, y(0)==0.5)
Solution =
 1/(t^2 + 2)
```

Here is another alternative syntax for newer versions of MATLAB:

```
>> syms y(t) t; Dy=diff(y, t); Equation = Dy ==-2*y^2*t; IC=y(0)==0.5;
Solution=dsolve(Equation, IC); pretty(Solution)
   1
 ------
   2
 t  + 2
```

The resulting analytical solution is in a symbolic formulation and thus can be plotted (Figure 1-2) with ezplot or the recommended fplot. (ezplot will not be supported in future releases of MATLAB.)

```
>> fplot(Solution, [-5, 5], 'r--o'), grid on
>>title('Solution of: $$ \frac{dy}{dt}+2y^2t=0, y_0 = 0.5 $$',
'interpreter', 'latex')
>> xlabel('\it t'), ylabel '\it Solution, y(t)'
```

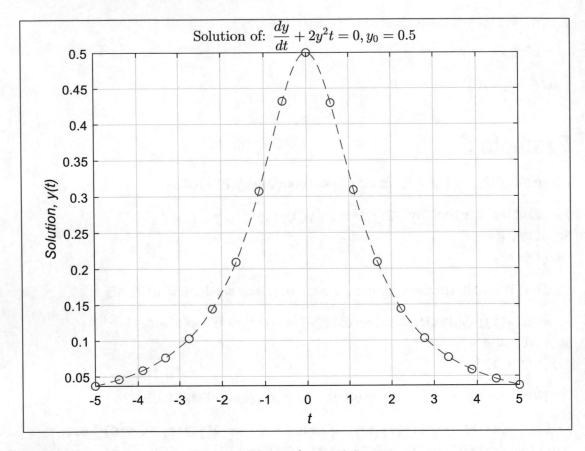

Figure 1-2. *Analytical solution of $\dot{y} + 2ty^2 = 0$, $y(0)=0.5$*

Numerical values of the resulting analytical solution (equation) can be computed by vectorizing (parameterizing) the resulting symbolic formulation (solution), as shown here:

```
>> ysol=vectorize(solution)
ysol =
1./(t.^2 + 2)
>> t=(-5:.1:5); ysol_values=eval(ysol);
```

An alternative way of computing numerical solution values is to use `fplot`, as shown here:

```
[t, yt]=fplot(Solution, [-5, 5]);
```

Example 6

Given $\dot{y} + kty^2 = 0, y(0) = 0.5$. Note that this exercise has one unspecified parameter, k.

```
>> syms k
>> solution=dsolve('Dy=-k*y^2*t', 'y(0)=0.5')
solution =
1/((k*t^2)/2 + 2)
```

Here is the alternative command syntax:

```
>> syms y(t) k; solution=dsolve(diff(y) == -k*y^2*t, y(0)==0.5)
solution =
1/((k*t^2)/2 + 2)
```

Example 7

Given $\dot{y} - |y|e^t = 2, y(0) = 2$. Let's solve this exercise in a MuPAD note. Figure 1-3 shows the commands used to compute an analytical solution for this exercise.

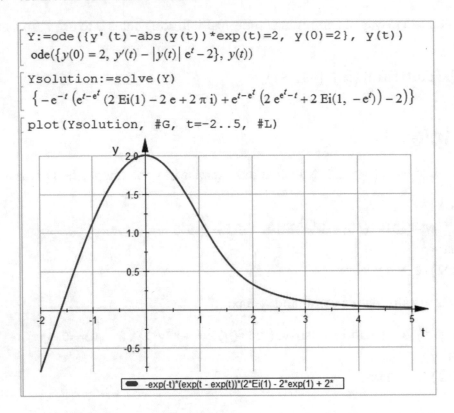

Figure 1-3. *MuPAD commands to solve* $\dot{y} - |y|e^t = 2,\, y(0) = 2$ *analytically*

Note   The options in `dsolve()` need to be set appropriately depending on the problem type. For MATLAB 2008–2010 or earlier versions, you should set `IgnoreAnalyticConstraints` to none to obtain all or any possible solutions.

Here's an example:

```
solution=dsolve('Dy=-k*y^2*t', 'y(0)=0.5', ... 'IgnoreAnalyticConstraints', 'none')
```

Note   For MATLAB 2012 or newer versions, you should set `IgnoreAnalyticConstraints` to `false` to get all the possible correct answers for all the argument values. Otherwise, `dsolve()` may output an incorrect answer because of its pre-algebraic simplifications.

Here's an example:

```
solution=dsolve(diff(y)==-k*y^2*t, y(0)==0.5, ... 'IgnoreAnalyticConstraints',
false)
```

Second-Order ODEs and a System of ODEs

There are a myriad of processes and phenomena that are expressed via second-order differential equations, for example, simple harmonic motions of a spring-mass system, motions of objects with some acceleration (Newton's second law), damped vibrations, current flows in resistor-capacitor-inductance circuits, and so forth. In a general form, second-order ODEs are expressed in two different forms: homogeneous, as shown in Equation (1-4), and nonhomogeneous, as shown in Equation (1-5).

$$\ddot{y} + p(x)\dot{y} + q(x)y = 0 \tag{1-4}$$

$$\ddot{y} + p(x)\dot{y} + q(x)y = g(x) \tag{1-5}$$

Note that the homogeneous ODEs in Equation 1-4 always have one trivial solution, which is $y(x) = 0$, that satisfies the given equations in Equation 1-4. With respect to the independent functions $p(x)$, $q(x)$, and $g(x)$, the ODEs might be linear or nonlinear. In some cases, the independent functions $p(x)$, $q(x)$, and $g(x)$ might be constant values or nonconstant values.

Let's consider several examples of using second-order ODEs to compute general and specific solutions with the Symbolic MATH Toolbox of MATLAB.

Example 8

Given an ODE of $\dfrac{d^2u}{dt^2} + 100u = 2.5\sin(10t)$, the initial conditions: $u(0) = 0$ and $\dfrac{du(0)}{dt} = 0$.

```
usol=dsolve('D2u+100*u=2.5*sin(10*t)', 'u(0)=0', 'Du(0) 0');
pretty(usol)

%% Alternative syntax
syms u(t)
Du = diff(u);
u(t) = dsolve(diff(u, 2)==2.5*sin(10*t)-100*u, u(0)==0, Du(0) == 0);
```

```
pretty(u(t))
```

The output from executing the two short scripts/commands is as follows:

```
3 sin(10 t)    sin(30 t)
----------- - --------- -
    320           320

              / t   sin(20 t) \
  cos(10 t) | - - --------- |
              \ 8      160    /
```

Example 9

Given the system of ODEs $\begin{cases} y_1' = y_2 \\ y_2' = -y_1 - 0.125y_2 \end{cases}$, the initial conditions: $y_1(0)=1, y_2(0)=0$;.

The given problem is a system of two first-order ODEs. This problem can be solved directly with dsolve() similar to the previous examples, as shown here:

```
%% System of two 1st order ODEs solved with dsolve
yt=dsolve('Dy1=y2', 'Dy2=-y1-0.125*y2','y1(0)=1', 'y2(0)=0');
pretty(yt.y1)
pretty(yt.y2)
```

```
%% Alternative syntax
syms y1(t) y2(t)
z=dsolve(diff(y1,1)==y2, diff(y2,1)==(-y1-0.125*y2), y1(0)==1, y2(0)==0);
pretty(z.y1), pretty(z.y2)
```

The computed analytical solutions of the problem displayed in the command window are as follows:

```
    /    1/2   \                    /    1/2   \
   | 255    t |        1/2         | 255    t |
  cos| -------- |      255      sin| -------- |
    \    16   /                    \    16   /
  --------------- + ----------------------
         1/16                      1/16
     exp(t)               255 exp(t)
```

```
                 /      1/2    \
        1/2      | 255      t |
   16 255    sin| -------- |
                 \     16     /
   - ------------------------
                   1/16
           255 exp(t)
```

The computed analytical solutions are as follows:

$$y_1(t) = \frac{\cos\left(\dfrac{\sqrt{255}\,t}{16}\right)}{\sqrt[16]{e^t}} + \frac{\sqrt{255}\,\sin\left(\dfrac{\sqrt{255}\,t}{16}\right)}{255\sqrt[16]{e^t}}$$

$$y_2(t) = -\frac{16\sqrt{255}\,\sin\left(\dfrac{\sqrt{255}\,t}{16}\right)}{255\sqrt[16]{e^t}}$$

Example 10

Given a second-order ODE of $2\ddot{y} + 3\dot{y}^3 - |y|\cos(100t) = 2$, the initial conditions are: $y(0) = 0$ and $\dot{y}(0) = 0$.

```
>> Solution_dsolve=dsolve('2*D2y+3*(Dy^3)-cos(100*t)*abs(y)-2=0', ...
'y(0)=0','Dy(0) =0');
Warning: Explicit solution could not be found.
> In dsolve at 194
```

This is a good example for many ODEs that cannot be solved analytically by using dsolve().

Let's consider several examples to compute analytical solutions of ODEs in MuPAD notebooks that are opened by typing a command in the command window, as shown here:

```
>> mupad
```

Example 11

Let's consider a second-order ODE, as shown in Figure 1-4:

$$\ddot{y} + \dot{y} = \sin(t),\, y(0) = 1,\; \dot{y}(0) = 2.$$

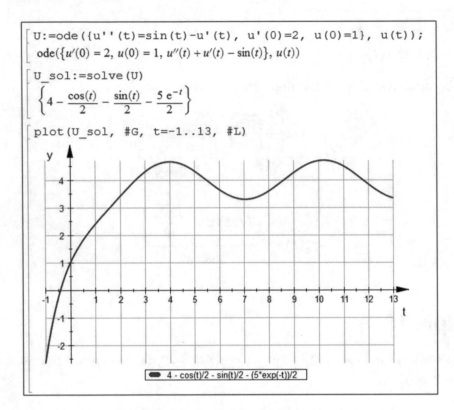

Figure 1-4. *MuPAD tools used to solve a second-order ODE*

Example 12

Given $\ddot{y} + 3\dot{y}^3 - |y|y = 2,\; y(0) = 0,\; \dot{y}(0) = 0$. This is a second-order nonhomogeneous and nonlinear ODE. Here is the solution syntax (solution command) of the given example in MuPAD notes.

[Y:=ode({y″(t)+3*(y′(t)^3)-y(t)*abs(y(t))=2,y(0)=0,y′(0)=0},y(t))

$$\text{ode}(\{y'(0) = 0,\, y(0) = 0,\, y''(t) + 3\, y'(t)^3 - |y(t)|\, y(t) - 2\},\, y(t))$$

[Ysolution:=solve(Y)

$$\text{solve}(\text{ode}(\{y'(0) = 0,\, y(0) = 0,\, y''(t) + 3\, y'(t)^3 - |y(t)|\, y(t) - 2\},\, y(t)))$$

No solution is computed in this exercise. When the MuPAD note (Symbolic MATH Toolbox) cannot compute an analytical solution of a given ODE problem, it rewrites a given problem formulation and produces no analytical solution expression.

Example 13

Given $A\ddot{y} + B\dot{y} + Cy = D$, the initial conditions are: $y(0) = a$ and $\dot{y}(0) = b$. This is a second-order and linear ODE. The $[A, B, C, D]$ and $[a, b]$ are scalars. Let's write Unicode code capable of solving the given ODE for any values of these scalars.

Here is the solution script (Unicode_2nd_ODE.m) with the dsolve() and syms functions of the Symbolic MATH Toolbox:

```
% Unicode_2nd_ODE.m
syms y(t) t
Dy=diff(y, t);
D2y=diff(y, t, 2);
fprintf('To solve: A*ddy+B*dy+C*y=D at ICs: y(0)=a, dy(0)=b \n')
fprintf('Enter the values of [A, B, C, D] and [a, b] \n');
A = input('Enter A = ');
B = input('Enter B = ');
C = input('Enter C = ');
D = input('Enter D = ');
a = input('Enter a = ');
b = input('Enter b = ');
% Given ODE equation:
Equation = D2y==(1/A)*(D - B*Dy-C*y);
ICs = [y(0)==a; Dy(0)==b];
Solution=dsolve(Equation, ICs);
% Display the computed anaytical solution in the command window:
pretty(Solution)
fplot(Solution, [0, 5], 'b-'), grid on
xlabel('\it t')
ylabel('\it Solution, y(t) ')
title('\it Solution of: $$\frac{A*d^2y}{dt^2}+\frac{B*dy}{dt}+C*y=D,
y(0)=a, dy(0)=b$$', 'interpreter', 'latex')
```

```
gtext(['\it A = '  num2str(A), '\it,  B = '  num2str(B), '\it,
C = '  num2str(C), '\it,  D = '  num2str(D)])
gtext(['\it Initial Conditions:  a = '  num2str(a), '\it,
b = '  num2str(b)])
grid on; shg
```

Now, let's test the code to solve this equation: $1.3\ddot{y} + \dot{y} + 2y = 3$, $y(0) = 1$, $\dot{y}(0) = 0$.

```
To solve: A*ddy+B*dy+C*y=D at ICs: y(0)=a, dy(0)=b
Enter the values of [A, B, C, D] and [a, b]
Enter A = 1.3
Enter B = 1
Enter C = 2
Enter D = 3
Enter a = 1
Enter b = 0
                    /   5 t \     / sqrt(235) t \
      sqrt(235) exp| - --- | sin| ----------- |
  3                 \  13 /     \     13      /
  - - ----------------------------------------
  2                              94

          /   5 t \     / sqrt(235) t \
      exp| - --- | cos| ----------- |
          \  13 /     \     13      /
  - -----------------------------
              2
```

In addition, we obtain the next plot figure, shown in Figure 1-5, which displays the numerical values of the computed analytical solution $y(t)$.

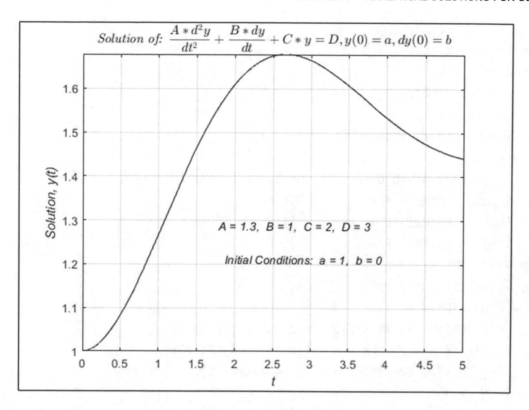

Figure 1-5. *Solution of the second-order ODE: 1.3ÿ+ẏ+2y=3, y(0) = 1, ẏ(0)= 0*

You can use this code to solve any second-order ODE of an IVP that has scalar [A, B, C, D] values. This code can be extended for a third-, fourth-, or higher-order ODE in a similar manner. Here is one alternative solution script (Unicode_4th_order_ODE.m) to solve any ODE up to the fourth order:

```
% Unicode_4th_order_ODE.m
%% Unicode to Solve up to 4-th order non-homogenuous linear ODEs of IVPs
%
% EQN:   Ax+B*dx + C*ddx + ... = F
% user entry (1) is ODE order #:   N
% user entry (2) values of:             A, B, C, ....
% user entry (3) initial conditions: IC1, IC2, IC3,...
syms x(t)
Dx=diff(x, t);
D2x=diff(x, t, 2);
```

```
D3x=diff(x, t, 3);
D4x=diff(x, t, 4);
D5x=diff(x, t, 5);
D6x=diff(x, t, 6);
D7x=diff(x, t, 7);
N = input('Order of the ODE:  N =  ');
if N == 1
disp('Solving: A*x+B*dx=F with one initial condtion: x(0) = ? ')
A = input(' Enter:  A =           ');
B = input(' Enter:  B =           ');
F = input(' Enter:  F =           ');
EQN = A*x+B*Dx==F;
IC1 = input(' Enter: x(0) =     ');
IC = x(0)==IC1;
Str0 ='Solution of: $$ A*x+B*\frac{dx}{dt}=F $$';
Str1=sprintf('A = %d,  B = %d, F = %d',  A, B, F);
Str2 =(['Initial Conditions: [ '  num2str(IC1), ' ]' ]);
elseif N == 2
disp('Solving: A*x+B*dx+C*ddx=E with two initial condtions: x(0) = ?,
dx(0) = ? ')
A = input(' Enter:  A =           ');
B = input(' Enter:  B =           ');
C = input(' Enter:  C =           ');
F = input(' Enter:  F =           ');
EQN = A*x+B*Dx+C*D2x==F;
IC1 = input(' Enter:    x(0) =     ');
IC2 = input(' Enter: dx(0) =     ');
IC = [x(0)==IC1, Dx(0)==IC2];
Str0 ='Solution of: $$ A*x+B*\frac{dx}{dt}+C*\frac{d^2x}{dt^2}=F $$';
Str1=sprintf('A = %d,  B = %d, C = %d, F = %d',  A, B, C, F);
Str2 =(['Initial Conditions: [' num2str(IC1), ', ' num2str(IC2), ' ]']);
elseif N == 3
disp('Solving: A*x+B*dx+C*ddx+D*dddx=F ')
disp('with three initial condtions: x(0) = ?, dx(0) = ? ddx(0) = ?')
A = input(' Enter:  A =           ');
```

```
B = input(' Enter:   B =           ');
C = input(' Enter:   C =           ');
D = input(' Enter:   D =           ');
F = input(' Enter:   F =           ');
EQN = A*x+B*Dx+C*D2x+D*D3x==F;
IC1 = input(' Enter:     x(0) =       ');
IC2 = input(' Enter:   dx(0) =      ');
IC3 = input(' Enter: ddx(0) =       ');
IC = [x(0)==IC1, Dx(0)==IC2, D2x(0)==IC3];
Str0 ='Solution of: $$ A*x+B*\frac{dx}{dt}+C*\frac{d^2x}{dt^2}+D*\
frac{d^3x}{dt^3}=F $$';
Str1=sprintf('A = %d,   B = %d, C = %d, D = %d, F = %d',  A, B, C, D, F);
Str2 =(['Initial Conditions: [ '  num2str(IC1), ', ' num2str(IC2), ', '
num2str(IC3),' ]']);
else
 disp('Solving: a*x+b*dx+c*ddx+d*dddx+e*ddddx=0 ')
disp('with four initial condtions: x(0) = ?, dx(0) = ? ddx(0) = ?,
dddx(0) = ?')
A = input(' Enter:  A =            ');
B = input(' Enter:  B =            ');
C = input(' Enter:  C =            ');
D = input(' Enter:  D =            ');
E = input(' Enter:  E =            ');
F = input(' Enter:  F =            ');
EQN = A*x+B*Dx+C*D2x+D*D3x+E*D4x==F;
IC1 = input(' Enter:     x(0)   =     ');
IC2 = input(' Enter:   dx(0)   =     ');
IC3 = input(' Enter: ddx(0)   =      ');
IC4 = input(' Enter: dddx(0) =     ');
IC = [x(0)==IC1, Dx(0)==IC2, D2x(0)==IC3, D3x(0)==IC4];
Str0 ='Solution of: $$ A*x+B*\frac{dx}{dt}+C*\frac{d^2x}{dt^2}+D*\
frac{d^3x}{dt^3} +E*\frac{d^4y}{dt^4}=F $$';
Str1=sprintf('A = %d,   B = %d, C = %d, D = %d,  E = %d, F = %d',
A, B, C, D, E, F);
```

```
Str2 =(['Initial Conditions: [ '  num2str(IC1),', ' num2str(IC2), ',
' num2str(IC3), ', ' num2str(IC3) ' ]']);
end
Solution = dsolve(EQN, IC);
fplot(Solution, [0, 3*pi], 'LineWidth', 2), grid on
xlabel('\it t')
ylabel('\it Solution, x(t) ')
title(Str0, 'Interpreter', 'latex')
gtext(Str1)
gtext(Str2)
grid on; shg
```

With this script (Unicode_4th_order_ODE.m), a user can obtain analytical and numerical solutions up to fourth-order nonhomogeneous linear ODEs with respect to their entries.

Laplace Transforms

Solutions of linear ordinary differential equations with constant coefficients can be evaluated by using the Laplace transformation. One of the most important features of the Laplace transforms in solving differential equations is that the transformed equation is an algebraic equation that will be used to define a solution for the given differential equation. In general, the Laplace transform application to solve differential equations can be formulated in the following way.

Let's consider that the given n-th order derivative of $y^n(x) = f(t)$. The Laplace transform of $y^n(x)$ is as follows:

$$-sy'(0) - y(0) = F(s) \tag{1-6}$$

or as follows:

$$s^n Y(s) - \sum_{i=1}^{n} s^{n-1} y^{i-1}(0) = F(s) \tag{1-7}$$

In Equation (1.6) or (1.7), if we substitute constant values for the initial conditions at $t = 0$ given as $y(0) = a_0$; $y'(0) = a_1$; $y''(0) = a_2$; ...; $y^{n-2}(0) = a_{n-2}$; $y^{n-1}(0) = a_{n-1}$.

Now, we can rewrite the expression in Equation (1-6 or 1-7) as follows:

$$\mathcal{L}\left\{\frac{d^n y}{dt^n} = f(t)\right\} => s^n Y(s) - s^{n-1} a_0 - s^{n-2} a_1 - s^{n-3} a_2 - ... - s\, a_{n-2} - a_{n-1} = F(s) \quad (1\text{-}8)$$

Subsequently, we first solve for $Y(s)$, take the inverse Laplace transform from $Y(s)$, and obtain the solution $y(t)$ of the n-th order differential equation.

The general steps to apply the Laplace and the inverse Laplace transforms to determine the solution of differential equations with constant coefficients are as follows:

1. Take the Laplace transforms from both sides of the given equation.

2. Solve for $Y(s)$, in terms of $F(s)$ and other unknowns.

3. Take the inverse Laplace transform of the found expression to obtain the final solution of the problem.

Note that in step 3 we should also decompose the found expression from step 2 into partial fractions to use tables of the inverse Laplace transform correspondences. The schematic view of the Laplace and inverse Laplace transforms is given in Figure 1-6.

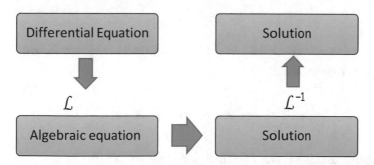

***Figure 1-6.** Flowchart of solving ODE with Laplace transform and its inverse*

Example 14

Let's consider a second-order nonhomogeneous differential equation, shown here:

$\dfrac{d^2 y}{dt^2} + A\dfrac{dy}{dt} + C = e^{nt}, \ dy(0) = k, y(0) = m.$

Now, by applying the steps depicted in the flowchart of the Laplace and inverse transforms, we can write the Laplace transform of the given problem in steps explicitly.

$$\mathcal{L}\left\{\frac{d^2 y}{dt^2} + A\frac{dy}{dt} + C = e^{nt}\right\} => L\left\{\frac{d^2 y}{dt^2}\right\} + \mathcal{L}\left\{A\frac{dy}{dt}\right\} + \mathcal{L}\{C\} = \mathcal{L}\{e^{nt}\} \tag{1-9}$$

$$\mathcal{L}\{e^{nt}\} = \frac{1}{s-n} \tag{1-10}$$

So,

$$\mathcal{L}\left\{\frac{d^2 y}{dt^2}\right\} = s^2 Y(s) - dy(0) - s*y(0) = s^2 Y(s) - k - s*m \tag{1-11}$$

$$\mathcal{L}\left\{A\frac{dy}{dt}\right\} = A*\mathcal{L}\left\{\frac{dy}{dt}\right\} = A*(s*Y(s) - y(0)) = A*s*Y(s) - m \tag{1-12}$$

$$\mathcal{L}\{C\} = C \tag{1-13}$$

Now plugging Equations (1-10), (1-11), (1-12), and (1-13) back into Equation (1-9), we obtain the next expression:

$$s^2 Y(s) - k - sm + AsY(s) - m + C = \frac{1}{s-n} \tag{1-14}$$

We solve Equation (1-14) for $Y(s)$ and determine the following:

$$Y(s) = \frac{1 + (sm + (k + m - C))(s - n)}{(s - n)(s^2 + As)} = -\frac{Cn - Cs - kn - mn + ks + ms + ms^2}{s(A+s)(n-s)} \tag{1-15}$$

From the expression $Y(s)$ in Equation (1-15), we decompose partial fractions, and then we take the inverse Laplace transform of both sides and obtain the solution in Equation (1-16), which is $y(t)$ of the given differential equation, as shown here:

$$y(t) = \frac{e^{nt}}{An + n^2} - \frac{Cn - kn - mn + 1}{An} + \frac{Cn - kn - mn - A(k - mn - C) + 1}{Ae^{At}(A + n)} \qquad (1\text{-}16)$$

A built-in function of the Symbolic MATH Toolbox, laplace(), is used to evaluate the Laplace transform of any algebraic expression or differential equation, and the ilaplace() function of the Symbolic MATH Toolbox is used to compute the inverse of the evaluated Laplace transformed s domain expression. These two functions handle all transformations by decomposing partial fraction procedures automatically and compute an analytical solution of a given ODE exercise.

laplace/ilaplace

laplace/ilaplace are based on the Laplace and inverse Laplace transforms that are built-in function tools of the Symbolic MATH Toolbox. The general syntaxes of laplace/ilaplace are as follows:

```
F=laplace(f)
F=laplace(f, t)
F=laplace(f, var1, var2)
```

   and

```
f=ilaplace(F)
f=ilaplace(F, s)
f=ilaplace(F, var1, var2)
```

Example 15

Given $x(t) = \sin(2t)$, the Laplace transform of $x(t)$ is computed with the following command syntaxes:

```
>> syms t
>> xt=sin(2*t); Xs=laplace(xt)
Xs =
2/(s^2 + 4)
```

Example 16

Given, $y(t) = \sin(Kt)$, we have the following:

```
>> syms t K
>> yt=sin(K*t); Ys=laplace(yt)
Ys =
K/(K^2 + s^2)
```

Example 17

Given $y(x) = ax^3 + b$, we have the following:

```
>> syms x a b
>> yx=a*x^3+b; Ys=laplace(yx)
Ys =
(6*a)/s^4 + b/s
```

Or, we can obtain the result in the t variable domain instead of s.

```
 >> yx=a*x^3+b; Yt=laplace(yx, x, t)
Yt =
(6*a)/t^4 + b/t
```

The `ilaplace()` function syntaxes and implementation are exactly the same for `laplace`. Let's look at several ODE exercises to demonstrate how to use `laplace` and `ilaplace` and compare their evaluated solutions with the ones obtained from `dsolve()`.

Example 18

Given $\dot{y} + 2y = 0, y(0) = 0.5$, let's solve it with `laplace/ilaplace` and `dsolve()`. The following script (`Laplace_vs_Dsolve_old_MATLAB_ex14.m`) is the solution. Note that this script works only on MATLAB 2017a or earlier versions.

```
% Laplace_vs_Dsolve_old_MATLAB_ex18.m
clearvars; clc; close all
% Step #1. Define symbolic variables' names
syms t s Y
```

```
ODE1='D(y)(t)=-2*y(t)';
% Step #2. Laplace Transforms
LT_A=laplace(ODE1, t, s);
% Step #3. Substitute ICs and initiate an unknown Y
LT_A=subs(LT_A,{'laplace(y(t),t, s)','y(0)'},{Y,0.5});
% Step #4. Solve for Y (unknown)
Y=solve(LT_A, Y);
display('Laplace Transforms of the given ODE with ICs'); disp(Y)
% Step #5. Evaluate Inverse Laplace Transform
Solution_Laplace=ilaplace(Y);
display('Solution found using Laplace Transforms: ')
pretty(Solution_Laplace)
% Step #6. Compute numerical values and plot them
t=0:.01:2.5; LTsol=eval(vectorize(Solution_Laplace));
figure, semilogx(t, LTsol, 'ro-')
xlabel('t'), ylabel('solution values')
title('laplace/ilaplace vs dsolve ')
grid on; hold on
% Compare with dsolve solution method
Y_d=dsolve('Dy=-2*y', 'y(0)=0.5'); display('Solution with dsolve')
pretty(Y_d); Y_sol=eval(vectorize(Y_d));
plot(t,Y_sol, 'b-', 'linewidth', 2), grid minor
legend('laplace+ilaplace', 'dsolve'), hold off; axis tight
```

Here is another version of the script (Laplace_vs_Dsolve_new_MATLAB_ex14.m) that works on newer versions of MATLAB (starting from MATLAB 2018b and on).

```
% Laplace_vs_Dsolve_New_MATLAB_ex18.m
clearvars; clc; close all
% Step #1. Define symbolic variables' names
syms t s Y y(t) Dy(t)
assume([t, Y]>0);
Dy=diff(y,t);
ODE1=Dy==-2*y(t);
% Step #2. Laplace Transform
LT_A=laplace(ODE1, t, s);
```

```
% Step #3. Substitute the arbitrary unknown Y and IC(s)
LT_A=subs(LT_A,laplace(y,t, s),Y);
LT_A=subs(LT_A, y(0), 0.5);
% Step #4. Solve for the arbitrary unknown Y
Y=solve(LT_A, Y);
disp('Laplace Transforms of the given ODE with ICs'); disp(Y)
% Step #5. Evaluate the Inverse Laplace Transform
Solution_Laplace=ilaplace(Y);
disp('Solution found using Laplace Transforms: ')
pretty(Solution_Laplace)
% Step #6. Compute numerical values and plot them
t=0:.01:2.5; LTsol=eval(vectorize(Solution_Laplace));
figure, semilogx(t, LTsol, 'ro-')
xlabel('t'), ylabel('solution values')
title('laplace/ilaplace vs. dsolve ')
grid on; hold on
% Compare with dsolve solution method
IC=y(0)==0.5;
Y_d=dsolve(ODE1, IC);
disp('Solution with dsolve')
pretty(Y_d); Y_sol=eval(vectorize(Y_d));
plot(t,Y_sol, 'b-', 'linewidth', 2), grid minor
legend('laplace+ilaplace', 'dsolve')
hold off; axis tight
```

Both versions of the script produce the same output. After executing the script (Laplace_vs_Dsolve_old_MATLAB_ex18.m or ODE_Laplace_New_MATLAB_vers.m), the following output is obtained:

```
Warning: Solutions are valid under the following conditions: -2 < s.
To include parameters and conditions in the solution, specify the
'ReturnConditions' value as 'true'.
Laplace Transforms of the given ODE with ICs
1/(2*s + 4)
```

Solution found using Laplace Transforms:

```
  exp(-2 t)
  ---------
      2
```
Solution with dsolve

```
  exp(-2 t)
  ---------
      2
```

Figure 1-7 shows the plotted solutions and clearly displays a perfect convergence of the solutions found with laplace/ilaplace and dsolve().

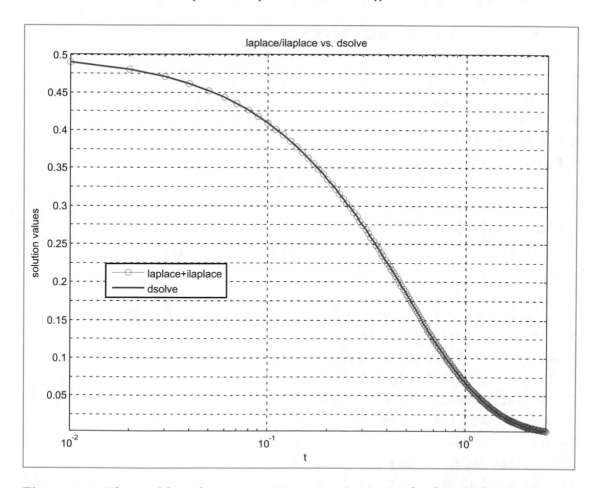

Figure 1-7. *The problem $\dot{y}+2y=0$, $y(0)=0.5$ solved with* laplace/ilaplace, *dsolve()*

Note   There are a few command syntaxes (functions) of the Symbolic Math Toolbox that are not forward compatible, such as laplace/ilaplace and subs employed while solving ODEs. This forward incompatibility issue was demonstrated in Examples 4 and 5.

Example 19

Given $\ddot{y} + \dot{y} = \sin t$, $y(0) = 1$, and $\dot{y}(0) = 2$, here is the solution script (Laplace_vs_ Dsolve_old_MATLAB_ex19.m) of this second-order nonhomogeneous ODE with laplace, ilaplace, and dsolve():

```
% Laplace_vs_Dsolve_old_MATLAB_ex19.m
% %%    LAPLACE TRANSFORMS
% Given: y"+y'=sin(t) with ICs: [1, 2] for y(0) & y'(0).
clearvars, clc, close all
syms t s Y
% Define the given ODE equation
ODE2nd='D(D(y))(t)+D(y)(t)-sin(t)';
% Step 1. Laplace Transforms
LT_A=laplace(ODE2nd, t, s);
% Step 2. Substitute ICs and initiate an unknown Y
LT_A=subs(LT_A,{'laplace(y(t),t, s)','y(0)','D(y)(0)'},{Y,1,2});
% Step 3.  Solve for Y unknown
Y=solve(LT_A, Y);
disp('Laplace Transforms of the given ODE with ICs'); disp(Y)
Solution_Laplace=ilaplace(Y);
disp('Solution found using ilaplace/laplace: ')
pretty(Solution_Laplace); t=0:.01:13;
LTsol=eval(vectorize(Solution_Laplace));
figure, plot(t, LTsol, 'ro-');
xlabel('\it t'),
ylabel('\it Solution y(t)')
ylabel('\it Solution y(t)')
```

```
title('\it laplace/ilaplace vs. dsolve Solutions: $$ \frac{d^2y}{dt^2}+\
frac{dy}{dt}=sin(t)$$', 'interpreter', 'latex');
hold on
% dsolve solution method
Y=dsolve('D2y+Dy=sin(t)', 'y(0)=1, Dy(0)=2', 't');
disp('Solution with dsolve:  ');
pretty(Y); Y_sol=eval(vectorize(Y));
plot(t,Y_sol, 'b-', 'linewidth', 2); grid minor
legend('laplace+ilaplace', 'dsolve', 'location', 'southeast'); hold off
```

The computed analytical solutions displayed in the command window are as follows (see Figure 1-8):

```
Laplace Transforms of the given ODE with ICs
(s + 1/(s^2 + 1) + 3)/(s^2 + s)

Solution found using Laplace Transforms:

      cos(t)    sin(t)    5 exp(-t)
  4 - ------ - ------ - ---------
        2         2         2
Solution with dsolve:

      cos(t)    sin(t)    5 exp(-t)
  4 - ------ - ------ - ---------
        2         2         2
```

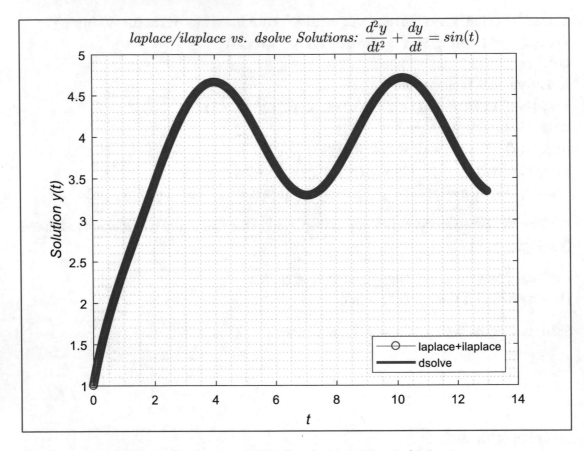

Figure 1-8. *Numerical solutions of* $\ddot{y}+\dot{y}=sin(t), y(0)=1, \dot{y}(0)=2$

Note that the previous script (Laplace_vs_Dsolve_old_MATLAB_ex19.m) works in older versions of MATLAB and will not run in later versions after MATLAB 2018a. This script (Laplace_vs_Dsolve_New_MATLAB_ex15.m) produces the same output:

```
% Laplace_vs_Dsolve_New_MATLAB_ex19.m
% %%   LAPLACE TRANSFORMS
% Given: y"+y'=sin(t) with ICs: [1, 2] for y(0) & y'(0).
clearvars, clc, close all
% Step 1. Define names of the symbolic variables
syms t s Y y(t) Dy(t)
Dy=diff(y, t);
D2y=diff(y, t, 2);
% Define the given ODE equation
```

```matlab
Equation=D2y+Dy==sin(t);
% Step 2. Laplace Transform
LT_A=laplace(Equation, t, s);
% Step 3. Substitute the arbitrary unknown Y and ICs
LT_A=subs(LT_A,laplace(y(t),t, s),Y);
LT_A=subs(LT_A,y(0),1);                              % IC: y(0)=1
LT_A=subs(LT_A,subs(diff(y(t), t), t, 0), 2);   % IC: dy(0)=2
% Step 4.  Solve for the arbitrary unknown Y
Y=solve(LT_A, Y);
disp('Laplace Transforms of the given ODE with ICs'); disp(Y)
% Step 5.  Compute the inverse of laplace Transform
Solution_Laplace=ilaplace(Y);
disp('Solution found using Laplace Transforms: ')
pretty(Solution_Laplace); t=0:.01:13;
LTsol=eval(vectorize(Solution_Laplace));
figure, plot(t, LTsol, 'ro-'); xlabel('\it t'),
ylabel('\it Solution y(t)')
title('\it laplace/ilaplace vs. dsolve Solutions: $$ \frac{d^2y}{dt^2}+\
frac{dy}{dt}=sin(t)$$', 'interpreter', 'latex');
hold on
% dsolve solution method
syms t y(t)
Dy=diff(y, t);
D2y=diff(y, t, 2);
% Define the given ODE equation
Equation=D2y+Dy==sin(t);
% Define Initial Conditions
ICs = [y(0)==1, Dy(0)==2];
Y=dsolve(Equation, ICs);
disp('Solution with dsolve:  ');
pretty(Y);
t=0:.01:13;
Y_sol=eval(vectorize(Y));
plot(t,Y_sol, 'b-', 'linewidth', 2); grid minor
legend('\it laplace+ilaplace', '\it dsolve', 'location', 'southeast') hold off
```

From the plot displayed in Figure 1-8, it is clear that the solutions found via the Laplace transforms' laplace()/ilaplace() and dsolve() functions converge perfectly well. Since both functions evaluate analytical solutions correctly.

Note   While employing the Symbolic MATH Toolbox functions (ilaplace()/laplace(), dsolve()) to solve ODEs, it is recommended that you introduce all the symbolic variables first and then any assumptions about the values of the independent variables, e.g., t, Y. Moreover, it is a good habit to specify all derivatives with respect to the independent variables and then the differential equation. Here's an example:

```
syms t s Y y(t) Dy(t)
Dy=diff(y, t);                  % First-order derivative
D2y=diff(y, t, 2);              % Second-order derivative
assume([t, Y]>0);               % Assumption
Equation=D2y==sin(t)-5*Dy-2*y;  % ODE formulation
```

Note   It is optional how to call the given ODE formulation, meaning independent variable t and arbitrary unknown Y, while using the ilaplace/laplace functions of the Symbolic MATH Toolbox.

It is not viable to compute the analytical solutions of all nonhomogenuous ODEs (particularly nonlinear types) with dsolve() and ilaplace/laplace of the Symbolic MATH Toolbox explicitly.

Example 20

Given a second-order nonhomogeneous and nonlinear ODE, $2\ddot{y} + 3\dot{y}^3 - |y|\cos(t) = 2$, $y(0) = 0$ and $\dot{y}(0) = 0$. Here is the solution script (Laplace_vs_Dsolve_Ex20.m) with Laplace and inverse Laplace transforms:

```
% Laplace_vs_Dsolve_Ex20.m
clearvars, clc, close all
```

```
% Step 1. Introduce symbolic variables
syms t s Y y(t)  Dy(t)
assume([t  Y]>0);
Dy=diff(y, t);
D2y=diff(y, t, 2);
% Define Equation
Equation=2*D2y+3*Dy^3==cos(100*t)*abs(y(t))-2;
% Step 2. Laplace Transforms
LT_Y=laplace(Equation, t, s);
% Step 3. Substitute the arbitrary unknown Y and ICs
LT_Y=subs(LT_Y,laplace(y(t),t, s),Y);
LT_Y=subs(LT_Y,y(0),1);
LT_Y=subs(LT_Y,subs(diff(y(t), t), t),0);
% Step 4.  Solve for Y unknown
Y=solve(LT_Y, Y);
% Step 5. Compute the inverse Laplace transform
disp('Laplace Transforms of the given ODE with ICs'); disp(Y)
Solution_Laplace=ilaplace(Y);
disp('Solution found using Laplace Transforms: ')
pretty(Solution_Laplace)
%  Solution with dsolve():
syms t y(t) Dy(t)
Dy=diff(y, t);
D2y=diff(y, t, 2);
% Define Equation
Equation=2*D2y+3*Dy^3==cos(100*t)*abs(y(t))-2;
ICs = [y(0)==0, Dy(0)==0];
Solution=dsolve(Equation, ICs);
pretty(Solution)
```

The script (Laplace_vs_Dsolve_Ex20.m) produces the following outputs (solution expressions of Example 20) in the command window:

```
Laplace Transforms of the given ODE with ICs
(4 + s*laplace(abs(y(t)), t, s - 100*i) + s*laplace(abs(y(t)), t,
s + 100*i) - 6*s*laplace(D(y)(t)^3, t, s))/(4*s^3)
```

Solution found using Laplace Transforms:

```
                /                   3              \
                |  laplace(D(y)(t) , t, s)         |
       3 ilaplace| -----------------------, s, t |
   2            |              2                  |
   t            \              s                  /
   -- - ------------------------------------------------- +
   2                               2

              / laplace(|y(t)|, t, s - 100 i)        \
    ilaplace| -----------------------------, s, t |
            |               2                       |
            \               s                       /
   ------------------------------------------------- +
                          4

              / laplace(|y(t)|, t, s + 100 i)        \
    ilaplace| -----------------------------, s, t |
            |               2                       |
            \               s                       /
   -------------------------------------------------
                          4
```

Warning: Unable to find explicit solution.

()

This output means that no analytical solution is computed explicitly with the laplace()/ilaplace() and dsolve() function tools.

Example 21

Given a second-order nonhomogeneous ODE, $2\ddot{y}+3\dot{y}-2y=g(t)$, $y(0)=0$ and $\dot{y}(0)=0$, where $g(t)$ is a forcing function that is discontinuous and defined by the following:

$$g(t)=u_2(t)-u_{10}(t)=\begin{cases} 5, & 2\le t <10, \\ 0, & 0\le t<2 \text{ and } t\ge 10. \end{cases}$$

the Laplace transform of the given equation is as follows:

$$\mathcal{L}\{2\ddot{y}+3\dot{y}-2y\}=\mathcal{L}\{u_2(t)-u_{10}(t)\}$$

$$2s^2Y(s)-2s*y(0)-2*\dot{y}(0)+3sY(s)-y(0)-2Y(s)=\frac{5\left(e^{-2s}-e^{-10s}\right)}{s}$$

$$Y(s)=\frac{5\left(e^{-2s}-e^{-10s}\right)}{s\left(2s^2+3s-2\right)}$$

Note that e^{-2s} and e^{-10s} are explained with time delays in the system output signals; in other words, -2 and -10 mean 2 and 10 seconds of time delays. 5 is the magnitude of the Heaviside (step) function.

The formulation $Y(s)$ is the solution of the differential equation in the s domain, but we need it in the time domain, and thus we need to compute its inverse Laplace transform, e.g., $\mathcal{L}^{-1}\{Y(s)\}=y(t)$. By employing ilaplace(), the next short script (iLaplace_Ex21.m) is created that solves the given problem and computes its analytical and numerical solutions:

```
% iLaplace_Ex21.m
syms t s
% Solution of a 2nd order nonhomogenuous ODE with discontinuous % forcing
function.

F=5*(exp(-2*s)-exp(-10*s))/s; % Forcing fcn
Y=2*s^2+s+2;                   % ODE
TF=F/Y;                        % Equation
TFt=ilaplace(TF);              % Inverse Laplace
pretty(TFt);                   % Display
Sol=vectorize(TFt);            % Vectorization
t=linspace(0, 20, 400);        % t space (independent variable)
S=eval(Sol);                   % Numerical solutions computed
plot(t, S, 'b-', 'linewidth', 2);
grid minor
title('\it Solution of: $$ 2*\frac{d^2y}{dt^2}+3*\frac{dy}{dt}-2*y=g(t)
$$', 'interpreter', 'latex')
Str='\it g(t) is a discontinuous Forcing Fcn';
```

```
text(7.5, 3.75, Str, 'backgroundcolor', [1 1 1]);
grid on, xlabel('\it t ')
ylabel('\it Solution, y(t) '), shg
```

By executing the script iLaplace_Ex21.m, we obtain the solution plot shown in Figure 1-9, along with the solution formulation in the command window.

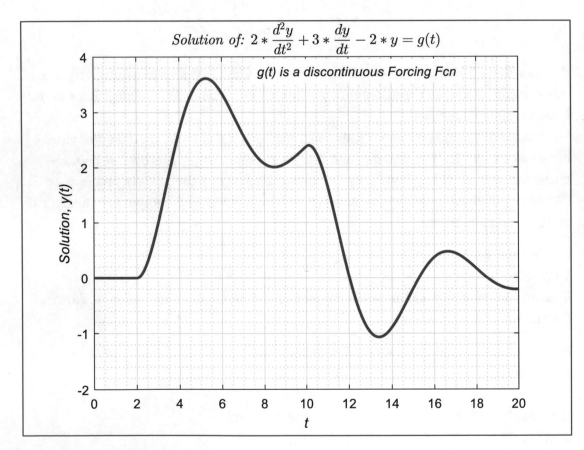

Figure 1-9. *Simulation of the second-order nonhomogeneous ODE subject to discontinuous forcing function*

```
>> pretty(TFt)
```

```
                        /                 /              1/2           \          \
                        |    /        t \ |               15    sin(#1) |          |
                        | exp| 5/2 - - | | | cos(#1) + ------------- |          |
                        |    \        4 / \                15        /          |
   5 heaviside(t - 10) | --------------------------------------------------- - 1/2 | -
                        \                         2                                 /

                        /                 /              1/2           \          \
                        |    /        t \ |               15    sin(#2) |          |
                        | exp| 1/2 - - | | | cos(#2) + ------------- |          |
                        |    \        4 / \                15        /          |
      5 heaviside(t - 2) | --------------------------------------------------- - 1/2 |
                        \                         2                                 /
```

where

```
            1/2
          15    (t - 10)
   #1 == --------------
                4

            1/2
          15    (t - 2)
   #2 == -------------
                4
```

As you can see from the previous exercises, using the Laplace transforms (laplace()/ilaplace()), computing the analytical solutions of nonhomogeneous ODEs subject to external forcing functions, which are discontinuous, is relatively easy and fast. Such exercises are often found within control engineering settings. Moreover, note that the Laplace and inverse Laplace transforms (laplace()/ilaplace()) are straightforward to implement when solving ODEs. The solutions of ODEs found with them match perfectly well with the ones found by dsolve(). As stated earlier, many ODEs cannot be solved analytically with the laplace()/ilaplace() and dsolve() functions. Thus, numerical methods are the only option.

References

[1] Website: `http://tutorial.math.lamar.edu/Classes/DE/Definitions.aspx`, viewed on Sept. 20, 2013.

[2] Gear, C.W., *Numerical Initial-Value problems in Ordinary Differential Equations*, Prentice-Hall (1971).

[3] Potter, M. C., Goldberg, J.L., Aboufadel, E. F., *Advanced Engineering Mathematics* (3rd ed.), Oxford University Press (2005).

[4] Boyce, W. E., Diprima, R.C., *Elementary Differential Equations and Boundary Value Problems* (7th ed.), John Wiley & Sons (2003).

[5] Hairer, E., Norsett, S. P., Wanner, G., *Solving ordinary differential equations I: Non-stiff problems* (2nd ed.), Springer Verlag (1993).

CHAPTER 2

Numerical Methods for First-Order ODEs

This chapter is dedicated to using numerical methods to solve initial-value problems (IVPs) of ODEs. The methods we'll use are Euler, Runge-Kutta, Adams, Taylor, Milne, and Adams-Moulton. Specifically, we will evaluate their accuracy and efficiency. Of these methods, the Euler method is the simplest one and actually has several types, such as forward, backward, and improved (modified).

Euler Method

As mentioned, there are three different types of Euler methods to compute numerical solutions of IVPs. They are the forward Euler, improved Euler, and backward Euler methods; the simplest one is the forward method.

The forward Euler method is formulated as follows:

$$\begin{cases} y_i = y_{i-1} + hk \\ k = f\left(t_{i-1}, y_{i-1}\right) \\ t_i = t_{i-1} + h \end{cases} \tag{2-1}$$

where $y_0 = y(t_0)$ and $i = 1, 2, 3, \ldots, N$. The forward method, shown in Equation (2-1), can be implemented in a numerical search solution via the following algorithm:

Step 1. Define $f(t, y)$.

Step 2. Input the initial conditions (ICs): t_0 and y_0.

Step 3. Input the step size, h, and the number of steps, N.

Step 4. Output t_0 and y_0.

Step 5. For i from 1 to N, do step 6 and step 7.

Step 6.

© Sulaymon L. Eshkabilov 2020

S. L. Eshkabilov, *Practical MATLAB Modeling with Simulink*, https://doi.org/10.1007/978-1-4842-5799-9_2

$$k = f(t,y)$$

$$y = y + hk$$

$$t = t + h$$

Step 7. Output t and y.

Step 8. End.

Sometimes step 6 is not easily understood, and there might be difficulties in its implementation. This step can be explained more explicitly with the following expressions:

$$y_1 = y_0 + f(t_0,y_0)(t_1 - t_0)$$

$$y_2 = y_1 + f(t_1,y_1)(t_2 - t_1)$$

$$y_3 = y_2 + f(t_2,y_2)(t_3 - t_2)$$

Therefore, we can formulate the following:

$$y_{i+1} = y_i + f(t_i,y_i)(t_{i+1} - t_i)$$

where $i = 1, 2, 3, ..., N$.

If we introduce an equally spaced step size of $h = t_1 - t_0 = t_2 - t_1 = ...$, then $y_{i+1} = y_i + f_i h$ with $i = 1, 2, 3, ..., N$.

Example 1

Here's the example problem: $\dot{y} + 2y - e^{-t} = 3$ with ICs: $y(0) = 1$.

To compute a numerical solution of the given first-order nonhomogeneous ODE problem for $t = 0...10$, we take $h = 0.05$ (time step size) and rewrite the given problem equation as $\dot{y} = 3 - 2y + e^{-t}$, which can be also written as $f(t,y) = 3 - 2y + e^{-t}$.

For $t_0 = 0$,

$$y_0 = 1$$

For $t_1 = 0.05$,

$$y_0 = 1, \, k = f(t_0, y_0) = 3 - 2 * 1 + e^0 = 2$$
$$y_1 = y_0 + h\,k = 1 + 0.05 * 2 = 1.1$$

For $t_2 = 0.1$,

$$y_1 = 1.1, \, k = f(t_1, y_1) = 3 - 2 * 1.1 + e^{-0.1} = 1.7512$$

$$y_2 = y_1 + kh = 1.1 + 1.7512 * 0.05 = 1.1876$$

...

Now, based on what we have derived, we put together a script (Example1.m) to obtain a complete numerical solution for the problem, as shown here:

```
% Example1.m
%% Euler's Forward Method.
% Part 1
clc; clearvars; close all;
t0=0;                   % Initial time
tend=10;                % End time
h=.05;                  % Time step size
t=t0:h:tend;            % Simulation time ranges
steps=length(t)-1;      % Number of steps
y0=1;                   % Initial Condition y0 at t0
f=zeros(size(t));       % Memory allocation
y1=[y0, f(1, end-1)];   % Memory allocation
for ii=1:steps
    f(ii)=3+exp(-t(ii))-2*y1(ii);
    y1(ii+1)=y1(ii)+f(ii)*h;
end
plot(t, y1, 'b-', 'linewidth', 1.5), grid on
xlabel('\it t')
ylabel('\it y(t)')
hold on
```

Improved Euler Method

The improved Euler method is called the *modified Euler method* or *Heun's method* in some literature. It is an explicit method implemented in a similar manner to the forward Euler method except for step 6. Its implementation in step 6 looks like this:

$$y_{i+1} = y_i + \left(f\left(t_i,y_i\right) + f\left(t_i + h, y_i + hf_i\right) \right)\frac{h}{2}; \qquad (2\text{-}2)$$

The improved Euler method, shown in Equation (2-2), is implemented in a numerical search solution via the following algorithm:

Step 1. Define $f(t,y)$.
Step 2. Input the ICs: t_0 and y_0.
Step 3. Input the step size, h, and the number of steps, N.
Step 4. Output t_0 and y_0.
Step 5. For i from 1 to N, do step 6 and step 7.
Step 6.

$$k_1 = f\left(t_i,y_i\right)$$

$$k_2 = f\left(t_i + h, y_i + hk_1\right)$$

$$y_i = y_i + \left(k_1 + k_2\right)\frac{h}{2}$$

$$t = t + h$$

Step 7. Output t and y.
Step 8. End.

In this method, step 6 is slightly different from the one given in the forward Euler method. This step can be explained more explicitly with the following expressions. This procedure can be rewritten via more explicit expressions in the example of the previous described ODE problem: $\dot{y} + 2y - e^{-t} = 3$, $y(0) = 1$.

$$k_{1,i} = f\left(t_i,y_i\right) = 3 + e^{-t_i} - 2y_i;$$

$$k_{2,i} = f\left(t_i + h, y_i + h*k_{1,i}\right) = 3 + e^{-(t_i + h)} - 2\left(y_i + hk_{1,i}\right);$$

$$y_{i+1} = y_i + \left(k_{1,i} + k_{2,i}\right)\left(\frac{h}{2}\right).$$

We don't necessarily need to have $[t_{i+1} = t_i + h]$, in other words, uniform time steps over the interval according to the reference source [1]. Variable step solvers are also considered in later sections. All depicted steps of the method are implemented in part 2 of the script, called Example1.m and shown here:

```
%% Euler's Improved method or Heun's method.
% Part 2
%{
y(n+1)=y(n)+( f(tn,y(n)) + f(tn+h, y(n)+h*fn) )*0.5*h;
or different expression (STEP 6 changed from Euler's method):
k1 =f(t,y) = 3+exp(-t(ii))-2*y(ii);
k2 =f(t+h, y+h*k1) = 3 + exp(-(t(ii)+h))-2*(y(ii)+h*k1);
y(ii+1) = y(ii)+(h/2)*(k1(ii)+k2(ii))
%}
f(1) = 3 + exp(-t(1))-2*y0;
f=[f(1), zeros(1, steps-1)];    % Memory allocation
y2=[y0, zeros(1, steps-1)];     % Memory allocation
for jj=1:steps
    f(jj+1)=3+exp(-(t(jj)+h))-2*(y2(jj)+h*f(jj));
    y2(jj+1)=y2(jj)+(f(jj)+f(jj+1))*.5*h;
end
plot(t,y2, 'k--', 'linewidth',1.5)
```

Backward Euler Method

The backward Euler method is a first-order implicit method. This method is expressed with the expression shown in Equation (2-3).

$$
\left\{
\begin{array}{c}
t_i = t_{i-1} + h \\
k = f(t_i, y_i) \\
y_i^* = y_i + hk \\
y_i = y_{i-1} + hf(t_i, y_i^*)
\end{array}
\right.
\tag{2-3}
$$

where $i = 1, 2, 3, ..., N$.

The only tricky question in the formulation of Equation (2-3) is how to compute y_i^*. There are several ways to compute it, and one of them is to use the forward Euler method to predict y_i^*, as shown in Equation (2-3), and then correct it by recomputing y_i according to the Equation (2-3) formulation. Another approach to find y_i^* is to use the Newton-Raphson technique or to use algebraic formulations based on the given IVP equation. The previously described backward Euler method can be implemented in a numerical search solution via the following algorithm:

Step 1. Define $f(t,y)$.

Step 2. Input the ICs: t_0 and y_0.

Step 3. Input the step size, h, and the number of steps, N.

Step 4. Output t_0 and y_0.

Step 5. For i from 1 to N, do step 6 and step 7.

Step 6.

$$t_{i+1} = t_i + h$$

$$k = f(t,y)$$

$$y_i^* = y_i + hk$$

$$y_i = y_{i-1} + hf(t_i, y_i^*)$$

Step 7. Output t and y.

Step 8. End.

In the backward Euler method, step 6 is found to be complicated due to the implicitly expressed formulation of y_i^*; thus, this approach cannot be understood easily, and sometimes users may find its implementation difficult. This step can be explained more explicitly with the following expressions:

$$t_1 = t_0 + h$$

$$k = f(t_1, y_0)$$

$$y_1^* = y_0 + kh$$

$$y_1 = y_0 + f(t_1, y_1^*)h$$

$$t_2 = t_1 + h$$

$$k = f\left(t_2, y_1\right)$$

$$y_2^* = y_1 + kh$$

$$y_2 = y_1 + f\left(t_2, y_2^*\right)h$$

Thus,

$$y_i = y_{i-1} + f\left(t_i, y_i^*\right)h$$

where i = 1, 2, 3, ..., N.

If we introduce an equally spaced step size $h = t_1 - t_0 = t_2 - t_1 = ...$, then $y_{i+1} = y_i + f_i h$ with i = 1, 2, 3, ..., N.

Example 2

Here's the example problem: $\dot{y} + 2y - e^{-t} = 3$, $y(0) = 1$. Let's compute a numerical solution to the problem for $t = 0...10$ with the time step of $h = 0.05$. First, we rewrite the given problem equation:

$$\dot{y} = 3 - 2y + e^{-t}$$

This can also be written as follows: $f(t, y) = 3 - 2y + e^{-t}$.

Now, based on what we have derived, we write part 3 of the script (Example2.m) to obtain a complete numerical solution for the problem, as shown here:

```
%% Example2.m: EULER's backward method (1st order implicit)
% Part 3
f=zeros(1,steps);               % Memory allocation
T=zeros(1, steps);              % Memory allocation
yNEW=zeros(1,steps+1);          % Memory allocation
y3=[y0, zeros(1,steps-1)];      % Memory allocation
for ii=1:steps
   T(ii+1)=T(ii)+h;
    f(ii)=3-2*y3(ii)+exp(-T(ii));
    yNEW(ii)=y3(ii)+h*f(ii);
    f(ii+1)=3-2*yNEW(ii)+exp(-T(ii+1));
    y3(ii+1)=y3(ii)+f(ii+1)*h;
end
```

```
plot(T, y3, 'm-.','linewidth', 1.5),grid on
title('\it Solutions with Euler methods: $$ \frac{dy}{dt}+2*y-e^{-t}=3,
y(0)=1 $$', 'interpreter', 'latex')
legend('\it Euler forward ', '\it Euler improved', '\it Euler backward')
xlabel '\it t'; ylabel('\it Solution, y(t)'), shg
```

From the plotted results shown in Figure 2-1, we can conclude that the three solution approaches, namely, forward, improved (Heun's), and backward (implicit), produce visibly diverged solutions. Among these methods, the backward Euler method produces the most accurate results, but it is the costliest one in terms of computation time because of the additional y_i^* computation.

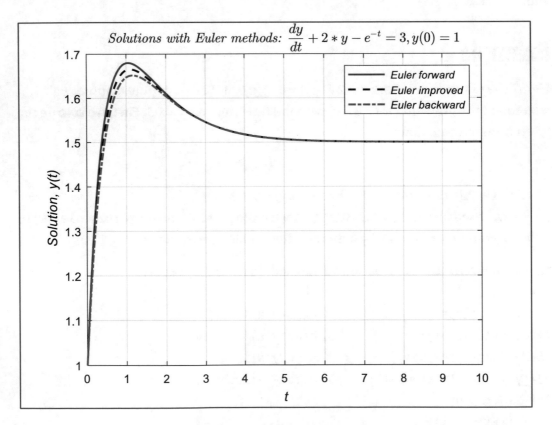

Figure 2-1. *Simulation results of the problem ($\dot{y}+2y-e^{-t}=3$, $y(0) = 1$) with Euler's forward, improved, and backward methods*

The backward Euler method is demonstrated again in the example of $\dot{y} + 2y - e^{-t} = 3$, $y(0) = 1$. The given problem can be reformulated as follows:

$$f(t_i, y_i) = 3 - 2y_i + e^{-t_i} \text{ for } i = 1, 2, 3, ..., N$$

Thus, the backward Euler method can be written with clearer indexes in the following way:

$$y_{i+1} = y_i + hf(t_{i+1}, y_{i+1})$$

In addition, this can be rewritten for the given problem as follows:

$$y_{i+1} = y_i + hf(t_{i+1}, y_{i+1}) = y_i + h\left(3 - 2y_{i+1} + e^{-t_{i+1}}\right)$$

$$y_{i+1} = y_i + h\left(3 - 2y_{i+1} + e^{-t_{i+1}}\right) \xrightarrow{yields} y_{i+1} = \frac{y_i + h\left(3 + e^{-t_{i+1}}\right)}{1 + 2h}$$

Now, the formulation $y_{i+1} = \dfrac{y_i + h\left(3 + e^{-t_{i+1}}\right)}{1 + 2h}$ is explicit and can be solved for $i = 1, 2, 3, ..., N$ for $t_1 = t_0 = 0$ and $y_1 = y(t_0) = 1$ and is implemented via Example2.m as part 4, as shown here:

```
%% Euler's Backward Method (Another approach). Example2.m
% Part 4
% Memory allocation to speed up computation
T=zeros(1, steps);          % Memory allocation
y4=[y0, zeros(1,steps-1)];  % Memory allocation
for ii=1:steps
    T(ii+1)=T(ii)+h;
    y4(ii+1)=(y4(ii)+h*(3+exp(-T(ii+1))))/(1+2*h);
end
```

Midpoint Rule Method

The midpoint rule method is a second-order explicit method and has two steps, which can be expressed with the expressions in Equation (2-4).

$$
\begin{cases}
y_i^* = y_i(t) + \dfrac{h}{2} f\left(t_i, y_i(t)\right) \\[2mm]
y_i(t+h) = y_i(t) + hf\left(t_i + \dfrac{h}{2}, y_i^*\right)
\end{cases}
\tag{2-4}
$$

where $i = 1, 2, 3, ..., N$.

This method is relatively simple and straightforward to implement via the following algorithm:

Step 1. Define $f(t, y)$.

Step 2. Input the ICs: t_0 and y_0.

Step 3. Input the step size, h, and the number of steps, N.

Step 4. Output t_0 and y_0.

Step 5. For i from 1 to N, do step 6 and step 7.

Step 6.

$$t_{i+1} = t_i + h$$

$$y_i^* = y_i + \frac{h}{2} f(t_i, y_i)$$

$$y(t_i + h) = y(t_i) + hf\left(t_i + \frac{h}{2}, y_i^*\right)$$

Step 7. Output t and y.

Step 8. End.

Like the previously described methods, we describe step 6 with more explicitly defined formulations.

$$t_1 = t_0 + \frac{h}{2}$$

$$y_1^* = y_0 + \frac{h}{2} f(t_0, y_0)$$

$$y_1 = y_0 + h f(t_1, y_1^*)$$

$$t_2 = t_1 + \frac{h}{2}$$

$$y_2^* = y_1 + \frac{h}{2} f\left(t_1, y_1\right)$$

$$y_2 = y_1 + h\, f\left(t_2, y_2^*\right)$$

Example 3

Here's the example problem: $\dot{y} + 2y - e^{-t} = 3$, $y(0) = 1$. Let's compute a numerical solution for the problem of $t = 0...10$ with a time step of $h = 0.05$. First, we rewrite the given problem equation as $\dot{y} = 3 - 2y + e^{-t}$. This also can be written as follows: $f(t, y) = 3 - 2y + e^{-t}$.

Now, based on what we have derived, we write part 5 of the script (Example3.m) to obtain numerical solutions for the problem, as shown here:

```
%%  2nd ORDER(explicit): Mid-Point Rule method
% Example3.m: dy/dt=3-2*y+exp(-t); with ICs: y(0)=1.0
% Part 5
yp=zeros(1,steps);          % Memory allocation
y5=[y0, zeros(1,steps-1)];  % Memory allocation
f1=3-2*y5(1)+exp(-t(1));
f=[f1, zeros(1, steps-1)];  % Memory allocation
for ii=1:steps
    yp(ii)=y5(ii)+(h/2)*f(ii);
    f(ii+1)=3-2*yp(ii)+exp(-(t(ii)+h/2));
    y5(ii+1)=y5(ii)+h*f(ii+1);
end
plot(t, y5, 'g:', 'linewidth', 2.2)
title('\it Solutions of: $$ \frac{dy}{dt}+2*y-e^{-t}=3, y(0)=1 $$',
'interpreter', 'latex')
xlabel('\it t'), ylabel('\it y(t)')
% Compute the analytical solution:
syms u(x) x
Du=diff(u, x);
Equation=Du==3+exp(-x)-2*u;
```

```
IC = u(0)==1;
u_sol=dsolve(Equation, IC);
Solution=vectorize(u_sol);
x=t;
Solution=eval(Solution);
plot(t, Solution, 'r-.', 'linewidth', 2)
legend('\it Euler forward ', '\it Euler improved', '\it
Euler  backward',...
    '\it Mid-point rule','\it Analytical Solution', 'location',
'SouthEast')
xlim([0, 5])
hold off
```

We also obtain numerical values from these simulations by combining all data via the next script included in our existing script (Example3.m) as an additional cell, as shown here:

```
%% Combine all data of simulations from four methods
DATA = [t; y1; y2; y3; y4; y5; Solution];
fid = fopen('Example1_Out.dat','w');
fprintf(fid,'%2.2f  %4.5f %6.5f %6.5f %6.5f %6.5f %6.5f\n',DATA);
fclose(fid);
type Example1_Out.dat  % View output data file
```

Time	Euler's Forward Method	Euler's Improved Method	Euler's Backward Method	Midpoint Rule	Euler's Backward (Another Way)	Analytical (Exact) Solution
0.00	1.00000	1.00000	1.00000	1.00000	1.00000	1.00000
0.05	1.10000	1.09378	1.08756	1.08869	1.09377	1.09397
0.10	1.18756	1.17612	1.16517	1.16721	1.17609	1.17674
0.15	1.26405	1.24853	1.23381	1.23659	1.24849	1.24948
0.20	1.33068	1.31204	1.29440	1.29775	1.31199	1.31325
0.25	1.38855	1.36760	1.34775	1.35154	1.36754	1.36900

(continued)

Time	Euler's Forward Method	Euler's Improved Method	Euler's Backward Method	Midpoint Rule	Euler's Backward (Another Way)	Analytical (Exact) Solution
0.30	1.43863	1.41605	1.39460	1.39871	1.41599	1.41760
0.35	1.48181	1.45816	1.43562	1.43995	1.45810	1.45981
0.40	1.51886	1.49461	1.47141	1.47588	1.49454	1.49633
0.45	1.55049	1.52603	1.50251	1.50705	1.52596	1.52777
0.50	1.57732	1.55296	1.52942	1.53398	1.55288	1.55471
0.55	1.59992	1.57590	1.55259	1.55712	1.57583	1.57764
0.60	1.61877	1.59531	1.57241	1.57687	1.59523	1.59702
0.65	1.63434	1.61158	1.58925	1.59361	1.61150	1.61325
0.70	1.64701	1.62507	1.60344	1.60767	1.62500	1.62669
0.75	1.65713	1.63612	1.61526	1.61936	1.63604	1.63767
0.80	1.66504	1.64499	1.62500	1.62893	1.64492	1.64648
0.85	1.67100	1.65197	1.63287	1.63664	1.65189	1.65339
0.90	1.67527	1.65727	1.63910	1.64270	1.65720	1.65862
0.95	1.67807	1.66111	1.64389	1.64730	1.66104	1.66239
1.00	1.67960	1.66368	1.64740	1.65063	1.66360	1.66488
1.95	1.61462	1.61168	1.60815	1.60889	1.61164	1.61191
2.00	1.61027	1.60766	1.60447	1.60515	1.60762	1.60786
...						
2.95	1.54814	1.54824	1.54816	1.54819	1.54823	1.54823
3.00	1.54594	1.54608	1.54605	1.54607	1.54607	1.54607
...						
4.95	1.50687	1.50702	1.50718	1.50715	1.50702	1.50701
5.00	1.50654	1.50668	1.50683	1.50680	1.50668	1.50667
...						
10.00	1.50004	1.50005	1.50005	1.50005	1.50005	1.50005

From the numerical simulations that are tabulated (written in the output file IVP_ODE.txt) and plotted in Figure 2-2, it is clear that Euler's improved (Heun's method) and backward methods converge much faster and more closely with the analytical solution than the other two methods (forward and midpoint rules). It is also worth mentioning that the accuracy of these methods depends much on a step size and a problem type, such as stiff or nonstiff ODEs. Besides, Euler's improved (Heun's method) and backward methods are costlier in terms of computation time than the other two methods.

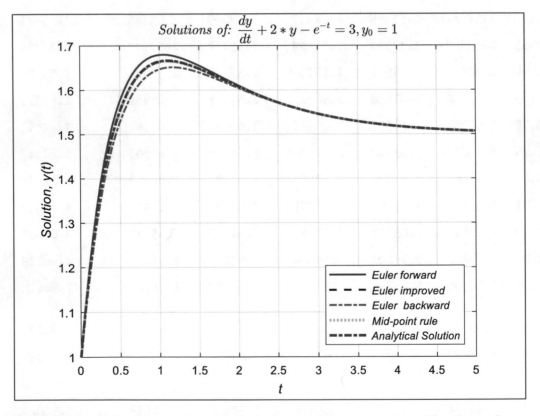

Figure 2-2. *Simulation results of Euler's forward, improved, and backward methods and the midpoint rule method versus the analytical solution of the problem*

Ralston Method

The Ralston method is like Euler's improved method (Heun's method) with small differences in coefficients. It is expressed in the form of the expressions in Equation (2-5).

$$
\begin{cases}
y_{i+1} = y_i + \left(\dfrac{k_1}{3} + \dfrac{2k_2}{3} \right) h \\[2mm]
k_1 = f(t_i, y_i) \\[2mm]
k_2 = f\left(t_i + \dfrac{3h}{4}, y_i + \dfrac{3hk_1}{4} \right)
\end{cases}
\qquad (2\text{-}5)
$$

From the expression, it is clear that this is an explicit method and can be implemented like the improved Euler method algorithm. We add one more additional cell to our existing script (Example3.m) as part 6, shown here:

```
%% Ralston Method
% Part 6
k1=zeros(1,steps);        % Memory allocation
k2=zeros(1,steps);        % Memory allocation
y6=[y0, zeros(1,steps)];  % Memory allocation
for jj=1:steps
    k1(jj) = 3 + exp(-t(jj))-2*y6(jj);
    k2(jj) = 3 + exp(-(t(jj)+3*h/4))-2*(y6(jj)+3*h*k1(jj)/4);
    y6(jj+1)=y6(jj)+(k1(jj)+2*k2(jj))*h/3;
end
figure
plot(t, y6, 'b-',t, y1, 'r--','linewidth', 2)
legend('\it Ralston method','\it Euler forward method'), grid on
title('\it Solutions of: $$ \frac{dy}{dt}+2*y-e^{-t}=3, y_0=1 $$',
'interpreter', 'latex')
xlabel('\it t'), ylabel('\it Solution, y(t)'), shg
```

We plot the results from Ralston's method against the forward Euler method and obtain the plot shown in Figure 2-3.

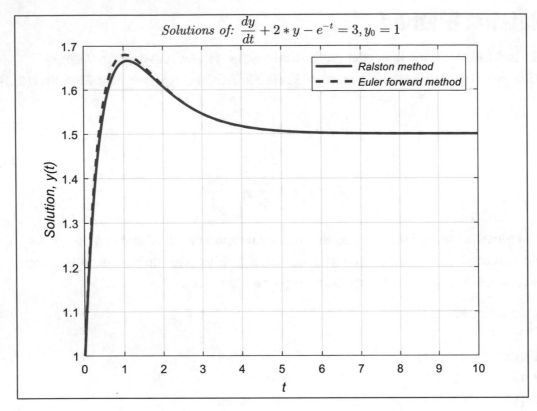

Figure 2-3. *Solutions from Ralston's method compared with the ones from Euler's forward method*

From the plot in Figure 2-3, we can see that the Ralston method, which is a two-step method, performs better than the forward Euler method that is a relatively crude method but more efficient in terms of computation time.

Runge-Kutta Method

The Runge-Kutta method can be implemented in different orders such as the 1st-order Runge-Kutta method (equivalent to the Euler method), the 2nd-order Runge-Kutta method (equivalent to Heun's method), and the classical 4th- and 5th-order Runge-Kutta methods. These methods can be also implemented in a similar algorithm, as shown with Euler's method algorithm, by introducing a few changes in step 6. So, in the Runge-

Kutta method, the following changes in step 6 of the forward Euler method algorithm are introduced by computing the function values from Equation (2-6).

$$y_{i+1} = y_i + \frac{h(k_1 + 2k_2 + 2k_3 + k_4)}{6} \tag{2-6}$$

The terms k_1, k_2, k_3, k_4 are computed from the following:

$$k_1 = f(t_i, y_i)$$

$$k_2 = f\left(t_i + \frac{h}{2}, y_i + \frac{k_1 h}{2}\right)$$

$$k_3 = f\left(t_i + \frac{h}{2}, y_i + \frac{k_2 h}{2}\right)$$

$$k_4 = f\left(t_i + \frac{h}{2}, y_i + k_3 h\right)$$

for $i = 0, 1, 2, ..., N - 1$. The sum $\dfrac{(k_1 + 2k_2 + 2k_3 + k_4)}{6}$ can be interpreted as an average slope.

Example 4

Here's the example problem: $\dot{y} + 2y - e^{-t} = 3$, $y(0) = 1$. Let's compute the numerical solution for the problem of $t = 0...10$ with a time step of $h = 0.05$ and rewrite the given problem equation as $\dot{y} = 3 - 2y + e^{-t}$. This can be also expressed by $f(t, y) = 3 - 2y + e^{-t}$ for $t = 0...10$.

For $t_0 = 0$,

$$y_0 = 1$$

For $t_1 = 0.05$,

$$y_0 = 1, \ k_1 = f(t_0, y_0) = 3 - 2 * 1 + e^0 = 2$$

$$k_2 = f\left(t_0 + \frac{h}{2}, y_0 + \frac{k_1 h}{2}\right) = 3 - 2*\left(1 + \frac{2 * 0.05}{2}\right) + e^{-\left(0 + \frac{0.05}{2}\right)} = 1.8753$$

$$k_3 = f\left(t_0 + \frac{h}{2}, y_0 + \frac{k_2 h}{2}\right) = 3 - 2*\left(1 + \frac{1.7620 * 0.05}{2}\right) + e^{-\left(0 + \frac{0.05}{2}\right)} = 1.8815$$

$$k_4 = f\left(t_0 + \frac{h}{2}, y_0 + k_3 h\right) = 3 - 2*(1 + 0.0938*0.05) + e^{-\left(0 + \frac{0.05}{2}\right)} = 1.7631$$

$$y_1 = y_0 + \frac{h(k_1 + 2k_2 + 2k_3 + k_4)}{6} = 1 + \frac{0.05(2 + 2*0.0985 + 2*0.0938 + 0.0897)}{6} = 1.094$$

For $t_2 = 0.1$,

$$y_1 = 1.094, \ k_1 = f(t_1, y_1) = 3 - 2*1.094 + e^{-0.1} = 1.7633$$

$$k_2 = f\left(t_1 + \frac{h}{2}, y_1 + \frac{k_1 h}{2}\right) = 3 - 2*\left(1.094 + \frac{1.7633*0.05}{2}\right) + e^{-\left(0.05 + \frac{0.05}{2}\right)} = 1.6516$$

$$k_3 = f\left(t_1 + \frac{h}{2}, y_1 + \frac{k_2 h}{2}\right) = 3 - 2*\left(1 + \frac{1.6516*0.05}{2}\right) + e^{-\left(0.05 + \frac{0.05}{2}\right)} = 1.6572$$

$$k_4 = f\left(t_1 + \frac{h}{2}, y_1 + k_3 h\right) = 3 - 2*(1 + 0.0938*0.05) + e^{-\left(0.05 + \frac{0.05}{2}\right)} = 1.5512$$

$$y_2 = y_1 + \frac{h(k_1 + 2k_2 + 2k_3 + k_4)}{6} = 1 + \frac{0.05(1.7633 + 2*1.6516 + 2*1.6572 + 1.5512)}{6} = 1.1767$$

...

Now, we implement all of the previously defined six-step procedures of the classical 4th-order Runge-Kutta method in our existing script (Example1.m) as an additional cell.

```
%% 4th-Order RUNGE-KUTTA Method. Example4.m
% Part 7
% EXAMPLE. y'+2*y-exp(-t)=3 with ICs: y(0)=1. t=0...10.
%{
Step 6 changed from Euler's Method
    k_n1=f(t(n), y(n))
    k_n2=f(t(n)+h/2, yn+k_n1*h/2)
    k_n3=f(t(n)+h/2, yn+k_n2*h/2)
    k_n4=f(t(n)+h, yn+k_n3*h)
    y(n+1)=y(n)+h*(k_n1+2*k_n2+2*k_n3+k_n4)/6;
%}
```

```
k1=zeros(1,steps);              % Memory allocation
k2=k1;                          % Memory allocation
k3=k1;                          % Memory allocation
k4=k1;                          % Memory allocation
y7=[y0, zeros(1,steps)];        % Memory allocation
for ii=1:steps
    k1(ii)=3+exp(-t(ii))-2*y7(ii);
    k2(ii)=3+exp(-(t(ii)+h/2))-2*(y7(ii)+k1(ii)*h/2);
    k3(ii)=3+exp(-(t(ii)+h/2))-2*(y7(ii)+k2(ii)*h/2);
    k4(ii)=3+exp(-(t(ii)+h))-2*(y7(ii)+k3(ii)*h);
    y7(ii+1)=y7(ii)+h*(k1(ii)+2*k2(ii)+2*k3(ii)+k4(ii))/6;
end
figure
plot(t, y7, 'k-.', 'linewidth', 1.5), grid on
hold on
plot(t, Solution, 'r-.', 'linewidth', 1.5)
legend('\it Runge-Kutta Method','\it Analytic solution')
title('\it Numerical Solutions of: $$ \frac{dy}{dt}+2*y-e^{-t}=3, y_0 = 1
$$', 'interpreter', 'latex')
xlabel('\it t'), ylabel('\it Solution, y(t)')
xlim([0, 5]), shg
hold off
```

After executing part 7 of the script Example4.m, we obtain the plot shown in Figure 2-4. From the plot, we can see that the classical 4th-order Runge-Kutta method for this given first-order ODE problem performs adequately well and converges nicely with the analytical solution.

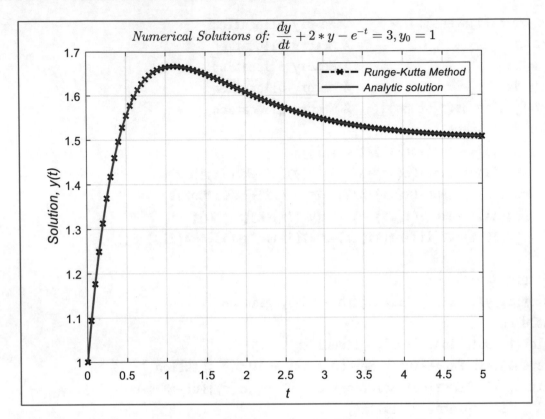

Figure 2-4. *Numerical solutions of first-order ODE found by 4th-order Runge-Kutta method versus analytical method (dsolve())*

Runge-Kutta-Gill Method

There are several modified versions of the Runge-Kutta method, and one of them is the Runge-Kutta-Gill method. This method is aimed at minimizing round-off errors and expressed by the expression in Equation (2-7).

$$y_{i+1} = y_i + \frac{1}{6}(k_1 + k_4) + \frac{1}{3}(bk_2 + dk_3) \qquad (2\text{-}7)$$

The terms k_1, k_2, k_3, k_4 in Equation (2-7) are computed from the following expressions:

$$k_1 = hf(t_i, y_i)$$

$$k_2 = hf\left(t_i + \frac{h}{2}, y_i + \frac{k_1}{2}\right)$$

$$k_3 = hf\left(t_i + \frac{h}{2}, y_i + ak_1 + bk_2\right)$$

$$k_4 = hf\left(t_i + h, y_i + cbk_2 + dk_3\right)$$

In general, the following values of terms a, b, c, d are used:

$$a = \frac{\sqrt{2}-1}{2}, \ b = \frac{2-\sqrt{2}}{2}, \ c = -\frac{\sqrt{2}}{2}, \ d = 1 + \frac{\sqrt{2}}{2}$$

for $i = 0, 1, 2, ..., N-1$.

Now, we implement the previously written expressions in our script (Example1.m) as an additional cell to compute numerical solutions of $\dot{y} + 2y - e^{-t} = 3$, $y(0) = 1$, as shown here:

```
%% Runge-Kutta-Gill method
% EXAMPLE. y'+2*y-exp(-t)=3 with ICs: y(0)=1. t=0...10.
% Part 8
a=(sqrt(2)-1)/2; b=(2-sqrt(2))/2; c=-sqrt(2)/2; d=1+sqrt(2)/2;
y8=[y0, zeros(1, steps-1)];
for ii=1:steps
    k1(ii)=h*(3+exp(-t(ii))-2*y8(ii));
    k2(ii)=h*(3+exp(-(t(ii)+h/2))-2*(y8(ii)+k1(ii)*h/2));
    k3(ii)=h*(3+exp(-(t(ii)+h/2))-2*(y8(ii)+a*k1(ii)+b*k2(ii)));
    k4(ii)=h*(3+exp(-(t(ii)+h/2))-2*(y8(ii)+c*k2(ii)+d*k3(ii)));
    y8(ii+1)=y8(ii)+(k1(ii)+k4(ii))/6+(b*k2(ii)+d*k3(ii))/3;
end
DATA1 = [t; y1; y7; y8; Solution];
% Data export
fid1 = fopen('Example1_Out2.dat', 'w');
fprintf(fid1, 'time  Euler-F  Runge-K  RK-Gill Analytic \n');
fprintf(fid1,'%2.2f  %4.5f %4.5f %4.5f %4.5f\n',DATA1);
fclose(fid1);
type Example1_Out2.dat    % View the imported data on the command window
```

Now, we compare numerical simulation results of the example
$\dot{y} + 2y - e^{-t} = 3, y(0) = 1$ from the forward Euler method, the classical Runge-Kutta
fourth-order method, the Runge-Kutta-Gill methods, and method analytical solution.

Time	Euler's Forward	Runge-Kutta Fourth Order	Runge-Kutta-Gill Fourth Order	Analytical Solution
0.00000	1.00000	1.00000	1.00000	1.00000
0.05000	1.10000	1.09397	1.09462	1.09397
0.10000	1.18756	1.17674	1.17791	1.17674
0.15000	1.26405	1.24948	1.25106	1.24948
0.20000	1.33068	1.31325	1.31516	1.31325
0.25000	1.38855	1.36900	1.37116	1.36900
0.30000	1.43863	1.41760	1.41993	1.41760
0.35000	1.48181	1.45981	1.46227	1.45981
0.40000	1.51886	1.49633	1.49887	1.49633
0.45000	1.55049	1.52777	1.53036	1.52777
0.50000	1.57732	1.55471	1.55731	1.55471
0.55000	1.59992	1.57764	1.58022	1.57764
0.60000	1.61877	1.59702	1.59956	1.59702
0.65000	1.63434	1.61325	1.61574	1.61325
0.70000	1.64701	1.62669	1.62911	1.62669
0.75000	1.65713	1.63767	1.64001	1.63767
0.80000	1.66504	1.64648	1.64874	1.64648
0.85000	1.67100	1.65339	1.65556	1.65339
0.90000	1.67527	1.65862	1.66069	1.65862
0.95000	1.67807	1.66239	1.66436	1.66239
1.00000	1.67960	1.66488	1.66675	1.66488

From these numerical simulations, we can conclude that the classical 4th-order Runge-Kutta method performs much better than the forward Euler method. Local truncation errors are as follows:

Forward Euler method: $error = \mathcal{O}(h)$

4th-order Runge-Kutta: $error = \mathcal{O}(h^5)$

For every step in the 4th-order Runge-Kutta method, there are four evaluations. Thus, when there is a need for extensive and large computations, this method will become computationally costlier than the Euler methods.

Runge-Kutta-Fehlberg Method

This is another modified version of the higher-order Runge-Kutta method that contains six coefficients instead of four. The method is formulated by the expression in Equation (2-8).

$$y_{i+1} = y_i + \left(\frac{16k_1}{135} + \frac{6656k_3}{12825} + \frac{28561k_4}{56430} - \frac{9k_5}{50} + \frac{2k_6}{55} \right) \tag{2-8}$$

The coefficient terms, such as k_1, k_2, k_3, k_4, k_5, k_6, are computed from the following expressions:

$$k_1 = hf(t_i, y_i)$$

$$k_2 = hf\left(t_i + \frac{h}{4}, y_i + \frac{k_1}{4} \right)$$

$$k_3 = hf\left(t_i + \frac{3h}{8}, y_i + \frac{3k_1}{32} + \frac{9k_2}{32} \right)$$

$$k_4 = hf\left(t_i + \frac{12h}{13}, y_i + \frac{1932k_1}{2197} - \frac{7200k_2}{2197} + \frac{7296k_3}{2197} \right)$$

$$k_5 = hf\left(t_i + h, y_i + \frac{439k_1}{216} - 8k_2 + \frac{3680k_3}{513} - \frac{845k_3}{4104} \right)$$

$$k_6 = hf\left(t_i + \frac{h}{2}, y_i - \frac{8k_1}{27} + 2k_2 - \frac{3544k_3}{2565} + \frac{1589k_3}{4104} - \frac{11k_4}{40} \right)$$

This has the following local error: $error = \mathcal{O}(h^6)$.

This has the following global error: $error = \mathcal{O}(h^5)$.

A formula for error estimation from the Runge-Kutta-Fehlberg method is as follows:

$$e = \frac{k_1}{360} - \frac{128k_3}{4275} - \frac{2197k_4}{7524} + \frac{k_5}{50} + \frac{2k_6}{55} \tag{2-9}$$

So now, we implement all of these procedures given in the expressions shown in Equations (2-8) and (2-9) to solve the previous example (Example4.m) and to compute error values in every computation step, as shown here:

```
%% Runge-Kutta-Fehlberg Method implemented in Example4.m
% EXAMPLE. y'+2*y-exp(-t)=3 with ICs: y(0)=1. t=0...10.
% Part 9
F=@(t,y)(3-2*y+exp(-t));
y9=[y0, ones(1,steps-1)];  % Memory allocation
e=zeros(1, steps);         % Memory allocation
for n=1:steps
    k_n1=h*F(t(n), y9(n));
    k_n2=h*F(t(n)+h/4, y9(n)+k_n1/4);
    k_n3=h*F(t(n)+3*h/8, y9(n)+3*k_n1/32+9*k_n2/32);
    k_n4=h*F(t(n)+12*h/13, y9(n)+1932*k_n1/2197-...
        7200*k_n2/2197+7296*k_n3/2197);
    k_n5=h*F(t(n)+h, y9(n)+439*k_n1/216-8*k_n2+...
        3680*k_n3/513-845*k_n4/4104);
    k_n6=h*F(t(n)+h/2,y9(n)-8*k_n1/27+2*k_n2-...
        3544*k_n3/2565+1859*k_n4/4104-11*k_n5/40);
    y9(n+1)=y9(n)+(16*k_n1/135+6656*k_n3/12825+...
        28561*k_n4/56430-9*k_n5/50+2*k_n6/55);
    e(n)=k_n1/360-128*k_n3/4275-2197*k_n4/7524+k_n5/50+2*k_n6/55;
end
subplot(211)
% Analytical Solution of the problem using dsolve
plot(t, y9, 'ko', t, Solution, 'r-', 'linewidth', 1.5)
legend('\it Runge-Kutta-Fehlberg Method','\it Analytic solution',
'location', 'southeast')
title('\it Numerical Solutions of: $$ \frac{dy}{dt}+2*y-e^{-t}=3, y_0 = 1
$$', 'interpreter', 'latex')
xlabel('\it t'), ylabel('\it Solution, y(t)')
```

```
xlim([0, 5]), grid on
subplot(212)
plot(e, 'r-', 'linewidth', 1.5), grid on
title('\it Global error: \epsilon')
xlabel('\it t'), ylabel('\it \epsilon(t)')
shg
```

From the simulation results (shown in Figure 2-5) of the script Example4.m using the Runge-Kutta-Fehlberg method to solve the given first-order differential equation, we see that after about 70 computation steps, the global error has become almost zero. In addition, it must be noted that this is a nonstiff ODE problem.

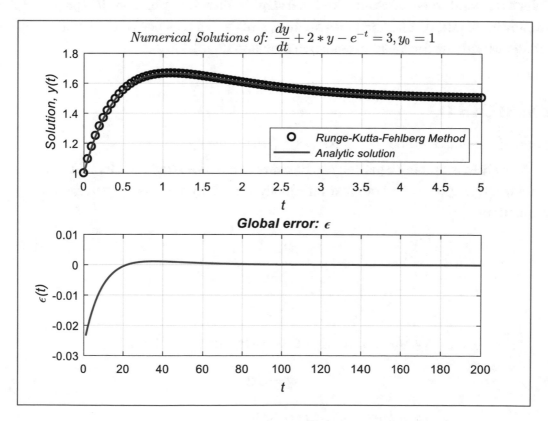

Figure 2-5. *Numerical simulation of the problem: $\dot{y} + 2y - e^{-t} = 3$, $y(0) = 1$*

Adams-Bashforth Method

Adams-Bashforth is a two-step method. This method is expressed in different literature sources in several different ways, one of which is in terms of the iteration indexes given here:

$$y_{i+2} = y_{i+1} + \frac{3}{2}hf(t_{i+1}, y_{i+1}) - \frac{1}{2}hf(t_i, y_i) \tag{2-10}$$

The expression in Equation (2-10) requires two values, y_{i+1} and y_i, to compute the value of y_{i+2}. For the IVP, y_0 (can be understood as y_1) is always given. That is not sufficient for this computation process, though. Hence, in this case, we can employ Euler's (forward) method as an additional step, for instance. Moreover, it is possible to employ in this step the Taylor series method, and we can advance with other iteration processes with the formulation shown in Equation (2-10).

Example 5

Here's the example problem: $\dot{y} + 2ty^2 = 0$, $y_0 = 0.5$

The given example's equation can be expressed by $\dot{y} = -2ty^2$, and its solution function is $f(t, y) = -2ty^2$. Let's take an equally spaced time step of $h = 0.01$ for our simulations.

$$t_0 = 0$$

$$y_0 = 0.5;$$

$$t_1 = 0.01$$

$$y_1 = y_0 + hf(t_0, y_0) = 0.5 + 0.01 * (-2 * 0 * 0.5^2) = 0.5;$$

$$t_2 = 0.02$$

$$y_2 = y_1 + \frac{3}{2}hf(t_1, y_1) - \frac{1}{2}hf(t_0, y_0)$$

$$= 0.5 + \frac{3}{2} * 0.01 * (-2 * 0.01 * 0.5^2) - \frac{1}{2} * 0.01 * (-2 * 0 * 0.5^2)$$

$$= 0.499925;$$

$$t_3 = 0.03$$

$$y_3 = y_2 + \frac{3}{2}hf(t_2, y_2) - \frac{1}{2}hf(t_1, y_1)$$

$$= 0.4999 + \frac{3}{2} * 0.01 * (-2 * 0.02 * 0.5^2) - \frac{1}{2} * 0.01 * (-2 * 0.01 * 0.5^2)$$

$$= 0.499875.$$

$$\cdots$$

The Adams-Bashforth method is implemented in the Example2.m script by using the same algorithm steps employed in the previous methods (e.g., the Euler method implementation algorithm) except for step 6, which is explained in the formulation of Equation (2-10). The solution is shown in Figure 2-6.

```
% Example5.m
%% ADAMS or ADAMS-Bashforth 2-step Method
% EXAMPLE 2. dy/dt=-2*t*(y^2); with ICs: y(0)=0.5
% Part 1.
Fn=0.5;                 % ICs: u0 at t0
tend=10;                % max. time limit
h=.01;                  % time step size
t=0:h:tend;             % time space
steps=length(t);        % number of steps
% NB: Dy = f(y,t)=-2*y^2*t defined via Function Handle
Fcn=@(Fn,t)(-2*(Fn.^2).*t);
% Memory allocation:
Fn=[Fn(1), zeros(1,steps-1)];
fn=[Fcn(Fn(1), t(1)),zeros(1,steps-1)];
for ii=1:steps-2
% EULER's method starts from here
Fn(ii+1)=Fn(ii)+h*Fcn(Fn(ii),t(ii));
% Adams-Bashforth method starts from here
fn(ii+2)=Fcn(Fn(ii+1),t(ii+1));
Fn(ii+2)=Fn(ii+1)+(h/2)*(3*fn(ii+1)-fn(ii));
end
figure
plot(t, Fn, 'r:+', 'linewidth', 2), hold on
syms u(T) T
```

```
Du = diff(u, T);
Equation=Du==-2*u^2*T;
IC=u(0)==0.5;
u=dsolve(Equation, IC);          % Analytical Solution:
U=vectorize(u);
T=t;
U=eval(U);
A =plot(T, U);
A.Color=[.2 .5 .5];
A.LineWidth=A.LineWidth+1.5;
title('Solution of: $$ \frac{dy}{dt}+2*y^2*t=0,   y_0=0.5 $$',
'interpreter', 'latex')
legend('\it Adams-Bashforth Method 2-step solver', '\it Analytical solution')
xlabel('\it t')
ylabel('\it Solution, y(t)')
grid on; shg
```

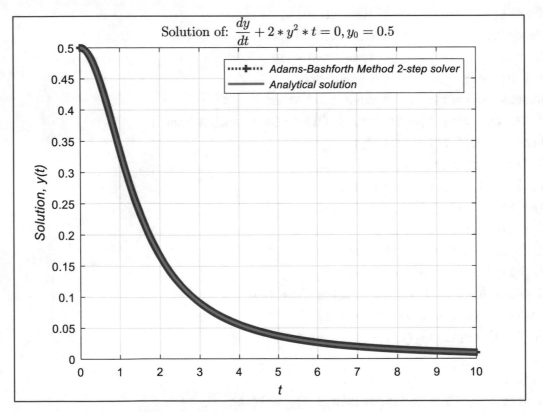

Figure 2-6. *Two-step Adams-Bashforth method*

This is the three-step Adams-Bashforth method [5]:

$$y_{i+3} = y_{i+2} + h\left(\frac{23}{12}f\left(t_{i+2},y_{i+2}\right) - \frac{4}{3}f\left(t_{i+1},y_{i+1}\right) + \frac{5}{12}f\left(t_i,y_i\right)\right) \qquad (2\text{-}11)$$

This is the four-step Adams-Bashforth method:

$$y_{i+4} = y_{i+3} + h\left(\frac{55}{24}f\left(t_{i+3},y_{i+3}\right) - \frac{59}{24}f\left(t_{i+2},y_{i+2}\right) + \frac{37}{24}f\left(t_{i+1},y_{i+1}\right) - \frac{3}{8}f\left(t_i,y_i\right)\right) \qquad (2\text{-}12)$$

This is the five-step Adams-Bashforth method:

$$y_{i+5} = y_{i+4} + h\left(\begin{array}{l}\dfrac{1901}{720}f\left(t_{i+4},y_{i+4}\right) - \dfrac{1387}{360}f\left(t_{i+3},y_{i+3}\right) + \dfrac{109}{30}f\left(t_{i+2},y_{i+2}\right) \\ -\dfrac{637}{360}f\left(t_{i+1},y_{i+1}\right) + \dfrac{251}{720}f\left(t_i,y_i\right)\end{array}\right) \qquad (2\text{-}13)$$

The implementation of these three-, four-, and five-step Adams-Bashforth methods expressed in Equations (2-11), (2-12), and (2-13) is like the two-step Adams-Bashforth method. First, the value of y_i is found with the forward Euler method, and the following values, such as the second and third values, y_{i+1} and y_{i+2}, are computed from the two- and three-step Adams-Bashforth methods. The other values, y_{i+3} and y_{i+4}, respectively, are computed from the four- and five-step Adams-Bashforth methods.

Here are the other parts of the script (Example5.m) written based on the three-, four-, and five-step Adams-Bashforth methods:

```
%% ADAMS-Bashforth 3-step Method
% EXAMPLE 5. dy/dt=-2*t*(y^2); with ICs: y(0)=0.5
% Part 2.
% NB: Dy = f(y,t)=-2*y^2*t defined via Function Handle
Fcn=@(Fn,t)(-2*(Fn.^2).*t);
% Memory allocation with ZERO matrices
Fn=[Fn(1), zeros(1,steps-1)];
fn=[Fcn(Fn(1),t(1)), zeros(1,steps-1)];
for ii=1:steps-3
% EULER's method is used here
Fn(ii+1)=Fn(ii)+h*Fcn(Fn(ii),t(ii));
% 2-step ADAMS-Bashforth method is used here
```

```
Fn(ii+2)=Fn(ii+1)+h*(1.5*Fcn(Fn(ii+1),t(ii+1))-0.5*Fcn(Fn(ii),t(ii)));
fn(ii+2)=Fcn(Fn(ii+2),t(ii+2));
% ADAMS-Bashforth method: 3-step starts from here
Fn(ii+3)=Fn(ii+2)+h*((23/12)*fn(ii+2)-(4/3)*fn(ii+1)+(5/12)*fn(ii));
end
plot(t, Fn, 'g:x', 'linewidth', 2), grid on
% ADAMS or ADAMS-Bashforth 4-step Method
Fcn=@(Fn,t)(-2*(Fn.^2).*t);
% Memory allocation with ZERO matrices
Fn=[Fn(1), zeros(1,steps-1)];
fn=[Fcn(Fn(1),t(1)), zeros(1,steps-1)];
for ii=1:steps-4
% EULER's method is used here
Fn(ii+1)=Fn(ii)+h*Fcn(Fn(ii),t(ii));
% ADAMS-Bashforth method: 2-step
Fn(ii+2)=Fn(ii+1)+h*(1.5*Fcn(Fn(ii+1),t(ii+1))-0.5*Fcn(Fn(ii),t(ii)));
% ADAMS-Bashforth method: 3-step
Fn(ii+3)=Fn(ii+2)+h*((23/12)*fn(ii+2)-(4/3)*fn(ii+1)+(5/12)*fn(ii));
fn(ii+3)=Fcn(Fn(ii+3),t(ii+3));
% ADAMS-Bashforth method: 4-step starts from here
Fn(ii+4)=Fn(ii+3)+h*((55/24)*fn(ii+3)-...
    (59/24)*fn(ii+2)+(37/24)*fn(ii+1)-(3/8)*fn(ii));
end
plot(t, Fn, 'm--', 'linewidth', 2), grid on
%   ADAMS-Bashforth 5-step Method
Fcn=@(Fn,t)(-2*(Fn.^2).*t);
% Memory allocation with ZERO matrices
Fn=[Fn(1), zeros(1,steps-1)];
fn=[Fcn(Fn(1), t(1)), zeros(1,steps-1)];
for ii=1:steps-5
% EULER's method is used here
Fn(ii+1)=Fn(ii)+h*Fcn(Fn(ii),t(ii));
% ADAMS-Bashforth method: 2-step
Fn(ii+2)=Fn(ii+1)+h*(1.5*Fcn(Fn(ii+1),t(ii+1))-0.5*Fcn(Fn(ii),t(ii)));
% ADAMS-Bashforth method: 3-step
```

```
Fn(ii+3)=Fn(ii+2)+h*((23/12)*fn(ii+2)-(4/3)*fn(ii+1)+(5/12)*fn(ii));
% ADAMS-Bashforth method: 4-step
Fn(ii+4)=Fn(ii+3)+h*((55/24)*fn(ii+3)-(59/24)*fn(ii+2)+...
    (37/24)*fn(ii+1)-(3/8)*fn(ii));
fn(ii+5)=Fcn(Fn(ii+4),t(ii+4));
% ADAMS-Bashforth method: 5-step starts from here
Fn(ii+5)=Fn(ii+4)+h*((1901/720)*fn(ii+4)-...
    (1387/360)*fn(ii+3)+(109/30)*fn(ii+2)-...
    (637/360)*fn(ii+1)+(251/720)*fn(ii));
end
plot(t, Fn, 'k-', 'linewidth', 2), grid on
title('Solution of: $$ \frac{dy}{dt}+2*y^2*t=0,  y_0=0.5 $$',
'interpreter', 'latex')
legend('AB Method 2-step', 'Analytical solution', 'AB Method 3-step',...
'AB Method 4-step', 'AB Method 5-step'), shg
```

By executing the script, we obtain the plot shown in Figure 2-7 displaying the computation results of the Adams-Bashforth three-step, four-step, and five-step methods versus the numerical results from the analytical solution of the problem computed with dsolve().

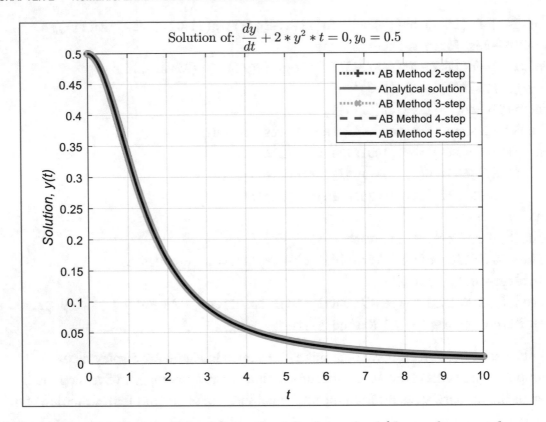

Figure 2-7. *Simulation of the Adams-Bashforth method (three-, four-, and five-step) versus analytical solution with dsolve()*

From the plots (Figures 2-6 and 2-7) of the simulations, the results of the Adams-Bashforth methods with two, three, four, and five steps, for the given first-order ODE, are converged with the analytic solution so well that an accuracy level of higher orders of the Adams-Bashforth method is not visible. In fact, the accuracy of the Adams-Bashforth method depends on a given ODE type (stiff or nonstiff, linear or nonlinear), and that can be improved considerably with the cost of computation time.

Milne Method

The Milne method is another kind of explicit and multistep ODE solver that is formulated with the following expression:

$$y_{i+3} = y_i + \frac{4h}{3}\left(2f\left(t_{i+2}, y_{i+2}\right) - f\left(t_{i+1}, y_{i+1}\right) + 2f\left(t_i, y_i\right)\right) \tag{2-14}$$

In implementing the Milne method in Equation (2-14), like with Adams-Bashforth, we should employ other methods to define the values of y_i, y_{i+1}, y_{i+2}. Here's an example:

Euler's forward: $y_{i+1} = y_i + hf(t_i, y_i)$

Adams-Bashforth: $y_{i+2} = y_{i+1} + \dfrac{3}{2}hf(t_{i+1}, y_{i+1}) - \dfrac{1}{2}hf(t_i, y_i)$

Example 6

Here's the example problem: $\dfrac{dy}{dt} = 1 - t\sqrt[3]{y}$, $y(0) = 1$.

Here is the Milne method implementation in the Example3.m script that solves the given problem:

```
% Example6.m
%% MILNE's Method
% First term of f(y,t) found by Euler's method
% Second term of f(y,t) by 2 step Adams-Bashforth
y0=1;                      % ICs: u0 at t0
tend=5;                    % max. time limit
h=.01;                     % time step size
t=0:h:tend;                % time space
steps=length(t);           % # of steps
fun = @(t, y)(1-t*y^(1/3));
% Memory allocation
y=[y0, zeros(1,steps-1)];
for ii=1:steps-3
% EULER's method used here
y(ii+1)=y(ii)+h*fun(t(ii), y(ii));
% Adams-Bashforth 2-step method used here
y(ii+2)=y(ii+1)+(3/2)*h*fun(t(ii+1),y(ii+1))-(1/2)*h*fun(t(ii),y(ii));
% MILNE's method starts from here
y(ii+3)=y(ii)+(h/3)*(8*fun(t(ii+2),y(ii+2))-4*fun(t(ii+1),...
    y(ii+1))+8*fun(t(ii),y(ii)));
end
```

```
plot(t, y, 'b-', 'linewidth', 1.5), grid on
title('\it Milne method solution of: $$ \frac{dy}{dt}=1-t*y^\frac{1}{3}
$$', 'interpreter', 'latex')
xlabel '\it t'; ylabel '\it Solution, y(t)'; shg
```

Numerical solutions of the exercise from the previous script (Example3.m) are shown in Figure 2-8.

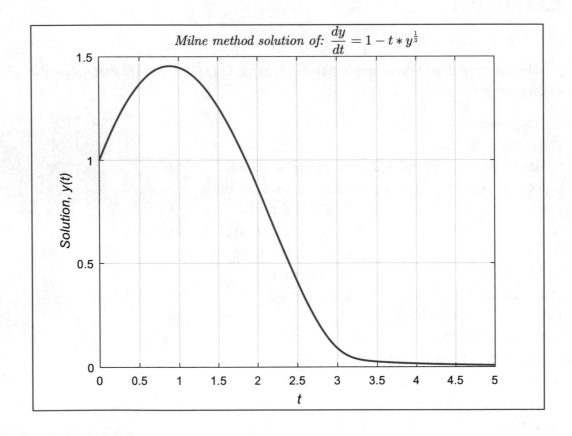

Figure 2-8. *Numerical solutions of $\frac{dy}{dt} = 1 - t\sqrt[3]{y}$, y(0) = 1 computed with the Milne method*

Taylor Series Method

The Taylor series method is one of the simplest techniques for solving first-order ODEs. This method can be expressed with the following formulation:

$$y_{i+1} = y_i + h\dot{y}_i + \frac{h^2}{2}\ddot{y}_i + \frac{h^3}{3}\dddot{y}_i + \frac{h^4}{4}y_i^{iv} + \dots \tag{2-15}$$

where h is the step size equal to $t_{i+1} - t_i$. This technique requires a few derivatives of the given problem equation at t_i depending on the order of the terms truncated. From the formula shown in Equation (2-15), it is clear that a first-order Taylor series expansion is equal to the forward Euler equation. By increasing the order of the Taylor series, we can attain more accurate numerical results. At the same time, we should note that the higher the order of the Taylor series, the higher the computation costs will be. The derivatives of $\dot{y}_i, \ddot{y}_i, \dots$ in Equation (2-15) are found by differentiating the given problem equation $\dot{y} = f(t, y)$. Since from the previously written formulation we know that the function f depends on two variables, y and t, and that y is a function of t, while differentiating the function with respect to t, we have to be careful. Now, look at a specific example to explain the Taylor series application more explicitly.

Example 7

Here's the example problem: $\dot{y} = 3 + e^{-t} - 2y$, $y(0) = 1$. Let's compute numerical solutions of the given first-order ODE using the Taylor series method with two terms. First, we compute the second and third derivatives of the given first-order ODE, as shown here:

$$\ddot{y} = -e^{-t} - 2\dot{y} \quad \text{and} \quad \dddot{y} = e^{-t} - 2\ddot{y} = e^{-t} - 2\left(-e^{-t} - 2\dot{y}\right).$$

By taking the found $\ddot{y}$ and $\dddot{y}$ expressions, we write formulations to compute y_{i+1} for the second- and third-order Taylor series expansion. A second-order Taylor series expansion of the problem is as follows:

$$y_{i+1} = y_i + h\left(3 + e^{-t_i} - 2y_i\right) + \frac{h^2}{2}\left(-e^{-t_i} - 2\left(3 + e^{-t_i} - 2y_i\right)\right).$$

A third-order Taylor series expansion of the problem is as follows:

$$y_{i+1} = y_i + h\left(3 + e^{-t_i} - 2y_i\right) + \frac{h^2}{2}\left(-e^{-t_i} - 2\left(3 + e^{-t_i} - 2y_i\right)\right)$$
$$+ \frac{h^3}{6}\left(e^{-t_i} - 2\left(-e^{-t_i} - 2\left(3 + e^{-t_i} - 2y_i\right)\right)\right).$$

Now, we implement all that in the script called `Example4.m` and obtain numerical solutions by executing it.

```
% Taylor series method
% Example7.m solves dy/dt=3+exp(-t)-2*y; y(0)=1
t0=0;
tend=10;
h=0.05;
t=t0:h:tend;
steps=length(t)-1;
y0=1;
y2=[y0, zeros(1, steps-1)];    % Memory allocation for 2nd-order
y3=[y0, zeros(1, steps-1)];    % Memory allocation for 3rd-order
for ii=1:steps
    % 2nd-order Taylor series:
    y2(ii+1)=y2(ii)+h*(3+exp(-t(ii))-2*y2(ii))+...
    (h^2/2)*(-exp(-t(ii))-2*(3+exp(-t(ii))-2*y2(ii)));
    % 3rd-order Taylor series:
    y3(ii+1)=y3(ii)+h*(3+exp(-t(ii))-2*y3(ii))+...
    (h^2/2)*(-exp(-t(ii))-2*(3+exp(-t(ii))-2*y3(ii)))+...
    (h^3/6)*(exp(-t(ii))-2*(-exp(-t(ii))-2*(3+exp(-t(ii))-2*y3(ii))));
end
plot(t, y2, 'bx--', t,y3, 'ko:', 'linewidth', 1.5), grid minor
xlabel '\it t', ylabel('\it Solution, y(t)')
syms u(x)  x
Du=diff(u, x);
Equation=Du==3+exp(-x)-2*u;
IC = u(0)==1;
u_sol=dsolve(Equation, IC);
U_sol=vectorize(u_sol); x=t;
Solution=eval(U_sol);
```

```
hold on
plot(x, Solution, 'r-', 'linewidth', 1.5)
title('\it Solutions of: $$ \frac{dy}{dt}+2*y-e^{-t}=3, y_0=1 $$',
'Interpreter', 'latex')
legend('\it 2nd order Taylor series','\it 3rd order Taylor series', '\it
Analytical solution'), hold off
```

By executing the previous script, we obtain the plot shown in Figure 2-9, which displays perfect convergences of second- and third-order Taylor series solutions with its analytical solution.

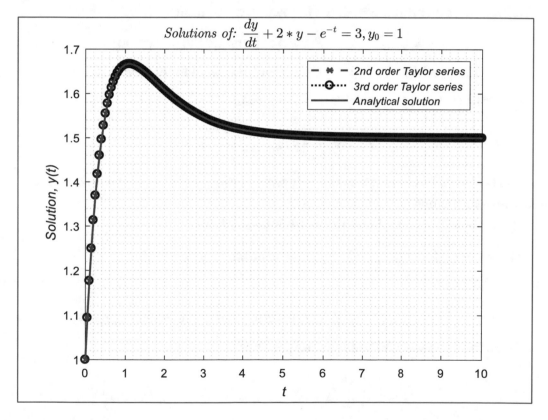

Figure 2-9. *Numerical solutions of $\dot{y} = 3 + e^{-t} - 2y$ with Taylor series versus analytical solution*

From Figure 2-9, we can see that any divergence of the solutions found via second-order and third-order Taylor series is hardly visible, and there is a good convergence of numerical solutions found with the Taylor series method and analytical solution. Thus, it can be concluded that, for the given problem, the accuracy of the second-order Taylor series is sufficient.

Adams-Moulton Method

The Adams-Moulton methods, unlike the Runge-Kutta and Adams-Bashforth methods, are an implicit type of predictor-corrector ODE solvers. A single-step Adams-Moulton method [2] is expressed in the following simple implicit formulation:

$$y_i = y_{i-1} + hf(t_i, y_i) \tag{2-16}$$

The expression in Equation (2-16) is also called the backward Euler's method, and we have covered it in the previous discussions of the Euler methods.

In implementing this method, the following can be used:

$y_{i+1}{}^P = y_i + hf(t_i, y_i)$ Predictor (forward Euler method)

$y_{i+1} = y_i + hf(t_{i+1}, y_{i+1}{}^P)$ Corrector (backward Euler method)

Here is the two-step Adams-Moulton method:

$$y_{i+1} = y_i + \frac{h}{2}\left(f(t_{i+1}, y_{i+1}) + f(t_i, y_i)\right) \tag{2-17}$$

This is also called the *trapezoidal rule method* that can be defined with one of the following expressions:

$y_{i+1}^P = y_i + hf(t_i, y_i)$ predictor (forward Euler method)

$y_{i+1}^P = y_i + \frac{3h}{2}f(t_i, y_i) - \frac{h}{2}f(t_{i-1}, y_{i-1})$, an alternative predictor for

$y_i = y_{i-1} + hf(t_{i-1}, y_{i-1})$

$y_{i+1} = y_i + \frac{h}{2}\left(f(t_{i+1}, y_{i+1}^P) + f(t_i, y_i)\right)$, corrector used for an iteration process

Here's the three-step Adams-Moulton method:

$$y_{i+2} = y_{i+1} + h\left(\frac{5}{12}f(t_{i+2}, y_{i+2}) + \frac{2}{3}f(t_{i+1}, y_{i+1}) - \frac{1}{12}f(t_i, y_i)\right) \tag{2-18}$$

The method in Equation (2-18) will be implemented with the following additional (predictor and corrector) expressions:

$y_2^P = y_1 + hf(t_1, y_1)$ predictor

$y_2 = y_1 + \frac{h}{2}\left(f(t_2, y_2^P) + f(t_1, y_1)\right)$ corrector

$$y_{i+2}^P = y_{i+1} + \frac{3h}{2} f(t_{i+1}, y_{i+1}) - \frac{h}{2} f(t_i, y_i) \text{ predictor used for iteration}$$

$$y_{i+2} = y_{i+1} + h\left(\frac{5}{12} f(t_{i+2}, y_{i+2}^P) + \frac{2}{3} f(t_{i+1}, y_{i+1}) - \frac{1}{12} f(t_i, y_i) \right) \text{ corrector}$$

used for iteration process

Here's the four-step Adams-Moulton method:

$$y_{i+3} = y_{i+2} + h\left(\frac{3}{8} f(t_{i+3}, y_{i+3}) + \frac{19}{24} f(t_{i+2}, y_{i+2}) - \frac{5}{24} f(t_{i+1}, y_{i+1}) + \frac{1}{24} f(t_i, y_i) \right) \quad (2\text{-}19)$$

The expression in Equation (2-19) should be implemented with the following additional (predictor and corrector) expressions:

$$y_2^P = y_1 + hf(t_1, y_1) \text{ predictor}$$

$$y_2 = y_1 + \frac{h}{2}\left(f(t_2, y_2^P) + f(t_1, y_1) \right) \text{ corrector}$$

$$y_3^P = y_2 + \frac{3h}{2} f(t_2, y_2) - \frac{h}{2} f(t_1, y_1) \text{ predictor}$$

$$y_3 = y_2 + h\left(\frac{5}{12} f(t_3, y_3^P) + \frac{2}{3} f(t_2, y_2) - \frac{1}{12} f(t_1, y_1) \right) \text{ corrector}$$

$$y_{i+3}^P = y_{i+2} + h\left(\frac{23}{12} f(t_{i+2}, y_{i+2}) - \frac{4}{3} f(t_{i+1}, y_{i+1}) + \frac{5}{12} f(t_i, y_i) \right) \text{ predictor}$$

used for iteration

$$y_{i+3} = y_{i+2} + h\left(\frac{3}{8} f(t_{i+3}, y_{i+3}^P) + \frac{19}{24} f(t_{i+2}, y_{i+2}) - \frac{5}{24} f(t_{i+1}, y_{i+1}) + \frac{1}{24} f(t_i, y_i) \right)$$

corrector used for iteration

Here's the five-step Adams-Moulton method:

$$y_{i+4} = y_{i+3} + h\left(\begin{array}{l} \dfrac{251}{720} f(t_{i+4}, y_{i+4}) + \dfrac{646}{720} f(t_{i+3}, y_{i+3}) - \dfrac{264}{720} f(t_{i+2}, y_{i+2}) + \\ \dfrac{106}{720} f(t_{i+1}, y_{i+1}) - \dfrac{19}{720} f(t_i, y_i) \end{array} \right) \quad (2\text{-}20)$$

Again, the expression in Equation (2-20) will be implemented with the following predictor-corrector steps:

$$y_2^P = y_1 + hf(t_1, y_1) \text{ predictor}$$

$$y_2 = y_1 + \frac{h}{2}\left(f(t_2, y_2^P) + f(t_1, y_1)\right) \text{ corrector}$$

$$y_3^P = y_2 + \frac{3h}{2}f(t_2, y_2) - \frac{h}{2}f(t_1, y_1) \text{ predictor}$$

$$y_3 = y_2 + h\left(\frac{5}{12}f(t_3, y_3^P) + \frac{2}{3}f(t_2, y_2) - \frac{1}{12}f(t_1, y_1)\right) \text{ corrector}$$

$$y_4^P = y_3 + h\left(\frac{23}{12}f(t_3, y_3) - \frac{4}{3}f(t_2, y_2) + \frac{5}{12}f(t_1, y_1)\right) \text{ predictor}$$

$$y_4 = y_3 + h\left(\frac{3}{8}f(t_4, y_4^P) + \frac{19}{24}f(t_3, y_3) - \frac{5}{24}f(t_2, y_2) + \frac{1}{24}f(t_1, y_1)\right) \text{ corrector}$$

$$y_{i+4}^P = y_{i+3} + h\left(\frac{55}{24}f(t_{i+3}, y_{i+3}) - \frac{59}{24}f(t_{i+2}, y_{i+2}) + \frac{37}{24}f(t_{i+1}, y_{i+1}) - \frac{3}{8}f(t_i, y_i)\right) \text{ predictor}$$

$$y_{i+4} = y_{i+3} + h\left(\begin{array}{l} \frac{251}{720}f(t_{i+4}, y_{i+4}^P) + \frac{646}{720}f(t_{i+3}, y_{i+3}) - \frac{264}{720}f(t_{i+2}, y_{i+2}) + \\ \frac{106}{720}f(t_{i+1}, y_{i+1}) - \frac{19}{720}f(t_i, y_i) \end{array}\right) \text{ corrector used for}$$

the iteration process

Example 8

Solve this: $\dot{y} + 2ty^2 = 0$, $y_0 = 0.5$. Now we will implement all of the previously written expressions via a script (Example8.m) to compute numerical solutions of the given problem: $\dot{y} + 2ty^2 = 0$, $y_0 = 0.5$.

```
% Example8.m
%% ADAMS-Moulton 1, 2, 3, 4 and 5 step Methods.
clc; close all; clearvars
Fn0=0.5;            % ICs: u0 at t0
tend=10;            % end time
h=.1;               % time step size
```

```matlab
t=0:h:tend;          % time space
steps=length(t);     % number of steps
% Part 1
% ADAMS-Moulton 1-step Method
% NB: Dy = f(y,t)=-2*y^2*t defined via Function Handle
Fcn=@(Fn,t)(-2*(Fn.^2).*t);
% Memory allocation with ZERO matrices
Fn=[Fn0, zeros(1,steps-1)];
for ii=1:steps-1
% EULER's forward method is used here
Fn(ii+1)=Fn(ii)+h*Fcn(Fn(ii), t(ii));
% ADAMS-Moulton method: 2-step starts from here
Fn(ii+1)=Fn(ii)+h*Fcn(Fn(ii+1), t(ii+1));
end
plot(t, Fn, 'bs-'), grid on
hold on
%%    ADAMS-Moulton 2-step Method
% Part 2
Fcn=@(Fn,t)(-2*(Fn.^2).*t);
% Memory allocation with ZERO matrices
Fn=[Fn(1), zeros(1,steps-1)];
for ii=1:steps-1
% EULER's forward method is used here
Fn(ii+1)=Fn(ii)+h*Fcn(Fn(ii), t(ii));
% ADAMS-Moulton method: 2-step starts from here
Fn(ii+1)=Fn(ii)+(h/2)*(Fcn(Fn(ii+1), t(ii+1))+Fcn(Fn(ii),t(ii)));
end
plot(t, Fn, 'mx--'), grid on
%%  ADAMS-Moulton 3-step Method
% Part 3
% Memory allocation with ZERO matrices
Fn=[Fn0, zeros(1,steps-1)];
for ii=1:steps-2
% Predicted value by Euler's forward method
Fn(ii+1)=Fn(ii)+h*Fcn(Fn(ii),t(ii));
```

```
% Corrected value by trapezoidal rule
Fn(ii+1)=Fn(ii)+(h/2)*(Fcn(Fn(ii+1),t(ii+1))+Fcn(Fn(ii), t(ii)));
% Predicted solution:
Fn(ii+2)=Fn(ii+1)+(3*h/2)*Fcn(Fn(ii+1), t(ii+1))-(h/2)*Fcn(Fn(ii), t(ii));
% ADAMS-Moulton method: 3-step starts from here
Fn(ii+2)=Fn(ii+1)+h*((5/12)*Fcn(Fn(ii+2),t(ii+2))+...
    (2/3)*Fcn(Fn(ii+1),t(ii+1))-(1/12)*Fcn(Fn(ii),t(ii)));
end
plot(t, Fn, 'ko-'), grid on
%%  ADAMS-Moulton 4-step Method
% Part 4
% Memory allocation with ZERO matrices
Fn=[Fn0, zeros(1,steps-1)];
for ii=1:steps-3
% Predicted value by Euler's forward method
Fn(ii+1)=Fn(ii)+h*Fcn(Fn(ii),t(ii));
% Corrected value by trapezoidal rule
Fn(ii+1)=Fn(ii)+(h/2)*(Fcn(Fn(ii+1), t(ii+1))+Fcn(Fn(ii), t(ii)));
% Predicted value
Fn(ii+2)=Fn(ii+1)+(3*h/2)*Fcn(Fn(ii+1),t(ii+1))-...
    (h/2)*Fcn(Fn(ii),t(ii));
% Corrected value
Fn(ii+2)=Fn(ii+1)+h*((5/12)*Fcn(Fn(ii+2),t(ii+2))+...
    (2/3)*Fcn(Fn(ii+1),t(ii+1))-(1/2)*Fcn(Fn(ii),t(ii)));
% Predicted solution:
Fn(ii+3)=Fn(ii+2)+h*((23/12)*Fcn(Fn(ii+2), t(ii+2))-...
    (4/3)*Fcn(Fn(ii+1), t(ii+1))+(5/12)*Fcn(Fn(ii),t(ii)));
% ADAMS-Moulton method: 4-step starts from here
Fn(ii+3)=Fn(ii+2)+h*((3/8)*Fcn(Fn(ii+3),t(ii+3))+...
  (19/24)*Fcn(Fn(ii+2),t(ii+2))-(5/24)*Fcn(Fn(ii+1),t(ii+1))+...
  (1/24)*Fcn(Fn(ii),t(ii)));
end
plot(t, Fn, 'gh--'), grid on
%%  ADAMS-Moulton 5-step Method
% Part 4
```

```
% Memory allocation with ZERO matrices
Fn=[Fn(1), zeros(1,steps-1)];
for k=1:steps-4
% Predicted value by Euler's forward method
Fn(k+1)=Fn(k)+h*Fcn(Fn(k),t(k));
% Corrected value by trapezoidal rule
Fn(k+1)=Fn(k)+(h/2)*(Fcn(Fn(k+1), t(k+1))+Fcn(Fn(k), t(k)));
% Predicted value
Fn(k+2)=Fn(k+1)+(3*h/2)*Fcn(Fn(k+1), t(k+1))- (h/2)*Fcn(Fn(k),t(k));
% Corrected value
Fn(k+2)=Fn(k+1)+h*((5/12)*Fcn(Fn(k+2),t(k+2))+...
    (2/3)*Fcn(Fn(k+1),t(k+1))-(1/2)*Fcn(Fn(k),t(k)));
% Predicted value
Fn(k+3)=Fn(k+2)+h*((23/12)*Fcn(Fn(k+2), t(k+2))-...
    (4/3)*Fcn(Fn(k+2), t(k+2))+(5/12)*Fcn(Fn(k),t(k)));
% Corrected value
Fn(k+3)=Fn(k+2)+h*((3/8)*Fcn(Fn(k+3),t(k+3))+...
    (19/24)*Fcn(Fn(k+2),t(k+2))-(5/24)*Fcn(Fn(k+1),t(k+1))+...
    (1/24)*Fcn(Fn(k),t(k)));
% Predicted solution:
Fn(k+4)=Fn(k+3)+h*((55/24)*Fcn(Fn(k+3), t(k+3))-...
    (59/24)*Fcn(Fn(k+2),t(k+2))+(37/24)*Fcn(Fn(k+1),t(k+1))-...
    (3/8)*Fcn(Fn(k),t(k)));
% ADAMS-Moulton method: 5-step starts from here
Fn(k+4)=Fn(k+3)+h*((251/720)*Fcn(Fn(k+4),t(k+4))+...
(646/720)*Fcn(Fn(k+3),t(k+3))-(264/720)*Fcn(Fn(k+2),t(k+2))+ ...
(106/720)*Fcn(Fn(k+1),t(k+1))-(19/720)*Fcn(Fn(k),t(k)));
end
plot(t, Fn, 'c--', 'linewidth', 2), grid on
title('\it Solutions of Adams-Moulton method: $$\frac{dy}{dt}= -2*u^2*t,
u_0=0.5$$ ', 'interpreter', 'latex')
legend('1-step', '2-step','3-step', '4-step','5-step')
xlabel '\it t', ylabel('\it Solution, y(t)'), shg
```

The simulation results are displayed in Figure 2-10; there is a good convergence, and the difference between the one- or two-step and higher-step approaches of the Adams-Moulton method is marginally small for this given problem. It should be noted that with higher-step approaches, we can always achieve higher accuracy; however, higher accuracy will be attained with higher computation costs, which is true for previously discussed multistep methods, such as Runge-Kutta, Adams-Bashforth, Milne, and Taylor series.

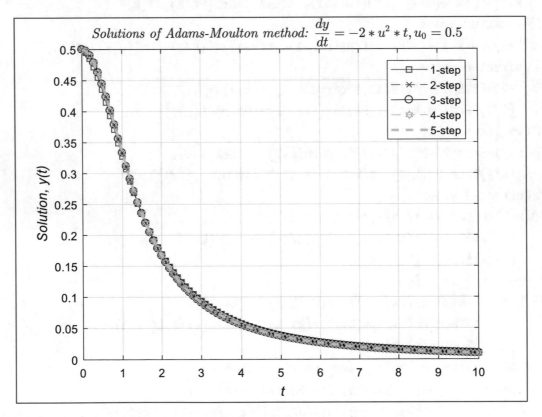

Figure 2-10. *Simulation results of the Adams-Moulton method with one, two, three, four, and five steps*

We have analyzed a few explicit and implicit and single-step and multistep numerical methods to solve ODEs. Within the explicit methods, we covered the Euler methods (forward and improved), Ralston, Runge-Kutta, Adams-Bashforth (two-, three-, four-, and five-step), Milne, and the Taylor series that are all explicitly defined, may provide good results for nonstiff IVPs of ODEs, and are easy to implement with relatively low computation costs. One of the biggest drawbacks of these methods is that they cannot provide accurate solutions for many stiff problems.

In addition, we have studied implicitly defined predictor-corrector methods that are employed to find numerical solutions of stiff ODE problems. They are the backward Euler method, the Adams-Moulton method, and Milne's method. The predictor-corrector methods are very stable in solution search but quite expensive to implement, and the simulation costs are much higher than explicit methods because of their additional computation steps. Predictor-corrector methods work according to their name; first they predict a candidate solution for a given problem, and then they correct a predicted solution. One of the simplest types of these methods is the backward Euler method, which first predicts a solution via the forward Euler method or Newton-Raphson's method and then corrects it. The same technique works for higher-order predictor-corrector methods. In many literature sources, these methods are grouped under the name of Adams-Moulton with different orders [2].

MATLAB's Built-in ODE Solvers

In MATLAB, there are a few built-in ODE solvers, such as ode15s, ode15i, ode23, ode23s, ode23t, ode23tb, ode45, and ode113, which are efficient at finding numerical solutions of many different types of initial value problems. These solvers are based on the explicit Runge-Kutta method and implicit Adams-Bashforth-Moulton method with different implementation algorithms and ODE solver methods, such as Dormand-Prince (ode45), Bogacki-Shampine (ode23), Rosenbrock (ode23s), trapezoidal rule (ode23t), Adams-Bashforth-Moulton (ode113), Gear's method (ode15s), and so forth. Using MATLAB's built-in ODE solvers is relatively simple. The following general syntaxes of the ODE solvers are shown here:

```
solver(odefun,tspan,y0)
[T,Y] = solver(odefun,tspan,y0)
[T,Y] = solver(odefun,tspan,y0,options)
[T,Y,TE,YE,IE] = solver(odefun,tspan,y0,options)
sol = solver(odefun,[t0 tf],y0...)
```

Any of these solvers can be chosen depending on the given problem type, depending on whether the given problem is stiff (how far stiff, e.g., very stiff or moderately stiff) or nonstiff and explicit or implicit: ode15s, ode15i, ode23, ode23s, ode23t, ode23tb, ode45, and ode113.

It is worth noting that an ODE solver type needs to be selected carefully. In selecting a solver type, the following recommendations and points should be considered. These are taken from the help library of the MATLAB package.

Solver Type	Problem Type	Accuracy	When to Apply
ode15i	Fully implicit	Medium	For only fully implicit IVPs
ode15s	Stiff	Low to medium	If ode45 is too slow in finding solutions of the problem due to its stiffness
ode23	Nonstiff	Low	For moderately stiff problems with crude error tolerances
ode23s	Stiff	Low	For stiff problems with crude error tolerances
ode23t	Moderately stiff	Low	For moderately stiff problems
ode23tb	Stiff	Low	For stiff problems with crude error tolerances
ode45	**Nonstiff**	**Medium**	**Recommended for most problems; a must for a first ODE solver to try**
ode113	Nonstiff	Low to high	For problems with tight error tolerances

Moreover, the efficiency of these solvers depends on the chosen step type (fixed or variable), the size, and the relative and absolute error tolerances that directly affect the accuracy of the simulation results and the efficiency of the computation processes. While using built-in ODE solvers, the step size can be chosen as variable (automatically chosen) or fixed/specified by a user. All built-in ODE solvers by default will take variable step sizes automatically depending on the type of a given IVP (e.g., a stiffness level) and a solution search space. Error tolerance can be controlled in ODE solvers via their setting options. Hereafter, we will study via real exercises all of these key aspects and settings of the ODE built-in solvers.

The ODEFUN function for the ODE solvers can be defined by using the following:

- Anonymous function with function (@)

- Function file (*.m file)

- matlabFunction, a function file (*.m file) by employing the Symbolic Math Toolbox

- Inline function (in future MATLAB versions, this feature will be removed)

Note   You need to be careful while calling the function named ODEFUN. If it is defined via an anonymous function (@) or an inline function, then we should use the following command syntax:

[T Y]=ODEx(my_Function, t, y0);

If we define a given problem (function/expression) via a function file, then we need to use one of the following command syntaxes:

[T Y]=ODEx(@Fun_File, t, y0);
[T Y]=ODEx('Fun_File', t, y0);

Time space can be predefined as a row or column vector of time values or with two elements, such as starting and end values, e.g., t = linspace(0, 13, 1000), t = 0:0.001:13, t = [0, 13].

ODEx solvers will automatically take a different number of steps or step size with respect to the nature of the given ODE (stiff or nonstiff, linear or nonlinear, etc.).

Example 9

Here's the example problem: $\dot{y} + 2ty^2 = 0$, $y_0 = 0.5$. In this case, our function file called Fun_File.m is defined via the next function file.

```
function F=Fun_File(t, y)
F=(-2*y^2*t);
```

We will look at several different problems to show how to implement these built-in tools and their options in defining ODEFUN. In the first example, we show how to use an anonymous function (@) to simulate a first-order ODE: $\dot{y} + 2ty^2 = 0$, $y_0 = 0.5$.

The following script (ExampleODE1.m) shows the implementation of the ode45, ode23, and ode113 solvers with an anonymous function (@) with a fixed step size of $h = 0.1$.

```
%%  ExampleODE1.m
% Part 1
% dy/dt=-2*t*(y^2); with ICs: y(0)=0.5
% We test methods for simulation time
clearvars; close all; clc;
F=@(t,y)(-2*y^2*t);
tmax=10;          % max. time limit
h=0.1;            % time step size is defined by a user!!!
t=0:h:tmax;       % time space
y0=0.5;           % ICs: y0 at t0
[time Yode45]=ode45(F, t, y0);
[time Yode23]=ode23(F, t, y0);
[time Yode113]=ode113(F, t, y0);
plot(time,Yode45,'ks-',time,Yode23,'ro-.',time, Yode113,'bx--') grid on
title('\it Solutions of: $$\frac{dy}{dt}+2*t^2=0, y_0 = 0.5$$',
'interpreter', 'latex')
legend ('ode45','ode23','ode113')
xlabel('\it t'), ylabel('\it Solution, y(t)'), shg
```

The output of the script is shown in Figure 2-11. From the plot, we can conclude that for the given problem, ode23, ode45, and ode113 perform well with the fixed step size of $h = 0.1$.

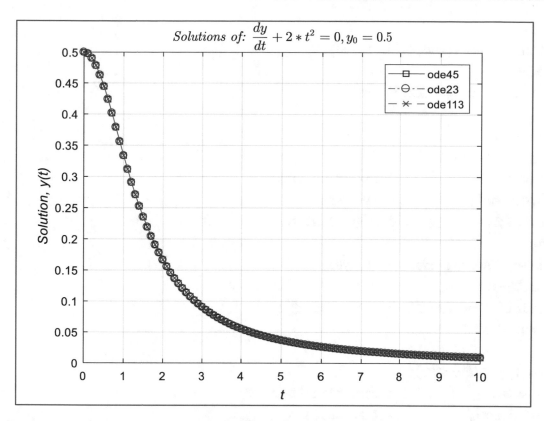

Figure 2-11. *Simulation results of ode23, ode45, and ode113*

Note   If we do not specify the output variable names (e.g., ode45(F, t, y0)), then the chosen solver displays computation results in a plot figure, and no numerical outputs are saved in the workspace.

Let's look at the issue of how MATLAB built-in solvers take variable steps in solving a given problem (Example 1: $\dot{y} + 2ty^2 = 0$, $y_0 = 0.5$) and how the step size will influence the accuracy of the simulations and computation (elapsed) time costs.

```
%% ExampleODE1.m
% Part 2
clearvars
F=@(t,y)(-2*y^2*t);
tmax=10;            % max. time limit
t=[0,tmax];         % time space
y0=0.5;             % ICs: y0 at t0
```

```
tic;
[time, Yode45]=ode45(F, t, y0);
Tode45=toc;
fprintf('Tode45= %2.6f  \n', Tode45)
%
clearvars
F=@(t,y)(-2*y^2*t);
tmax=10;                % max. time limit
t=[0,tmax];             % time space
y0=0.5;                 % ICs: y0 at t0
tic
[time, Yode23]=ode23(F, t, y0);
Tode23=toc;
fprintf('Tode23= %2.6f  \n', Tode23)
clearvars
%
F=@(t,y)(-2*y^2*t);
tmax=10;                % max. time limit
t=[0,tmax];             % time space
y0=0.5;                 % ICs: y0 at t0
tic
[time, Yode113]=ode113(F, t, y0);
Tode113=toc;
fprintf('Tode113= %2.6f \n', Tode113)
```

In part 2 of the script, the time space ($[t_0, t_{end}]$) is defined by the initial and end time values. Thus, in this case, each solver has taken variable steps while performing simulations. The simulations were performed on a laptop computer with Windows 10 (Intel Core i7: 6500 CPU at 2.50 GHz, with 8 GB RAM). The script outputs the following data that is the computational time of the solvers ode45, ode23, and ode113:

```
Tode45 = 0.003818
Tode23 = 0.004526
Tode113 = 0.008329
```

Note that the computation time (Tode45) of ode45 (in seconds) is the shortest.

Let's compare the simulation results of ode45 with the forward Euler method with a fixed step size and the analytical solution, as shown here:

```
%%  ExampleODE1.m
% Part 3
% Forward EULER method (1st order explicit)
% EXAMPLE #2: dy/dt=-2*t*(y^2); with ICs: y(0)=0.5
clearvars; close all
tic
tmax=10;                % max. time limit
h=0.001;                % time step size
t=0:h:tmax;             % time space
steps=length(t)-1;      % # of steps
y0=0.5;                 % ICs: y0 at t0
f=zeros(1,length(t));   % Memory allocation
y=[y0, zeros(1,steps)];    % Memory allocation
for ii=1:steps
    f(ii)=-2*y(ii).^2*t(ii);
    y(ii+1)=y(ii)+h*f(ii);
end
TEf=toc;
fprintf('TEf= %2.6f \n', TEf)
clearvars f y0 h t TEf steps tmax ii
%%
tic
F=@(t,y)(-2*y^2*t);
tmax=10;                % max. time limit
h=.001;                 % time step size
t=0:h:tmax;             % time space
y0=0.5;                 % ICs: y0 at t0
[time, Yode23]=ode23(F, t, y0);
Tode23=toc;
fprintf('Tode23= %2.6f \n', Tode23)
% Analytical solution:
syms u(T)  T
Du = diff(u, T);
```

```
Equation=Du==-2*u^2*T;
IC = u(0)==0.5;
u=dsolve(Equation,IC);
U=double(subs(u,'T',t));
plot(t, y, 'b-', t, Yode23, 'r-.', t, U, 'm:', 'linewidth',1.5)
xlim([1.399 1.3996])
grid on
legend('\it Euler forward', '\it ode23', '\it Analytical solution')
title('\it Solutions of: $$frac{dy}{dt}+2*t*y^2=0, y_0=0.5 $$',
'interpreter', 'latex')
shg
```

The outputs of the script are as follows:

```
T_Euler = 0.000312
Tode45 = 0.120490
```

From the computation results, we can conclude that the forward Euler method script solution is about 386 times faster than the ode45 solver in the solution search, which means it is 386 times more cost efficient in terms of computation time, given the nonstiff problem. At the same time, its accuracy is considerably high, and the error is marginal, as shown in Figure 2-12. This example is a good demonstration to show the importance of choosing an appropriate numerical search solution method to save a considerable amount of computation time.

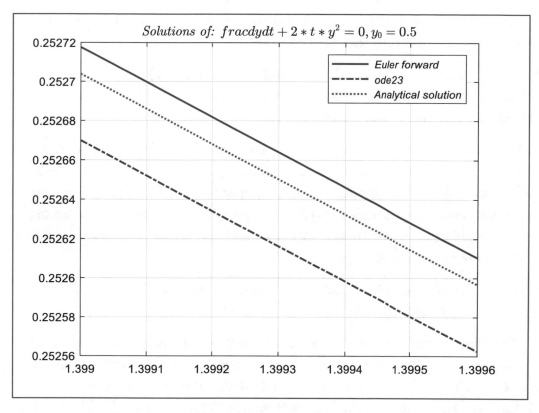

Figure 2-12. *Comparative studies of forward Euler method and MATLAB ode45 built-in solver*

The OPTIONS, ODESET, and ODEPLOT Tools of Solvers

OPTIONS settings are very important to get what we need while obtaining numerical solutions. For instance, via ODESET, we can set up the relative accuracy tolerance (RelTol), which controls the number of correct decimal digits in the solution ($RelTol < = 10^{-p}$, where p is correct digits), and the absolute tolerance (error margins: AbsTol), which determines the accuracy when the solution approaches zero. In every iteration step, the inequality equation shown in Equation (2-21) is computed and verified by relative and absolute error tolerances for an error signal of E_i.

$$|E_i| <= max\left(RelTol * |y_i|, AbsTol(i)\right) \qquad (2\text{-}21)$$

For instance, by default in MATLAB, the relative error tolerance (RelTol) is set to 0.001, which is roughly 0.1 percent of accuracy.

```
options = odeset('name1', value1, 'name2', value2,...)
options = odeset(oldopts, 'name1', value1,...)
options = odeset(oldopts, newopts)
```

ODESET is used to set up/adjust ODEx solver options. There are a number of options that can be adjusted according to a given problem type, accuracy level (absolute and relative error tolerances), computation time, and extra input data. For instance, for the previous example (Example 1: $\dot{y} + 2ty^2 = 0$, $y_0 = 0.5$), we can make the following adjustments in its options to improve accuracy and obtain output-function values and plot-computed values as an output:

```
options=odeset('OutputFcn',@odeplot,'RelTol',1e-5,'AbsTol',1e-7);
ode45(@(t,y)(-2*y^2*t), [0, 5], 0.5, options);
```

These two lines of commands simulate the problem and output simulation results in an interactively built plot by which we can observe how solutions converge over the specified solution space, e.g., $t \in [0,5]$. If we do not need an output function ('OutputFcn') or want to skip it, then we should put empty matrices ([ ], [ ]) instead of ['OutputFcn', @odeplot] or omit ['OutputFcn', @odeplot], as shown here:

```
options =odeset([],[], 'RelTol', 1e-5, 'AbsTol', 1e-7);
[time Yode45]=ode45(@(t,y)(-2*y^2*t), [0, 5], 0.5, options);
```

or

```
options =odeset('RelTol', 1e-5, 'AbsTol', 1e-7);
[time Yode45]=ode45(@(t,y)(-2*y^2*t), [0, 5], 0.5, options);
```

In addition, we can adjust some other options accordingly if we are not satisfied with the default values of the ODESET options. With the ODESET options, we can also print out statistics of performed simulation iterations (e.g., successful steps, failures, and simulation process components) of our selected ODE solver (e.g., ode23, ode45, ode15s, ode113). Statistics information might be of great use for benchmarking purposes when we are dealing with stiff problems with a large solution search space.

Example 10

Here's the example problem: $\dot{y} + 200ty^2 = 0$, $y_0 = 2.5$. Here is the solution script (ODEx_ Compare.m) that computes and compares several ODE solvers with the options set by the ODESET function:

```
clearvars; clc
disp('ODE45 performance: ')
tic;
options=odeset('stats','on','RelTol',1e-5,'AbsTol',1e-7);
ode45(@(t,y)(-200*t*y^2), [0, 2], 2.5, options);
toc;
disp('%-------------------------------%    ')
clearvars
disp ('ODE113 performance: ')
tic;
options=odeset('stats','on','RelTol',1e-5,'AbsTol',1e-7);
ode113(@(t,y)(-200*t*y^2), [0, 2], 2.5, options);
toc;
disp('%------------------------------%    ')
clearvars
disp('ODE23s performance: ')
tic;
options=odeset('stats','on','RelTol',1e-5,'AbsTol',1e-7);
ode23s(@(t,y)(-200*t*y^2), [0, 2], 2.5, options);
toc;
disp('%------------------------------%    ')
clearvars
disp('ODE15s performance: ')
tic;
ode15s(@(t,y)(-200*t*y^2),[0,2], 2.5, odeset('stats','on', ... 'RelTol',
1e-5,'AbsTol',1e-7));
toc;
disp('% ---------------------------%    ')
clearvars
disp('ODE23 performance: ')
```

```
tic;
ode23(@(t,y)(-200*t*y^2), [0, 2], 2.5,... odeset('stats','on','RelTol',1e-
5,'AbsTol',1e-7));
toc;
disp('%--------------------------------%     ')
```

Here is a comparative analysis of the solvers with their statistics data:

```
ODE45 performance:
37 successful steps
3 failed attempts
241 function evaluations
Elapsed time is 0.284619 seconds.
%-------------------------------%
ODE113 performance:
83 successful steps
4 failed attempts
171 function evaluations
Elapsed time is 0.172783 seconds.
%-------------------------------%
ODE23s performance:
204 successful steps
1 failed attempts
821 function evaluations
204 partial derivatives
205 LU decompositions
615 solutions of linear systems
Elapsed time is 0.298206 seconds.
%-------------------------------%
ODE15s performance:
122 successful steps
4 failed attempts
212 function evaluations
1 partial derivatives
24 LU decompositions
208 solutions of linear systems
```

```
Elapsed time is 0.212456 seconds.
% -----------------------------%
ODE23 performance:
152 successful steps
7 failed attempts
478 function evaluations
Elapsed time is 0.194447 seconds.
%-----------------------------%
```

From the comparative simulations in the script ODEx_Compare.m, it is clear that for the given first-order ODE simulation, the solver ode113 is the fastest solver of these five solvers. In fact, all of the ODEx solvers of MATLAB spend quite a large portion of their computation time at the start of a simulation. Therefore, to reduce the computation time, the initial step size must be introduced only if we can guess an appropriate step size for it. If we specify the wrong initial step size, it may lead to inaccurate solutions. The initial step size can be introduced using the ODESET options, as shown in the following command syntax. For instance, if we want to introduce that the initial step, e.g., 0.0025, then this ODESET command is appropriate:

```
options=odeset('InitialStep', 2.5e-3);
```

Now if we test the previous example with a specified initial step size of 0.0025 and adjusted relative and absolute error tolerances, then we see that ode23 is the most efficient solver of all the tested ODEx solvers and that it performs about 10 times faster than any of the other employed solvers.

Note   Specify the initial step size of a solver only if you know or can guess an appropriate or well-fit initial step size that would reduce computation/simulation time substantially. If not, then it is better to leave for a chosen solver to choose it automatically.

ODEXTEND is another tool of ODE solvers that might be useful when we are dealing with large and costly computations. The ODEXTEND tool is used to compute additional (extended) solution points after a computation process is completed. For example 2, this function tool can be used as follows:

```
% Already computed points:
SOL=ode23(@(t,y)(-200*t*y^2), [0, 2], 2.5, []);
%% Extended solution points are computed up to 5 seconds:
Tf =5;
Fun=@(t,y)(-200*t.*y.^2);
solext = odextend(SOL, Fun, Tf);
plot(solext.x, solext.y(:,:), 'ko'), shg
```

In this way, additional solution points can be computed without recomputing already computed data points.

Let's look at several examples to show how to employ and set up ODE options with ODESET.

Example 11

Given a first-order ODE, here's the example problem: $\dfrac{dy}{dx} + nty = 0$, $y(0) = 3$, $n = 1, 2, 3, 4, 5, 6$. We need to compute numerical values of the given first-order ODE for five different values of n. This ODE can be solved with dsolve() to find its analytical solution(s). Here we show how to employ ODEx solver options for loop-based problems and the importance of preset relative and absolute error tolerance values of ODE solvers. The solution script (ODE1sim.m) solves the given problem for different values of n within a [for, end] loop.

```
function ODE1sim
% Numerical solutions of: y'+nty=0; y(0)=3; for t=[0, 3.5];
% For n=1, 2, 3, 4, 5, 6.
y0=3;                   % Initial Condition
t=[0, 3.5];             % Simulation time
% RelTol is relative tolerance
% AbsTol is absolute tolerance
options =odeset('RelTol', 1e-2, 'AbsTol', 1e-1);
for n=1:6
   [T,Y] = ode23(@DY1, t, y0, options, n);
if n==1
      plot(T,Y, 'b-'); hold on
end
```

```
if n==2
        plot(T, Y, 'kx-')
end
if n==3
        plot(T, Y, 'ro--')
end
if n==4
        plot(T, Y, 'kv-.')
end
if n==5
        plot(T, Y, 'mp:')
end
if n==6
        plot(T, Y, 'cd-')
end
end
legend('n=1','n=2','n=3', 'n=4','n=5','n=6')
title('\it Solution of: $$ \frac{dy}{dt}+n*t*y=0, n=1,2...6, y_0=3 $$',
'interpreter', 'latex')
xlabel('\it t'), ylabel('\it Solution, y(t)'), grid on
hold off
function dy = DY1(t,y, n)
dy= -n.*t.*y;
end
end
```

In fact, this script is too long and cannot be efficient for larger values of n. Besides, the plot in Figure 2-13) is not smooth because of the poorly adjusted values of relative and absolute error tolerances, which are default values chosen by ODESET.

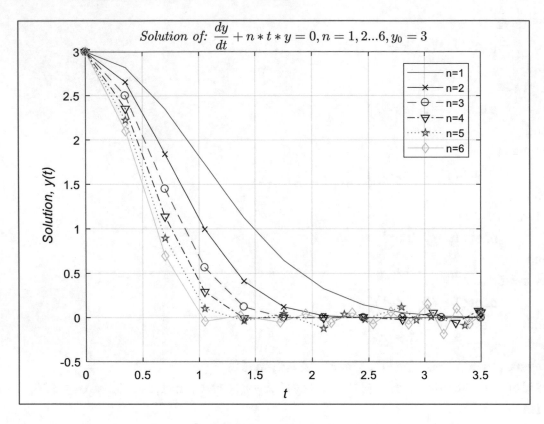

Figure 2-13. *Simulation of* $\dfrac{dy}{dx} + nty = 0$, *y(0) = 3, and n = 1, 2, 3, 4, 5, 6*

Here is one of the possible compact solutions, ODE1simOK.m, to the given problem, with the ODEPLOT option preset within ODESET as OuputFcn.

```
function ODE1simOK
% Numerical solutions of: y'+nty=0; y(0)=3; for t=[0, 3.5];
% For n=1, 2, 3, 4, 5 and 6.
%       color (color type): b-blue; g-green; r-red; m-magenta;
%       c-cyan; y-yellow; k-black.
%       lines (line type): - solid line; : colon; -- dashed line;
%       mark (marker type): o-circle; d-diamond; x-cross;
%       s-square; h-hexagon; + plus sign; * asterisk; etc.
y0=1;          % Initial Condition
t=[0, 3.5];    % Simulation time
% Pre-define plot marker labels:
labels = {};
```

```
color = 'bgrckmbgrckm';
lines = '--:-:-.--:-:-.';
mark  = 'od+xhsp*^>cs<d+xh';
% RelTol is relative tolerance
% AbsTol is absolute tolerance
options =odeset('RelTol', 1e-5, 'AbsTol', 1e-7);
labels=cell(1,6);
for n=1:6
   [T,Y] = ode23(@DY1, t, y0, options, n);
   style = [color(n) lines(n) mark(n)];
   labels{n} = ['n = ' num2str(n)];
   plot(T,Y,style), hold on
end
legend(labels{:},'Best'), grid on
title('Simulation of: $$ \frac{dy}{dt}+n*t*y=0, n=1,2...6 $$',
'interpreter', 'latex')
xlabel('\it t'), ylabel('\it  Solution, y(t)')
hold off

function dy = DY1(t,y, n)
dy= -n.*t.*y;
end
end
```

The script is considerably compact and produces the next interactively built plot, as shown in Figure 2-14.

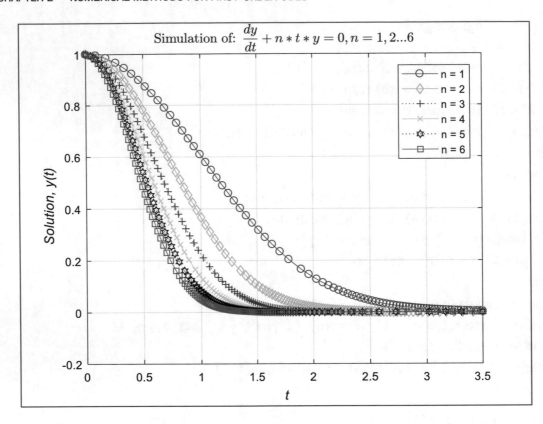

Figure 2-14. *Simulation of* $\dfrac{dy}{dx} + nty = 0$ *with y(0) = 3 and n = 1, 2, 3, 4, 5, 6*

The solutions, shown in Figure 2-14, are more accurate than the ones displayed in Figure 2-13. This accuracy has been attained with slightly tighter tuned relative and absolute error tolerances that can be tightened even more if higher accuracy is a concern. Besides, the latter script is more compact thanks to the use of ODEPLOT options preset with ODESET. Moreover, it has higher accuracy and tighter control over relative and absolute error tolerances.

In this section, we briefly covered the ODE built-in solvers and how to set up their settings such as relative and absolute error tolerances, how to plot outputs with ODESET, and how to compute extended additional values of already solved ODEs with ODEEXTEND with the example of several first-order ODEs. All these tools and procedures are also applicable for the second- and higher-order ODEs, which are discussed in the next chapter.

Simulink Modeling

In this chapter, we are limiting ourselves to explanations of how to employ standard and basic Simulink blocks to solve and simulate several ODE problems, with some explanatory notes and recommendations on tools and options (see [3] for more details on Simulink modeling). We start by modeling some simple first-order ODEs to demonstrate the simplicity of modeling ODEs in a Simulink environment.

Example 12

Here's the example problem: $\dot{y} + 2ty^2 = 0$, $y_0 = 0.5$. (Note that this problem was already solved previously by a few different methods.) The given problem is modeled in Simulink in the following steps:

Step 1. Rewrite the given problem: $\dot{y} = -2ty^2$.

Step 2. Start building a solution model.

To obtain $y(t)$ from $\dot{y} = \dfrac{dy}{dt}$, we need to integrate $\dfrac{dy}{dt}$. Another explanation of this procedure is in the s domain (with Laplace transforms): if we perform Laplace transforms of $\dfrac{dy}{dt}$, we will get $s\,Y(s) - y(t_0)$. To obtain $Y(s)$ from $s\,Y(s)$, we need to multiply $s\,Y(s)$ with $\dfrac{1}{s}$, which means integration. An integrator block $\left(\dfrac{1}{s}\right)$ can be taken from the Simulink Library Browser, which can be launched by clicking ▦ and then selecting Libraries ➤ Simulink ➤ Continuous. From the right pane, we take the [Integrator] block and drag it into our model area. Alternatively, the recent versions of Simulink starting from 2018a have a user-friendly novel option that with a single click on the blank area of the model window, the search box pops up as shown here: ▭. By typing **in**, the drop-down options shown in Figure 2-15 pop up.

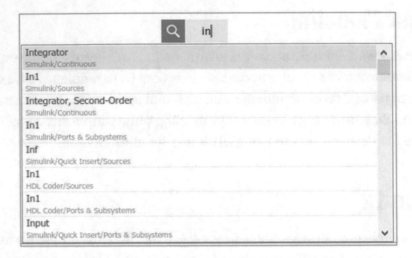

Figure 2-15. *Simulink model search box's drop-down options*

In the drop-down options, the first one is Integrator shown in Figure 2-16; hit Enter on the keyboard. That creates the integrator block with the initial condition prompt, as shown here: [image]. Similarly, all other blocks can be brought into the blank model window without jumping in the Simulink library [image]. For instance, the Gain block can be obtained by typing in **g**, as shown here: [image] and the first block is the Gain block. Next, hit Enter, and the Gain block pops up: [image]. This has an option to specify the Gain value. In a similar manner, we can locate all other necessary blocks for the model.

Note   This new feature of the search option in the model window area is available starting in MATLAB/Simulink 2018a.

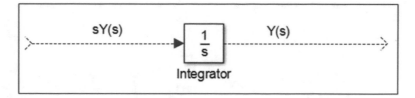

Figure 2-16. *Simulink Integrator block*

In this step, the given initial condition can also be defined within the integrator block parameters that can be opened by double-clicking the block. The initial conditions can be set as internal or external in the integrator parameters, as shown in Figure 2-17.

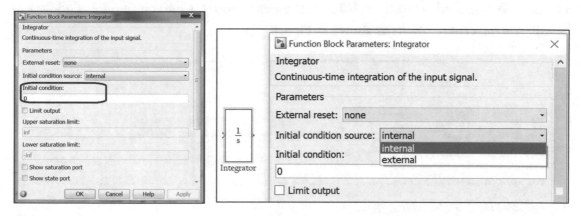

Figure 2-17. *Initial condition setup options for the Integrator block*

When doing Simulink modeling of ODEs, we always need to start with the [Integrator] block. This is a first fundamental step in building simulation models of differential equations or problems expressed via differential equations in a Simulink environment. If a given problem has zero initial conditions, then we should not worry about assigning initial conditions. By default, the initial condition of the integrator is set to zero.

In this given example, the initial condition value is 0.5, which we type in; then click Apply and OK. In the given algebraic expression $\dot{y} = -2ty^2$, we have a product $(-2t * y^2)$, constant gain (-2), input signal (time, t), and mathematical operation (y^2). In addition, other blocks, such as the [Out] and [Scope] blocks, to view and obtain computation results can be added to the model. Four out of the seven blocks used in this model can be taken from Libraries ➤ Simulink ➤ Commonly Used Blocks: [Gain], [Product], [Out1], and [Scope].

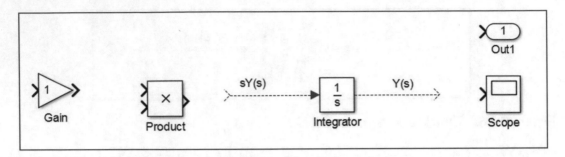

Figure 2-18. *Blocks of the model*

In addition, we need an input signal (time), which is a [Clock] block, located at Libraries ➤ Simulink ➤ Sources: [Clock]. Drag and drop this into our model window and obtain the next set of blocks as shown in Figure 2-19.

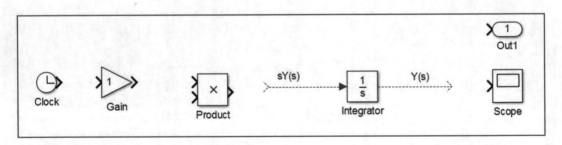

Figure 2-19. *Blocks of the model*

Finally, the last block, [Math Function], is taken from Simulink/Math operations. It has menu options from which math functions, such as exponential, logarithmic, square, power, etc., can be chosen. Finally, the list of all necessary blocks for our model will be complete. The block [Math Function] (Figure 2-20) is used to compute y^2 from a signal coming from the [Integrator] block's output port.

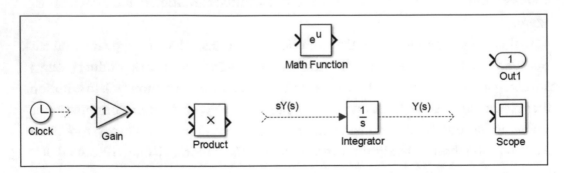

Figure 2-20. *Blocks of the model*

Now that we have collected all necessary blocks, we still need to make some adjustments and changes in the parameters of the chosen blocks. We can change the value of the constant [Gain] from 1 to -2, which can be done by double-clicking the block, changing its gain value in a pop-up window of the [Gain] block, and hitting OK. We can change the [Math Function] block function by double-clicking it and choosing the function type [square], which is u^2, and hitting OK. To get a better or more readable view of the model, we can flip/rotate some of the blocks (e.g., [Math Function]) by 180 degrees by clicking a block and pressing Ctrl+I or by right-clicking and choosing the Rotate & Flip option of the block. By performing these operations, we obtain the set of blocks shown in Figure 2-21.

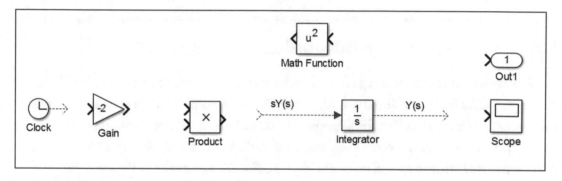

Figure 2-21. *Blocks of the model*

For the third step, we connect all blocks by dragging a signal line from an output port of a block to an input port of the next block or by clicking a block and holding the Ctrl key while clicking to the next block. The blocks are connected in the following order:

- For the input signal, [Clock] is linked with the [Gain] block.

- The [Gain] block is linked with the [Product] block.

- The [Product] block is linked with the [Integrator] block.

- The [Integrator] block is linked with the [Scope], [Out1], and [Product] block.

Finally, we get the following complete model, Example_2_1ODE.mdl, as shown in Figure 2-22.

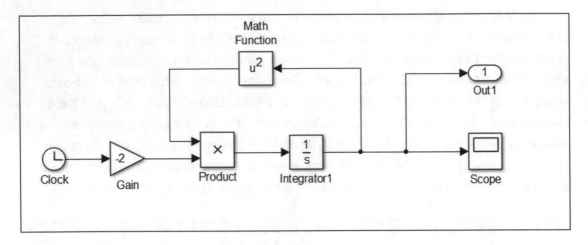

Figure 2-22. *Complete Simulink model of* $\dot{y} + 2ty^2 = 0$, $y_0 = 0.5$

For step 4, after connecting all the blocks, a created model can be saved with the `*.mdl` or `*.slx` file extension depending on which Simulink version is used. All unknown parameters of a created Simulink model can be defined in the command window or model properties/callbacks/InitFcn within Simulink. After completing all four steps, we can save our Simulink model as `Example_2_10DE.mdl`, as shown in Figure 2-22.

We simulate the created Simulink model by clicking the Run icon or pressing Ctrl+T. Then we can view the simulation results from the [Scope] block by double-clicking it. Moreover, we obtain simulation results from the [Out1] block in the MATLAB Workspace under two variables, `tout` and `yout`, which are named by default for the output data sets from the [Out1] block. This is one of the few possible ways to obtain simulation results from the Simulink model into MATLAB workspace. Another way, probably the easier one, is to use Configuration Properties: [Scope] block. To use this option (Figure 2-23), click the icon and select Configuration Properties A Scope window pops up. Go to the Logging tab and select the "Log data to workspace" option. If necessary, change the variable name and save format type. A third option to save simulation data or any signal values from a Simulink model is to use To Workspace block by clinking a signal from any block with it that collects a signal and saves in the MATLAB workspace. This block is located under the Simulink library and is called Sinks. A default variable name in this block is `simout`, which can be renamed, and its data structure type can be altered if necessary.

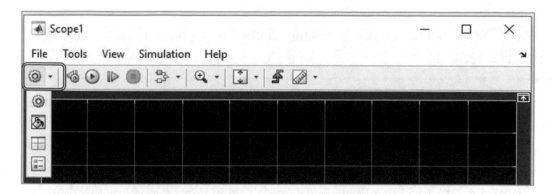

Figure 2-23. *Changing the [Scope] block display options interactively*

Finally, the simulation results are displayed in the [Scope] block, as shown in Figure 2-24, with the changed properties.

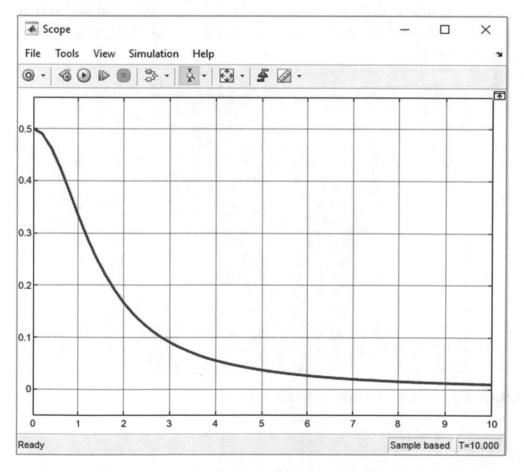

Figure 2-24. *Simulation resuts in the [Scope] block with the changed display options of the [Scope] block*

Note   Parameter icons and the options of the [Scope] block differ for different Simulink versions.

By employing the [Scope] block's parameters options, a few things in the plotted data can be adjusted such as the background, axis color, line type and marker type, and color of plotted data points. Also, the plot title and legend features can be added, and the save data options can be tuned.

Model parameters, such as solver type, solver, step size, relative and absolute tolerances, zero-crossing detection, and so forth., can be tuned via the configuration parameters. They can be accessed by clicking [⚙▾] and selecting - Model Configuration Parameters Ctrl+E from its drop-down options [🔲], pressing Ctrl+E, or selecting Tools ➤ Model Explorer ➤ Model Hierarchy ➤ Configuration, as shown in Figure 2-25.

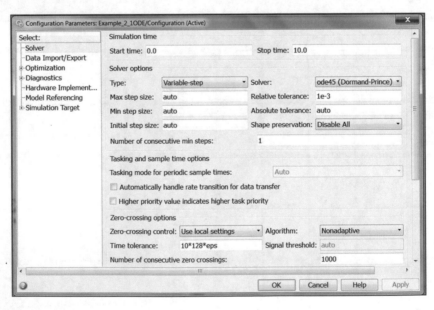

Figure 2-25. *Adjusting configuration parameters*

There are several parameters in the solver options. We may need to adjust them to achieve the desired accuracy of numerical solutions and computation time efficiency. They are the type (variable step or fixed step), solver, step size, relative and absolute error tolerances, and zero-crossing detection. The following are variable step solvers: ode45 (Dormand-Prince) by default, ode23 (Bogacki-Shampine), ode113 (Adams), ode15s (stiff/NDF), ode23t (Trapezoidal rule), ode23tb (stiff/TR-BD2). The fixed step solvers are discrete (no continuous states), ode1 (Euler), ode2 (Heun), ode3 (Bogacki-Shampine), ode4 (Runge-Kutta), etc.

Simulink models can be associated with MATLAB and can be simulated from the command window, M-files, or function files without launching the Simulink package. To execute a Simulink (e.g., MODELname.mdl or MODELname.slx) model, we need to use the following standard command syntaxes with sim():

```
sim('MODELname');
sim('MODELname', [time0, timeEND]);
```

That simulates a Simulink model called MODELname.mdl for specified time ranges of [time0, timeEND]; additionally, we can specify output data names while simulating the Simulink model.

```
[Tout, Uout]=sim('MODELname', [time0, timeEND]);
```

After obtaining the output data [Tout, Uout], we can perform further numerical analyses and plot the obtained results. In the example of our developed model (Example_2_ODE.mdl), we can simulate the model with the following command:

```
[Tout, Uout]=sim('Example_2_1ODE', [0, 10]);
```

SIMSET

The Simulink model solver options (settings), such as solver type, relative and absolute error tolerances, zero-crossings, and so forth, can be adjusted with the command simset(). Here's an example:

```
OPTIONS = simset('solver','ode3', 'reltol', 1e-3, 'abstol', 1e-5,
'zerocross', 'off');
[Tout, Uout]=sim('Example_2_1ODE', [0, 10], OPTIONS);
```

Moreover, there are other optimization options used from MATLAB to make adjustments in the Simulink model options, which are `optimset, sdo. OptimizeOptions`, etc.

```
opt = sdo.OptimizeOptions(Name, Value)
```

This can be applied to the previously mentioned example to employ a nonlinear least squares optimization with the following commands:

```
opts = sdo.OptimizeOptions('Method', 'lsqnonlin');
```

That sets optimization to be a nonlinear least squares method in the Simulink model solver.

The previously depicted steps and procedures are common and key procedures for Simulink model development and simulation. It is important to point out that in the previously created model, there are two blocks, such as [Out1] and [Scope], that do not directly influence on model performance simulation results. Another important point is that some blocks are interchangeable. For instance, the [Integrator] block with an internal initial condition option can be altered to the [Integrator] block with an external initial condition source. Another example is the [Divide] block that can be also changed and converted to the [Product] block. Any model can be developed in a few different ways by using different blocks and combining mathematical and algebraic operations.

References

[1] Boyce, W. E., Diprima, R. C., *Elementary Differential Equations and Boundary Value Problems* (7th ed.), John Wiley & Sons (2003).

[2] Hairer E., Wanner G., *Solving Ordinary Differential Equations II: Stiff and Differential Algebraic Problems* (2nd ed.), Springer-Verlag (1996).

[3] Eshkabilov, S., *Beginning MATLAB and Simulink*, Apress, 2019.

CHAPTER 3

Numerical Methods for Second-Order ODEs

Second-order or higher-order ODEs are vital for a broad range of problems in physics, engineering, and natural sciences. For instance, wave equations, electromagnetic phenomena, heat conduction, and fluid mechanics are modeled with second-order or higher-order ODEs. In general, second-order ODEs take the form shown in Equation (3-1).

$$\frac{d^2 y}{dt^2} = f\left(t, y, \frac{dy}{dt}\right),\qquad(3\text{-}1)$$

In general, the independent variable is denoted by t, which is often an independent variable (e.g., with respect to time) in many physical systems. There are also many systems where x, y, and z, or other variables, are used.

An n-th order linear differential equation can be expressed in the general form shown here:

$$P_0(t)\frac{d^n y}{dt^n} + P_1(t)\frac{d^{n-1} y}{dt^{n-1}} + \cdots + P_{n-1}(t)\frac{dy}{dt} + P_n(t)y = H(t)\qquad(3\text{-}2)$$

where $P_0(t)$, $P_1(t)...P_n(t)$, and $H(t)$ are continuous real-valued functions on some interval $\alpha < t < \beta$ and where $P_0(t)$ is not near zero in this interval. If we divide Equation (3-2) by $P_0(t)$, we obtain the following equation:

$$\frac{d^n y}{dt^n} + p_1(t)\frac{d^{n-1} y}{dt^{n-1}} + \cdots + p_{n-1}(t)\frac{dy}{dt} + p_n(t)y = h(t)\qquad(3\text{-}3)$$

Equation (3-3) can be also rewritten as follows:

© Sulaymon L. Eshkabilov 2020
S. L. Eshkabilov, *Practical MATLAB Modeling with Simulink*, https://doi.org/10.1007/978-1-4842-5799-9_3

$$\frac{d^n y}{dt^n} = h(t) - \left(p_1(t)\frac{d^{n-1}y}{dt^{n-1}} + \cdots + p_{n-1}(t)\frac{dy}{dt} + p_n(t)y \right) \tag{3-4}$$

and also as follows:

$$\frac{d^n y}{dt^n} = f\left(t, \frac{d^{n-1}y}{dt^{n-1}}, .. \frac{dy}{dt}, y(t) \right) \tag{3-5}$$

Equation (3-5) is similar to Equation (3-1) of the second-order ODE.

Let's look at a few examples of second-order ODEs to show how to use the Euler, Runge-Kutta, Adams-Bashforth, and Adams-Moulton methods, as well as others. In addition, we demonstrate how to use MATALB's built-in ODEx solvers and tools including dsolve() to compute numerical and analytic solutions.

The approaches and the previously employed methods for first-order ODEs are similar and valid for second-order and higher-order ODEs. There is one (or more) additional step required, which is to rewrite a given second-order or higher-order ODE as a system of first-order ODEs for all explicitly defined ODEs. For implicitly defined ODEs, we demonstrate a separate approach by using MATLAB's built-in solver ode15i and Simulink.

Euler Method

In previous chapters, we covered the formulation and implementation algorithm of the Euler methods in detail to show how to solve first-order ODEs numerically. The implementation steps of the Euler methods for higher-order ODEs are the same except for some additional steps in between. Thus, in this section, we will limit ourselves to the implementation of the Euler methods via MATLAB scripts to directly compute numerical solutions of second-order and higher-order ODEs.

Example 1

Given a second-order nonhomogeneous ODE, here's the example problem:

$$\frac{1}{2}\ddot{u} + \frac{2}{5}\dot{u} + u = 2t, \quad u(0) = 1, \dot{u}(0) = 2.$$

Before writing a script based on the Euler methods, we need to rewrite the given second-order ODE as a system of two first-order ODE equations by introducing new variables.

$$\ddot{u} = 4t - \frac{4}{5}\dot{u} - 2u \text{ needs to be rewritten as follows: } \begin{cases} \dot{u}_1 = u_2 \\ \dot{u}_2 = 4t - \frac{4}{5}u_2 - 2u_1 \end{cases}.$$

Note that $u_1 = u$ and $\dot{u}_2 = \ddot{u}$.

The previously rewritten system of first-order ODEs can be implemented in scripts in several ways, for instance, such as by writing a system of equations directly within the script, by using an anonymous function (@), by using a function, or by using a function file. Note that implementing a function within scripts is valid starting from MATLAB 2018a.

Here is the script (named Example1_2ndODE.m) that embeds the Euler forward method to compute numerical solutions (shown in Figure 3-1) of the given problem:

```
% EXAMPLE 1. (1/2)*u"+(2/5)*u'+u=2*t with ICs: u(0)=1 & u'(0)=2
% solved by EULER's Forward Method and DSOLVE.
clearvars, close all; clc
h=0.2;              % time step size
tmax=7;             % End of simulations
t0=0;               % Start of simulations
t=t0:h:tmax;
steps=length(t);    % # of steps of evaluations
u(1,:)=[1 2];       % Initial Conditions
%{
NB: Size of the u(t) is 2-dimensional that is due to the fact
 there are two variables used to rewrite 2nd order ODE as 2 1st order ODEs.
%}
% Way # 1. Direct way of expressing a system of ODEs
for k=2:steps
    f=[u(k-1,2), 4*t(k)-(4/5)*u(k-1,2)-2*u(k-1,1)];
    u(k,:)=u(k-1,:)+f*h;
end
% Way # 2. Anonymous Function (@)
%{
F = @(t, u1, u2)([u2, 4*t-(4/5)*u2-2*u1]);
```

```
for k=2:steps
    F1=F(t(k-1), u(k-1, 1), u(k-1, 2));
    u(k,:)=u(k-1,:)+F1*h;
end
%}
% Way # 3.   Function File called: FuFu.m
%{
function dudt = FuFu(t, u1, u2)
 dudt = [u2, 4*t-(4/5)*u2-2*u1]
 end
%}
%{
for k=2:steps
    F1=feval(@FuFu,t(k-1), u(k-1, 1), u(k-1, 2));
    u(k,:)=u(k-1,:)+F1*h;
end
%}
% Way #4
%{
 syms T u1 u2
  f=[u2, 4*T-(4/5)*u2-2*u1];
  matlabFunction(f, 'file', 'F');
for k=2:steps
    F1=feval(@F, t(k-1), u(k-1, 1), u(k-1, 2));
    u(k,:)=u(k-1,:)+F1*h;
end
%}
plot(t, u(:,1), 'bo'), grid minor, hold on
% Analytical solution of the problem:
syms x(T) Dx T
Dx=diff(x, T);
D2x=diff(x, T, 2);
Equation=D2x==-(4/5)*Dx-2*x+(4*T);
ICs = [x(0)==1, Dx(0)==2];
y=dsolve(Equation, ICs);
```

```
Y=double(subs(y,'T',t));
% or use: U=vectorize(u); T=t; U=eval(U);
plot(t, Y, 'g-', 'linewidth', 2)
legend('EULER forward method','Analytical Solution', 'location',
'southeast')
title('\it Solutions of: $$\frac{1}{2}*\frac{d^2u}{dt^2}+\frac{2}{5}*\
frac{du}{dt}+u=2*t $$', 'interpreter', 'latex')
xlabel('\it t'), ylabel('\it Solution, u(t)'), hold off
axis tight; shg
```

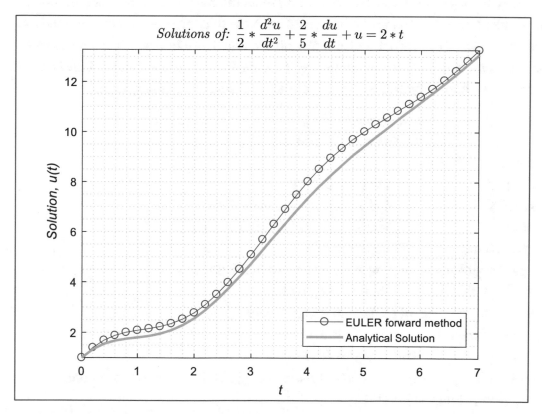

Figure 3-1. *The forward Euler method versus analytical solution with dsolve() to solve $\frac{1}{2}\ddot{u}+\frac{2}{5}\dot{u}+u=2t$, $u(0)=1$, $\dot{u}(0)=2$*

It should be noted in the previous example that by decreasing the step size (h), the accuracy of the Euler method improves considerably like with first-order ODEs.

Example 2

Given a second-order homogeneous ODE, here's the example problem: $\ddot{u} + 6\dot{u} + 9u = 0$, $u(0) = 2, \dot{u}(0) = 0$.

In this case, like the previous problem, we rewrite the given second-order differential equation with two first-order ODEs by introducing new variables.

$\ddot{u} = -6\dot{u} - 9u$ should be rewritten as two first-order ODEs: $\begin{cases} \dot{u}_1 = u_2 \\ \dot{u}_2 = 6u_2 - 9u_1 \end{cases}$.

Note that $u_1 = u$ and $\dot{u}_2 = \ddot{u}$.

Here is the script (Example2_2ndODE.m) that embeds the forward Euler method to compute numerical solutions of the problem. Note that the given system, defined by two first-order ODEs, is defined as a two-dimensional function within a loop. In this example, we demonstrate the influence of the step size on the accuracy of the computed numerical solutions with respect to the analytical solution of the system.

```
% EXAMPLE 2.  u"+6*u'+9*u=0with ICs: u(0)=2 & u'(0)=0
% solved by EULER's Forward Method and DSOLVE.
clearvars, close all
h=0.1;              % time step size
tend=2.5;           % End of calc's/simulations
t0=0;               % Start of calc's/simulations
t=t0:h:tend;        % Time space
steps=length(t);    % How many steps of evaluations
u(1,:)=[2 0];       % Initial Conditions
% NB: Size of the u(t) is two-dimensional
for k=1:steps-1
    f1=[u(k,2), -9*u(k,1)-6*u(k,2)];
    % f1(k,:)=[u(k,2), -9*u(k,1)-6*u(k,2)]; % If we need f1 values.
    u(k+1,:)=u(k,:)+f1*h;
end
plot(t, u(:,1), 'b--x', 'linewidth', 2), grid on, hold on
% Study step size change effects on errors
h1=0.025;           % time step size
t=t0:h1:tend;
step=length(t);     % How many steps of evaluations
y(1,:)=[2 0];       % Initial Conditions
```

```
for k=1:step-1
    ff1=[y(k,2), -9*y(k,1)-6*y(k,2)];
    y(k+1,:)=y(k,:)+ff1*h1;
end
plot(t, y(:,1), 'g--', 'linewidth', 2)
% Analytical solution of the problem:
syms x(T) Dx T
Dx=diff(x, T);
D2x=diff(x,T,2);
Equation=D2x==-6*Dx-9*x;
ICs=[x(0)==2, Dx(0)==0];
Y=dsolve(Equation, ICs);
Y=double(subs(Y,'T',t));
plot(t, Y, 'k-', 'linewidth', 1.5)
title('\it Solution of: $$ \frac{d^2u}{dt^2}+6*\frac{du}{dt}+9*u=0, u_0=2,
du_0=0 $$', 'interpreter', 'latex')
xlabel '\it t', ylabel '\it Solution, u(t)'
legend('EULER Method: h= 0.1', 'EULER Method: h = 0.05', 'Analytic
Solution')
err1= abs((Y(end)-u(end,1))/Y(end))*100;
err2= abs((Y(end)-y(end,1))/Y(end))*100;
error1=['Error (h=0.1): ', num2str(err1),      '%'];
error2=['Error (h=0.025): ', num2str(err2),  '%'];
text(1.0, .75, error1, 'color', 'r', 'fontsize', 14)
text(1.0, .6, error2, 'color', 'r', 'fontsize', 14)
axis([0, 3.5 -0.1 2.25])
hold off
axis tight
```

After executing the previous script, the plot of numerical solutions (shown in Figure 3-2) is displayed.

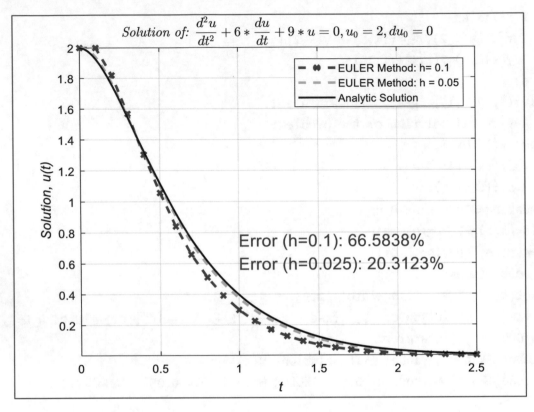

Figure 3-2. *Simulation results obtained from the forward Euler method for*
$\ddot{u} + 6\dot{u} + 9u = 0, \ u(0)=2, \dot{u}(0)=0$

From these simulations (shown in Figure 3-2), we can see that by decreasing the step size to four times, we can improve the accuracy by more than three times. This is a substantial improvement, but it is achieved with the cost of a four fold increase of computation time.

Example 3

Given a second-order nonhomogeneous ODE, here's the example problem:
$\ddot{u} + \sin(t) = 0, \ u(0)=0, \dot{u}(0)=0$.

Again, before writing a script, we rewrite the given problem as two first order ODEs.

$$\ddot{u} = -\sin(t) \ \text{is re-written as:} \ \begin{cases} \dot{u}_1 = u_2 \\ \dot{u}_2 = -\sin(t) \end{cases}$$

Note that $u_1 = u$ and $\dot{u}_2 = \ddot{u}$.

Here is a script (Example3_2ndODE.m) that embeds the forward Euler method to find numerical solutions (shown in Figure 3-3) of the problem. Note that, in our script, the system of first-order ODEs is expressed directly, as shown in the previously rewritten formulation.

```
% Example3_2ndODE.m
%% EXAMPLE 3.  u"=-sin(t) with ICs: u(0)=0 & u'(0)=0
% solved by forward EULER Method and DSOLVE
clc, clearvars; close all
h=0.1;                     % Time step size
tend=13;                   % End of calc's/simulations
t0=0;                      % Start of calc's/simulations
t=t0:h:tend;               % Time space
steps=length(t);           % How many steps of evaluations
u(1,:)=[0 0];              % Initial Conditions
for k=1:steps-1
f1=[u(k,2), -sin(t(k))];
u(k+1,:)=u(k,:)+f1*h;
end
plot(t', u(:,1),'bo')
hold on, axis tight
clearvars u
% Analytical solution of the problem:
syms u(T) T
Du=diff(u, T);
D2u=diff(u, T, 2);
Equation=D2u==-sin(T);
ICs = [u(0)==0, Du(0)==0];
u=dsolve(Equation, ICs);
U=vectorize(u);
T=t;
U=eval(U);
plot(T, U, 'k-', 'linewidth', 2), grid on
legend('Euler forward', 'Analytic (dsolve)')
```

121

```
title('\it Solutions of: $$ \frac{d^2u}{dt^2}+sin(t)=0, u_0=0, du(0)=0 $$',
'interpreter', 'latex')
xlabel('\it t'), ylabel('\it Solution, u(t)')
shg
```

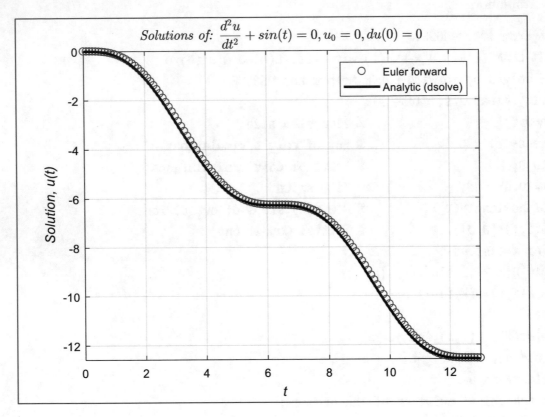

Figure 3-3. *Simulation results of the forward Euler method versus analytical solution (dsolve). Example 3: $\ddot{u}+sin(t)=0$, $u(0)=0$, $\dot{u}(0)=0$.*

Example 4

Given a second-order nonhomogeneous ODE, here's the example problem:
$\ddot{u}+\dot{u}=\sin(t)$, $u(0)=1,\dot{u}(0)=2$.

Again, before creating a script, we rewrite the given problem as two first-order ODEs.

$$\ddot{u}=-\dot{u}+\sin(t) \text{ is re-written as: } \begin{cases} \dot{u}_1 = u_2 \\ \dot{u}_2 = -u_2 + \sin(t) \end{cases}$$

Note that $u_1 = u$ and $\dot{u}_2 = \ddot{u}$.

Here is a script (Example4_2ndODE.m) that embeds the forward Euler method to find numerical solutions (shown in Figure 3-4) of the problem. Note that in our script, the system of first-order ODEs is expressed directly within a [for ... end] loop, as shown in the previous rewritten formulation.

```
% Example4_2ndODE.m
%% EXAMPLE 4. u"+u'=sin(t) with ICs: [1, 2]
% by EULER Forward Method& DSOLVE (analytical solution)
clearvars; clc; close all
u=[1, 2];              % ICs: u0 at t0
tmax=15;               % max. time limit
h=0.25;                % time step size
t=0:h:tmax;            % time space
steps=length(t);       % # of steps
f = @(t, u)([u(2), -u(2)+sin(t)]);
for k=1:steps-1
    u(k+1,:)=u(k,:)+f(t(k), u(k,:)).*h;
end
plot(t, u(:,1),'k--o'), grid minor; hold on
% Analytical solution of the problem:
syms u(T) T
Du=diff(u, T);
D2u=diff(u, T, 2);
Equation=D2u==sin(T)-Du;
ICs = [u(0)==1, Du(0)==2];
u=dsolve(Equation, ICs);
U=double(subs(u,'T',t));
plot(t, U, 'g-', 'linewidth', 1.5)
title('\it Solutions of: $$ \frac{d^2u}{dt^2}=sin(t)-\frac{du}{dt}, u_0=1,
du_0=2 $$', 'Interpreter', 'latex')
legend('EULER forward', 'DSOLVE (Analytic)', 'location', 'southeast')
xlabel '\it t', ylabel('\it Solution, u(t)')
axis tight, hold off
```

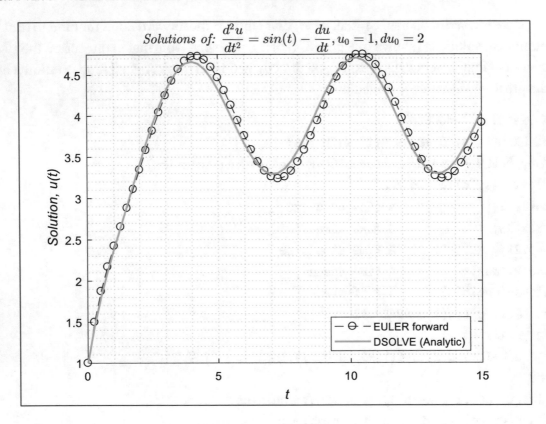

Figure 3-4. *Simulation results of Example 4:* $\ddot{u} = -\dot{u} + \sin(t),\ \ u(0) = 1,\ \dot{u}(0) = 2$

Example 5

Given a second-order nonhomogeneous ODE, here's the example problem:
$\ddot{u} + \dot{u} = \sin(ut),\ u(0) = 1, \dot{u}(0) = 2$.

Note that this is a nonlinear and nonhomogeneous second-order ODE. Let's test again the importance of choosing the step size for the Euler method.

Again, before creating a script, we rewrite the given problem with two first-order ODEs.

$\ddot{u} = -\dot{u} + \sin(ut)$ is expressed with this: $\begin{cases} \dot{u}_1 = u_2 \\ \dot{u}_2 = -u_2 + \sin(u_1 t) \end{cases}$

Note that $u_1 = u$ and $\dot{u}_2 = \ddot{u}$.

Here is a script (Example5_2ndODE.m) that embeds the forward Euler method to find numerical solutions (shown in Figure 3-3) of the problem. Note that in our script, the system of first-order ODEs is implemented in the script directly within a [for ... end] loop, an anonymous function (@), symbolic `matlabFunction`, and a function file. The chosen step sizes are 0.025, 0.05, 0.10, and 0.20.

```
% Example5_2ndODE.m
%% EXAMPLE 5.  u"=-u'+sin(t*u) with ICs: u(0)=1 & u'(0)=2
% solved by Forward EULER  Method
clearvars; clc; close all
h1=0.025;              % Time step size
tmax=13;               % End of simulations
t0=0;                  % Start of simulations
t=t0:h1:tmax;          % Time space
steps=length(t);       % How many steps of evaluations
u(1,:)=[1 2];          % Initial Conditions
% NB: Size of the u(t) is two-dimensional
% Direct method (direct expression)
for k=1:steps-1
    f1=[u(k,2), -u(k,2)+sin(t(k)*u(k,1))];
    u(k+1,:)=u(k,:)+f1*h1;
end
plot(t, u(:,1),'b-', 'linewidth', 2), grid minor
hold on
clearvars t u
h2=0.05;               % Time step size
t=t0:h2:tmax;
steps=length(t);
u(1,:)=[1 2];          % Initial Conditions
% Way # 2. Anonymous function (@)
F = @(t, u1, u2)([u2, -u2+sin(t*u1)]);
for k=1:steps-1
    u(k+1,:)=u(k,:)+F(t(k), u(k,1), u(k,2))*h2;
end
plot(t, u(:,1),'g-.', 'linewidth', 2), grid on
clearvars t u
h3=0.1;                % Step size
t=t0:h3:tmax;          % Time space
steps = numel(t);      % Number of steps
```

```matlab
u(1,:)=[1 2];           % Initial Conditions
% Way # 3. MatlabFunction
syms T u1 u2
 F=[u2, -u2+sin(T*u1)];
 matlabFunction(F, 'file', 'F5');
for k=1:steps-1
    u(k+1,:)=u(k,:)+h3*feval(@F5, t(k), u(k,1), u(k,2));
end
plot(t, u(:,1),'r:', 'linewidth', 2), grid on
clearvars t u
h4=0.2;                 % Step size
t=t0:h4:tmax;           % Time space
steps = numel(t);       % Number of steps
u(1,:)=[1 2];           % Initial Conditions
% Way # 4. Function File: FuFu5.m
for k=1:steps-1
    u(k+1,:)=u(k,:)+h4*feval(@FuFu5, t(k), u(k,1), u(k,2));
end
plot(t, u(:,1),'k--', 'linewidth', 2), grid on
title('\it Numerical Solutions of: $$ \frac{d^2u}{dt^2}=sin(t*u)-\frac{du}
{dt}, u_0 = 1, du_0 = 2 $$', 'interpreter', 'latex')
xlabel('\it t'), ylabel('\it Solution, u(t)')
xlim([0, 23])
legend('h = 0.025', 'h = 0.05', 'h = 0.1', 'h = 0.2', 'location',
'southeast')
axis tight
 function du = FuFu5(T, u1, u2)
du = [u2, -u2+sin(T*u1)];
end
```

From the simulation results shown in Figure 3-5, it is clear that if the chosen step size is not appropriate, the found numerical solutions will diverge from the true solutions substantially.

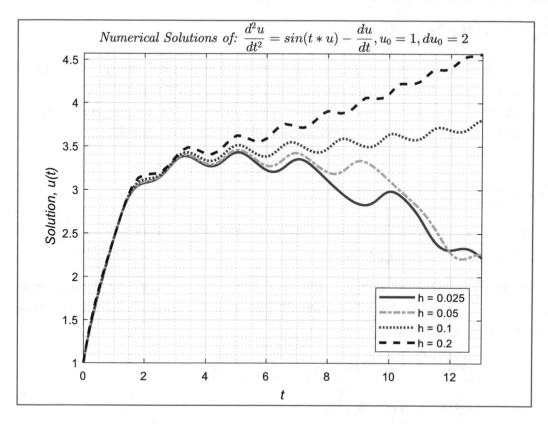

Figure 3-5. *Simulation results of example:* $\ddot{u} = -\dot{u} + sin(ut)$, $u(0)=1$, $\dot{u}(0)=2$

Based on our studied examples, we can make the following generalized remarks about Euler methods:

- They are simple and easy to implement via scripts. Thus, it is logical to start with them before using other higher-order and more complicated or computationally costly methods.

- They are good to compute crude and approximate solutions. However, their efficiency and accuracy depend on a chosen step size and a given ODE type, such as linear or nonlinear, homogeneous or nonhomogeneous, stiff or nonstiff.

- The accuracy of computed solutions with Euler methods can be assessed by comparing them with other methods.

Runge-Kutta Method

We have already formulated Runge-Kutta methods of different orders and gone through their implementation algorithm steps to solve first-order ODEs. Therefore, in this section, we will focus on how to implement the classical 4/5-order Runge-Kutta method via scripts to find numerical solutions of second- and higher-order ODEs.

Example 6

Given a second-order non-homogeneous ODE, here's the example problem:

$$\frac{1}{2}\ddot{u}+\frac{2}{5}\dot{u}+u=2t, \ \ u(0)=1, \dot{u}(0)=2.$$

First in implementing Runge-Kutta method, we need to rewrite the given second-order ODE as a system of two first-order ODE equations by introducing new variables similar to Euler's method.

$$\ddot{u}=4t-\frac{4}{5}\dot{u}-2u \ \text{ is converted to:} \ \begin{cases} \dot{u}_1=u_2 \\ \dot{u}_2=4t-\dfrac{4}{5}u_2-2u_1 \end{cases}$$

Note that $u_1=u$ and $\dot{u}_2=\ddot{u}_1$.

There are a few different ways in which the previously written system of first-order ODEs can be implemented. They are anonymous function (@), function file, symbolic matlabFunction, and direct method. The following script, RK_2ndODE_ex6.m, computes numerical solutions of the given problem based on the 4th-order Runge-Kutta method. In this example, we take different step sizes, where $h = 0.1, 0.2, 0.3$, and 0.4, and compare the found numerical solutions with the analytical solution computed with dsolve().

```
% RK_2ndODE_ex6.m
% EXAMPLE 6. (1/2)*u"+(2/5)*u'+u=2*t with ICs: u(0)=1 & u'(0)=2
% by 4th Order RUNGE-KUTTA Method and DSOLVE
clearvars; close all
h=0.1;                  % Time step size
t0=0;                   % Start of calc's/simulations
tmax=5;                 % End of calc's/simulations
t=t0:h:tmax;            % Time space
steps=length(t);        % Number of steps
u(1,:)=[1 2];           % Initial Conditions
```

```
% NB: Size of u(t) is two-dimensional.
% Way #1. Anonymous Function (@)
FuFu=@(t, u1, u2)([u2, 4*t-2*u1-(4/5)*u2]);
%%% for Way #1 Anonymous Function (@) FuFu:
for ii=1:steps-1
k1=FuFu(t(ii), u(ii,1), u(ii,2));
k2=FuFu(t(ii)+h/2, u(ii,1)+h*k1(:,1)/2,u(ii,2)+h*k1(:,2)/2);
k3=FuFu(t(ii)+h/2,u(ii,1)+h*k2(:,1)/2,u(ii,2)+h*k2(:,2)/2);
k4=FuFu(t(ii)+h,u(ii,1)+h*k3(:,1),u(ii,2)+h*k3(:,2));
u(ii+1,:)=u(ii,:)+h*(k1+2*k2+2*k3+k4)/6;
end
plot(t(:)', u(:,1), 'b:o', 'markersize', 9), grid on, hold on
clearvars t u
h=0.2;       % time step size
t=t0:h:tmax; steps=length(t);     u(1,:)=[1 2];
% Way #2. Function file called: F_RK_ex1.m
%%% with the function file called F.m  % Way #2
for ii = 1:steps-1
k1=feval(@RK_ex1,t(ii), u(ii,1), u(ii,2));
k2=feval(@RK_ex1,t(ii)+h/2,u(ii,1)+h*k1(:,1)/2,u(ii,2)+h*k1(:,2)/2);
k3=feval(@RK_ex1,t(ii)+h/2,u(ii,1)+h*k2(:,1)/2,u(ii,2)+h*k2(:,2)/2);
k4=feval(@RK_ex1,t(ii)+h,u(ii,1)+h*k3(:,1),u(ii,2)+h*k3(:,2));
u(ii+1,:)=u(ii,:)+h*(k1+2*k2+2*k3+k4)/6;
end
plot(t(:)', u(:,1), 'g-p')
clearvars  t u
h=0.3;        % time step size
t=t0:h:tmax; steps=length(t);     u(1,:)=[1 2];
% Way #3. matlabFunction
 syms T u1 u2
 f=[u2, 4*T-2.0*u1-(4/5)*u2];
 matlabFunction(f, 'file', 'F_ex1');
%%% with a function file F
for ii = 1:steps-1
k1 = feval(@F_ex1, t(ii), u(ii,1), u(ii,2));
k2 = feval(@F_ex1, t(ii), u(ii,1)+h*k1(:,1)/2,u(ii,2)+h*k1(:,2)/2);
```

```matlab
k3 = feval(@F_ex1, t(ii), u(ii,1)+h*k2(:,1)/2,u(ii,2)+h*k2(:,2)/2);
k4 = feval(@F_ex1, t(ii), u(ii,1)+h*k3(:,1),u(ii,2)+h*k3(:,2));
u(ii+1,:)=u(ii,:)+h*(k1+2*k2+2*k3+k4)/6;
end
plot(t(:)', u(:,1), 'r--h', 'linewidth', 2)
clearvars  t u
h=0.4;        % time step size
t=t0:h:tmax; steps=length(t);     u(1,:)=[1 2];
% Way # 4. Direct Way
for ii = 1:steps-1
k1 = [u(ii,2), 4*t(ii)-2*u(ii,1)-(4/5)*u(ii,2)];
k2 = [u(ii,2)+h*k1(:,1)/2, 4*(t(ii)+h/2)-...
    2*(u(ii,1)+h*(k1(:,1))/2)-(4/5)*(u(ii,2)+h*k1(:,2)/2)];
k3 = [u(ii,2)+h*k2(:,1)/2,    4*(t(ii)+h/2)- ...
    2*(u(ii,1)+h*k2(:,1)/2)-(4/5)*(u(ii,2)+h*k2(:,2)/2)];
k4 = [u(ii,2)+h*k3(:,1), 4*(t(ii)+h)-2*(u(ii,1)+h*k3(:,1))- ...
    (4/5)*(u(ii,2)+k3(:,2)*h)];
u(ii+1,:)=u(ii,:)+h*(k1+2*k2+2*k3+k4)/6;
end
plot(t(:)', u(:,1), 'm-.d', 'linewidth', 1.5)
% Analytical solution of the problem:
y=dsolve('D2x=-(4/5)*Dx-2*x+(4*t)', 'x(0)=1, Dx(0)=2','t');
Y=double(subs(y,'t',t));
plot(t, Y, 'k-', 'linewidth', 1.5)
legend('h = 0.1','h = 0.2','h = 0.3','h = 0.4', 'Analytical Solution',
'location', 'southeast')
title('\it Runge-Kutta Solutions vs. Analytical Solutions: $$ \frac{1}{2}*\
frac{d^2u}{dt^2}+\frac{2}{5}*\frac{du}{dt}=2*t$$', 'interpreter', 'latex')
xlim([0, 5])
xlabel('\it t')
ylabel('\it Solution, u(t)')
hold off

function f=RK_ex1(t, u1, u2)
f = [u2,4*t-2.0*u1-(4/5)*u2];
end
```

This example (shown in Figure 3-6) demonstrates clearly the importance of the chosen step size, which influences the computed solutions considerable like with the Euler method.

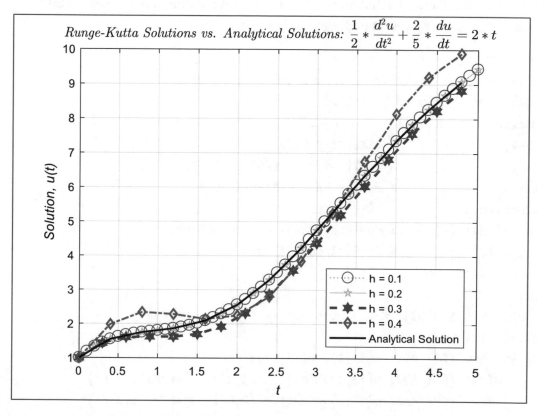

Figure 3-6. *Runge-Kutta (4/5-order) method versus analytical solution with dsolve()*

Example 7

Here's the example problem: $\ddot{u} + 6\dot{u} + 9u = 0$, $u(0) = 2$, $\dot{u}(0) = 0$.

We express the given equation via two first-order ODEs by introducing new variables.

$\ddot{u} = -6\dot{u} - 9u$ is rewritten as follows: $\begin{cases} \dot{u}_1 = u_2 \\ \dot{u}_2 = 6u_2 - 9u_1 \end{cases}$

Note that $u_1 = u$ and $\dot{u}_2 = \ddot{u}_1$.

Here is a script called RK_2ndODE_ex7.m that embeds the 4th-order Runge-Kutta and forward Euler methods and compares their numerical solutions with the analytical solution found with dsolve(). Note that the given system is expressed for Runge-Kutta with the matlabFunction that automatically generates a function file. For the Euler method, the given system is implemented directly.

```
% RK_2ndODE_ex7.m
% EXAMPLE 7. x"+9x'+6x=0 with ICs: x(0)=2, Dx(0)=0
% 4th Order RUNGE-KUTTA Method
clearvars; close all
h=0.2;                 % time step size
t0=0;                  % Start of calc's/simulations
tmax=3;                % End of calc's/simulations
t=t0:h:tmax;           % Time space
steps=length(t);       % How many steps of evaluations
u(1,:)=[2 0];          % Initial Conditions
% Runge-Kutta method
syms u1 u2
f=[u2, -6.*u1-9*u2]; matlabFunction(f, 'file', 'RK_ex2');
for ii=1:steps-1
    k1=feval(@RK_ex2, u(ii,1), u(ii,2));
    k2=feval(@RK_ex2, u(ii,1)+h*k1(:,1)/2,u(ii,2)+h*k1(:,2)/2);
    k3=feval(@RK_ex2, u(ii,1)+h*k2(:,1)/2,u(ii,2)+h*k2(:,2)/2);
    k4=feval(@RK_ex2, u(ii,1)+h*k3(:,1),u(ii,2)+h*k3(:,2));
    u(ii+1,:)=u(ii,:)+h*(k1+2*k2+2*k3+k4)/6;
end
plot(t(:)', u(:,1), 'ko'), hold on
% Euler forward method
for k=1:steps-1
    f1=[u(k,2), -6*u(k,1)-9*u(k,2)];
    u(k+1,:)=u(k,:)+f1*h;
end
plot(t, u(:,1), 'b--*')
% Analytical solution:
y=dsolve('D2x=-6*x-9*Dx', 'x(0)=2, Dx(0)=0','t');
Y=double(subs(y,'t',t)); plot(t, Y, 'g-', 'linewidth', 2)
```

```
grid on, xlabel('\it t'), ylabel('\it Solution, u(t)')
legend('RK: h = 0.2 ','EM: h = 0.2',  'Analytical (dsolve)')
title('Runge-Kutta, Euler Forward vs. Analytical Solutions of: $$ \
frac{d^2}{dt^2}+9*\frac{du}{dt}+6*u=0 $$', 'interpreter', 'latex')
xlim([0, 3])
hold off; shg
```

This is a moderately stiff problem for that forward Euler method with a large step size ($h=0.2$). It performs less well than the Runge-Kutta method, as shown in Figure 3-7.

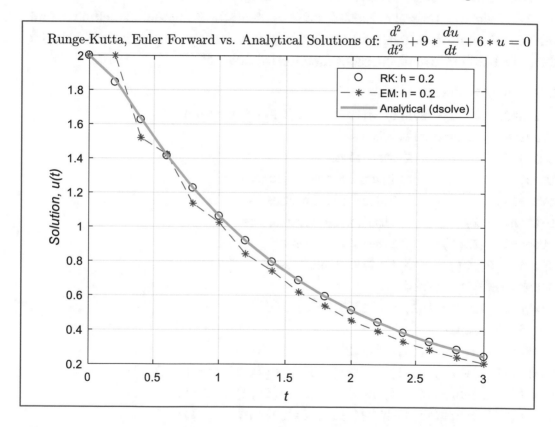

Figure 3-7. *Simulation of the problem: $\ddot{u}+6\dot{u}+9u=0$, $u(0)=2, \dot{u}(0)=0$*

Example 8

Given a second-order homogeneous ODE, here's the example problem: $\ddot{u} + \sin(t) = 0$, $u(0) = 0, \dot{u}(0) = 0$.

Again, before starting to create a script, we rewrite the given problem via two first-order ODEs.

$\ddot{u} = -\sin(t)$ is written as: $\begin{cases} \dot{u}_1 = u_2 \\ \dot{u}_2 = -\sin(t) \end{cases}$

Note that $u_1 = u$ and $\dot{u}_2 = \ddot{u}_1$.

The following script, RK_2ndODE_ex8.m, embeds the Runge-Kutta method to find numerical solutions of the problem. Note that, in this script, the system of first-order ODEs is defined via an anonymous function (@) called FF.

```
% RK_2ndODE_ex8.m
% EXAMPLE 8. u"=-sin(t) with ICs: u(0)=0 & u'(0)=0
clearvars; close all; clc
h=0.3;              % Step size
t0=0;               % Start of simulations
tmax=13;            % End of simulations
t=t0:h:tmax;        % Simulation time space
steps=length(t);    % Simulation steps
u(1,:)=[0, 0];      % Initial Conditions
% Runge-Kutta Method:
FF=@(t, u1, u2)([u2,-sin(t)]);
for ii=1:steps-1
k1=FF(t(ii), u(ii,1), u(ii,2));
k2=FF(t(ii)+h/2, u(ii,1)+h*k1(:,1)/2,u(ii,2)+h*k1(:,2)/2);
k3=FF(t(ii)+h/2,u(ii,1)+h*k2(:,1)/2,u(ii,2)+h*k2(:,2)/2);
k4=FF(t(ii)+h,u(ii,1)+h*k3(:,1),u(ii,2)+h*k3(:,2));
u(ii+1,:)=u(ii,:)+h*(k1+2*k2+2*k3+k4)/6;
end
plot(t, u(:,1), 'k--+'), grid on; hold on
% Euler forward method:
for ii=1:steps-1
u(ii+1,:)=u(ii,:)+h*[u(ii,2), -sin(t(ii))];
end
```

```
plot(t, u(:,1), 'r-.p')
% Analytical solution:
y=dsolve('D2y=-sin(t)', 'y(0)=0, Dy(0)=0','t');
Y=double(subs(y,'t',t));
plot(t, Y, 'm-', 'linewidth', 1.5)
title('Runge-Kutta, Euler Forward vs. Analytical Solutions of: $$ \
frac{d^2}{dt^2}+sin(t)=0 $$', 'interpreter', 'latex')
xlabel '\it t'; ylabel '\it Solution, u(t)'
legend('RK: h = 0.3', 'EM: h = 0.3', 'DSOLVE')
hold off; axis tight; shg
```

Again, the simulation results displayed in Figure 3-8 show that numerical solutions computed with the Runge-Kutta method converge with the analytical solution well and the ones computed with the forward Euler method have some offset from the analytical solution values.

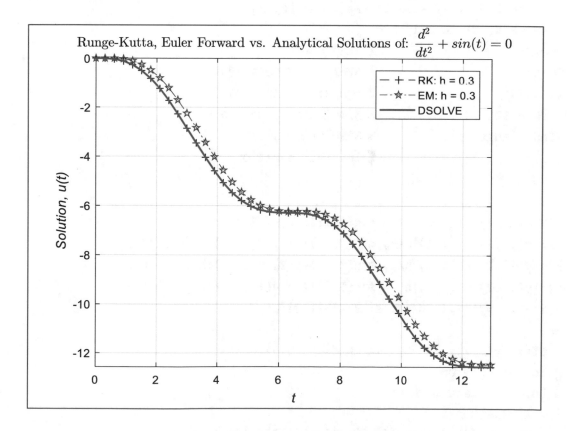

Figure 3-8. *Simulation of the problem* $\ddot{u} + sin(t) = 0$, $u(0) = 0, \dot{u}(0) = 0$

Example 9

Given a second-order nonhomogeneous ODE, here's the example problem:
$\ddot{y} + \dot{y} = \sin(t)$, $y(0)=1$, $\dot{y}(0)=2$.

Again, before starting to write a script, we rewrite the given problem with two first order ODEs.

$\ddot{y} = -\dot{y} + \sin(t)$ is expressed with this: $\begin{cases} \dot{y}_1 = y_2 \\ \dot{y}_2 = -y_2 + \sin(t) \end{cases}$

Note that $y_1 = y$ and $\dot{y}_2 = \ddot{y}_1$.

The following script (RK_2ndODE_ex9.m) embeds the Runge-Kutta method versus the forward Euler methods to compute numerical solutions of the problem. Note that in this script, the system of first-order ODEs is embedded directly as defined in the previously rewritten two first-order ODEs for the Runge-Kutta and forward Euler methods.

```
% RK_2ndODE_ex9.m
% EXAMPLE 9. y"+y'=sin(t) with ICs: [1, 2]
clearvars; close all; clc
h=0.25;                   % Step size
t0=0;                     % Start of simulations
tmax=13;                  % End of simulations
t=t0:h:tmax;              % Simulation time space
steps=length(t);          % Simulation steps
y(1,:)=[1, 2];            % Initial Conditions
% Runge-Kutta method:
for ii=1:steps-1
k1=[y(ii,2),                sin(t(ii))-y(ii,2)];
k2=[y(ii,2)+k1(:,1)*h/2, sin(t(ii)+h/2)-y(ii,2)+k1(:,2)*h/2];
k3=[y(ii,2)+k2(:,1)*h/2, sin(t(ii)+h/2)-y(ii,2)+k2(:,2)*h/2];
k4=[y(ii,2)+k3(:,1)*h,   sin(t(ii)+h)-y(ii,2)+k3(:,2)*h];
y(ii+1,:)=y(ii,:)+h*(k1+2*k2+2*k3+k4)/6;
end
plot(t, y(:,1), 'k--', 'linewidth', 1.5)
grid on; hold on
% Euler forward method:
for ii=1:steps-1
y(ii+1,:)=y(ii,:)+h*[y(ii,2), sin(t(ii))-y(ii,2)];
end
```

136

```
plot(t, y(:,1), 'r-.+', 'linewidth', 1.5)
% Analytical solution:
y=dsolve('D2y=-Dy+sin(t)', 'y(0)=1, Dy(0)=2','t');
Y=double(subs(y,'t',t));
plot(t, Y, 'm-', 'linewidth', 1.5)
title('\it Runge-Kutta, Euler Methods vs. Analytical Solution of: $$ \
frac{d^2u}{dt^2}=sin(t)-\frac{du}{dt} $$', 'Interpreter', 'latex')
xlabel '\it t'; ylabel '\it Solution, u(t)'
legend('Runge-Kutta: h = 0.25', 'Euler: h = 0.25', 'Analytical Solution',
'location', 'southeast')
hold off; axis tight; shg
```

From the simulation results shown in Figure 3-9 of the problem $\ddot{y} + \dot{y} = \sin(t)$, $y(0) = 1$, $\dot{y}(0) = 2$, it is clear that with a considerably big step size ($h = 0.25$), the Runge-Kutta method performs less well than the forward Euler method even though it is four times costlier in terms of computation time and simulation steps.

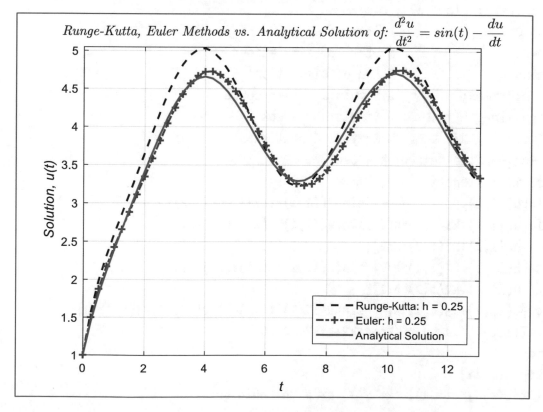

Figure 3-9. *Simulation of the problem:* $\ddot{y} + \dot{y} = \sin(t)$, $y(0) = 1$, $\dot{y}(0) = 2$

Example 10

Given a second-order non-homogeneous and nonlinear ODE, here's the example problem: $\ddot{u} + \dot{u} = \sin(ut)$, $u(0)=1, \dot{u}(0)=2$.

Again, before writing a script, we express the given problem with two first-order ODEs.

$\ddot{u} = -\dot{u} + \sin(ut)$ is expressed with this: $\begin{cases} u_2 = \dot{u}_1 \\ \dot{u}_2 = -u_2 + \sin(u_1 t) \end{cases}$

Note that $u_1 = u$ and $\dot{u}_2 = \ddot{u}$.

Here is a script called RK_2ndODE_ex10.m that embeds the Runge-Kutta and Euler forward methods and compares their results with two different step sizes: $h = 0.05$ and $h = 0.01$. The system of first-order ODEs for both methods is expressed directly within a for ... end loop.

```
% RK_2ndODE_ex10.m
% EXAMPLE 10.  u"=-u'+sin(t*u) with ICs: u(0)=1 & u'(0)=2
clearvars; close all; clc
h=0.05;                % Step size
t0=0;                  % Start of simulations
tmax=15;               % End of simulations
t=t0:h:tmax;           % Simulation time space
steps=length(t);       % Simulation steps
u(1,:)=[1, 2];         % Initial Conditions
% Runge-Kutta Method: h = 0.05
for ii=1:steps-1
k1=[u(ii,2),               sin(u(ii,1)*t(ii))-u(ii,2)];
k2=[u(ii,2)+k1(:,1)*h/2, sin(u(ii,1)*t(ii)+h/2)-...
    u(ii,2)+k1(:,2)*h/2];
k3=[u(ii,2)+k2(:,1)*h/2, sin(u(ii,1)*t(ii)+h/2)- ...
    u(ii,2)+k2(:,2)*h/2];
k4=[u(ii,2)+k3(:,1)*h,   sin(u(ii,1)*t(ii)+h)-u(ii,2)+k3(:,2)*h];
u(ii+1,:)=u(ii,:)+h*(k1+2*k2+2*k3+k4)/6;
end
subplot(211)
plot(t(:)', u(:,1), 'k--x'), grid on; hold on
```

```
title('Runge-Kutta vs. Euler Solutions of: $$ \frac{d^2u}{dt^2}+\frac{du}
{dt}=sin(t*u) $$', 'interpreter', 'latex')
xlabel '\it t'; ylabel '\it Solution, u(t)'

clearvars t u k1 k2 k3 k4
h=0.01;              % Step size
t=t0:h:tmax; steps=length(t);     u(1,:)=[1, 2];
% Runge-Kutta Method: h = 0.01
for ii=1:steps-1
k1=[u(ii,2),                sin(u(ii,1)*t(ii))-u(ii,2)];
k2=[u(ii,2)+k1(:,1)*h/2, sin(u(ii,1)*t(ii)+h/2)-...
    u(ii,2)+k1(:,2)*h/2];
k3=[u(ii,2)+k2(:,1)*h/2, sin(u(ii,1)*t(ii)+h/2)-...
    u(ii,2)+k2(:,2)*h/2];
k4=[u(ii,2)+k3(:,1)*h,   sin(u(ii,1)*t(ii)+h)-u(ii,2)+k3(:,2)*h];
u(ii+1,:)=u(ii,:)+h*(k1+2*k2+2*k3+k4)/6;
end
subplot(212)
plot(t(:)', u(:,1), 'k--x'), hold on
clearvars -except t0 tmax
h=0.05;             % Step size
t=t0:h:tmax;
steps=length(t);
u(1,:)=[1, 2];
% Euler forward method: h = 0.05
for ii=1:steps-1
u(ii+1,:)=u(ii,:)+h*[u(ii,2), sin(u(ii,1)*t(ii))-u(ii,2)];
end
subplot(211)
plot(t(:)', u(:,1), 'r-', 'linewidth', 1.5), grid on
legend('RK: h = 0.05', 'EM: h = 0.05')
clearvars t u
h=0.01;             % Step size
t=t0:h:tmax;
steps=length(t);
u(1,:)=[1, 2];
```

139

```
% Euler forward method: h = 0.01
for ii=1:steps-1
u(ii+1,:)=u(ii,:)+h*[u(ii,2), sin(u(ii,1)*t(ii))-u(ii,2)];
end
subplot(212)
plot(t(:)', u(:,1), 'r-', 'linewidth',1.5), grid on
xlabel '\it t'; ylabel '\it Solution, u(t)'
legend( 'RK: h = 0.01',  'EM: h = 0.01')
hold off; axis tight; shg
```

From the simulation results (shown in Figure 3-10) of this example, we can see that both methods are very sensitive to the chosen step size. With a considerably small step size ($h = 0.01$), both methods (shown in Figure 3-10, lower subplot) converge very well.

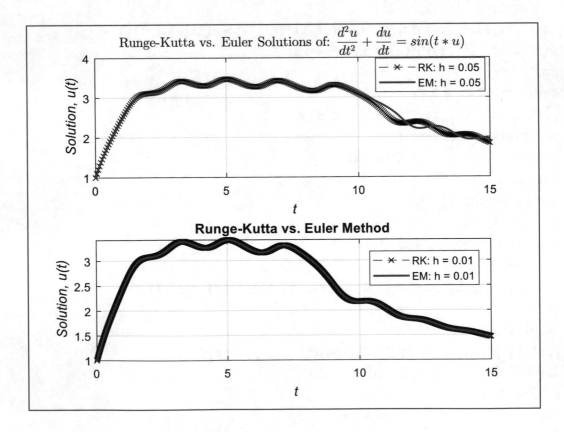

Figure 3-10. *Simulation results of example:* $\ddot{u} = -\dot{u} + sin(ut)$, $u(0) = 1, \dot{u}(0) = 2$.

As you can see from the exercises of second-order ODEs, it is clear that the Runge-Kutta method outperforms the forward Euler method in terms of accuracy when solving linear and nonhomogeneous ODEs. On the other hand, the forward Euler method performs considerably well when solving nonlinear and nonhomogeneous ODEs.

Adams-Moulton Method

In the previous sections of first-order ODEs, we formulated the Adams-Moulton method with first-, second-, and fifth-order order and described its implementation algorithm to solve the IVPs of ODEs. This method, unlike the Euler method or the Runge-Kutta method, is a predictor-corrector method defined with implicit formulations. In this context, we will focus on how to implement the first-, second-, third-, fourth-, and fifth-order Adams-Moulton method via MATLAB scripts to compute the numerical solutions of second-order ODEs.

Example 11

Here's the example problem: $\ddot{y} = -\dot{y} + \sin(t)$, $y(0) = 1$, $\dot{y}(0) = 2$.

Again, before writing a script, we express the given problem with two first-order ODEs.

$\ddot{y} = -\dot{y} + \sin(t)$ is expressed with this: $\begin{cases} y_2 = \dot{y}_1 \\ \dot{y}_2 = -y_2 + \sin(t) \end{cases}$

Note that $y_1 = y$ and $\dot{y}_2 = \ddot{y}_1$.

Here is a script (Adams_Moulton_2ndODE_ex11.m) that embeds the Adams-Moulton method of different orders to compute numerical solutions of the problem. Note that in our script, the system of first-order ODEs is defined as a two-dimensional function via an anonymous function (@).

```
% ADAMS-Moulton 1st order method
% Adams_Moulton_2ndODE_ex11.m
% Part 1
clearvars; close all
Fn(1,:)=[1, 2];        % ICs: u0 at t0
tend=15;               % max. time limit
h=0.10;                % time step size
```

```
t=0:h:tend;            % time space
steps=length(t);       % # of steps
% Function is defined via anonymous function:
Fcn =@(t, y1, y2)([y2, -y2+sin(t)]);
for ii=1:steps-1
% Predicted solution by EULER's forward method here:
Fn(ii+1,:)=Fn(ii,:)+h*Fcn(t(ii), Fn(ii, 1),Fn(ii,2));
% ADAMS-Moulton method: 2-step starts from here
Fn(ii+1,:)=Fn(ii,:)+h*Fcn(t(ii+1),Fn(ii, 1),Fn(ii,2));
end
plot(t, Fn(:,1), 'bs-'), grid on; hold on

% ADAMS-Moulton 2-step Method
% Part 2
clearvars Fn
Fn(1,:)=[1, 2];        % ICs: u0 at t0
Fcn =@(t, y1, y2)([y2, -y2+sin(t)]);
for ii=1:steps-1
% EULER's forward method is used here
Fn(ii+1,:)=Fn(ii,:)+h*Fcn(t(ii), Fn(ii, 1),Fn(ii,2));
% ADAMS-Moulton method: 2-step starts from here
Fn(ii+1,:)=Fn(ii,:)+(h/2)*(Fcn(t(ii+1),Fn(ii+1,1),Fn(ii+1,2))+...
Fcn(t(ii),Fn(ii,1),Fn(ii,2)));
end
plot(t, Fn(:,1), 'mx--')

% ADAMS-Moulton 3-step Method
% Part 3
clearvars Fn
Fn(1,:)=[1, 2];
Fcn =@(t, y1, y2)([y2, -y2+sin(t)]);
for k=1:steps-2
% 1st: Predicted value by EULER's forward method
Fn(k+1,:)=Fn(k,:)+h*Fcn(t(k), Fn(k, 1),Fn(k,2));
% 2nd: Corrected value by trapezoidal rule
Fn(k+1,:)=Fn(k,:)+(h/2)*(Fcn(t(k+1),Fn(k+1,1),Fn(k+1,2))+ ...
```

```
Fcn(t(k),Fn(k,1),Fn(k,2)));
% Predicted solution:
Fn(k+2,:)=Fn(k+1,:)+(3*h/2)*Fcn(t(k+1),Fn(k+1,1),Fn(k+1,2))-...
    (h/2)*Fcn(t(k),Fn(k,1),Fn(k,2));
% ADAMS-Moulton method: 3-step starts from here
Fn(k+2,:)=Fn(k+1,:)+h*((5/12)*Fcn(t(k+2),Fn(k+2,1),Fn(k+2,2))+...
    (2/3)*Fcn(t(k+1),Fn(k+1,1),Fn(k+1,2))-...
    (1/12)*Fcn(t(k),Fn(k,1),Fn(k,2)));
end
plot(t, Fn(:,1), 'ko:')

% ADAMS-Moulton 4-step Method
% Part 4
clearvars Fn
Fn(1,:)=[1, 2];
Fcn =@(t, y1, y2)([y2, -y2+sin(t)]);
for k=1:steps-3
% 1st Predicted value: by Euler's forward method
Fn(k+1,:)=Fn(k,:)+h*Fcn(t(k),Fn(k,1),Fn(k,2));
% 1st Corrected value: by trapezoidal rule
Fn(k+1,:)=Fn(k,:)+(h/2)*(Fcn(t(k+1),Fn(k+1,1),Fn(k+1,2))+...
Fcn(t(k),Fn(k,1),Fn(k,2)));
% 2nd Predicted value
Fn(k+2,:)=Fn(k+1,:)+(3*h/2)*Fcn(t(k+1),Fn(k+1,1),Fn(k+1,2))-...
(h/2)*Fcn(t(k),Fn(k,1),Fn(k,2));
% 2nd Corrected value
Fn(k+2,:)=Fn(k+1,:)+h*((5/12)*Fcn(t(k+2),Fn(k+2,1),Fn(k+2,2))+...
(2/3)*Fcn(t(k+1),Fn(k+1,1),Fn(k+1,2))-...
(1/2)*Fcn(t(k),Fn(k,1),Fn(k,2)));
% Predicted solution:
Fn(k+3,:)=Fn(k+2,:)+h*((23/12)*Fcn(t(k+2),Fn(k+2,1),Fn(k+2,2))-...
(4/3)*Fcn(t(k+1),Fn(k+1,1),Fn(k+1,2))+...
(5/12)*Fcn(t(k),Fn(k,1),Fn(k,2)));
% Corrected solution by ADAMS-Moulton method 4-step from here:
Fn(k+3,:)=Fn(k+2,:)+h*((3/8)*Fcn(t(k+3),Fn(k+3,1),Fn(k+3,2))+...
(19/24)*Fcn(t(k+2),Fn(k+2,1),Fn(k+2,2))-...
```

```
(5/24)*Fcn(t(k+1),Fn(k+1,1),Fn(k+1,2))+...
(1/24)*Fcn(t(k),Fn(k,1),Fn(k,2)));
end
plot(t, Fn(:,1), 'k-.*')

% ADAMS-Moulton's 5th-order method
% Part 5
clearvars Fn
Fn(1,:)=[1, 2];
Fcn =@(t, y1, y2)([y2, -y2+sin(t)]);
for k=1:steps-4
% 1st Predicted value: by Euler's forward method
Fn(k+1,:)=Fn(k,:)+h*Fcn(t(k),Fn(k,1),Fn(k,2));
% 1st Corrected value: by trapezoidal rule
Fn(k+1,:)=Fn(k,:)+(h/2)*(Fcn(t(k+1),Fn(k+1,1),Fn(k+1,2))+...
Fcn(t(k),Fn(k,1),Fn(k,2)));
% 2nd Predicted value
Fn(k+2,:)=Fn(k+1,:)+(3*h/2)*Fcn(t(k+1),Fn(k+1,1),Fn(k+1,2))-...
(h/2)*Fcn(t(k),Fn(k,1),Fn(k,2));
% 2nd Corrected value
Fn(k+2,:)=Fn(k+1,:)+h*((5/12)*Fcn(t(k+2),Fn(k+2,1),Fn(k+2,2))+...
(2/3)*Fcn(t(k+1),Fn(k+1,1),Fn(k+1,2))-...
(1/2)*Fcn(t(k),Fn(k,1),Fn(k,2)));
% 3rd Predicted solution:
Fn(k+3,:)=Fn(k+2,:)+h*((23/12)*Fcn(t(k+2),Fn(k+2,1),Fn(k+2,2))-...
(4/3)*Fcn(t(k+1),Fn(k+1,1),Fn(k+1,2))+...
(5/12)*Fcn(t(k),Fn(k,1),Fn(k,2)));
% Corrected solution by ADAMS-Moulton method 4-step from here:
Fn(k+3,:)=Fn(k+2,:)+h*((3/8)*Fcn(t(k+3),Fn(k+3,1),Fn(k+3,2))+...
(19/24)*Fcn(t(k+2),Fn(k+2,1),Fn(k+2,2))-...
(5/24)*Fcn(t(k+1),Fn(k+1,1),Fn(k+1,2))+...
(1/24)*Fcn(t(k),Fn(k,1),Fn(k,2)));
% Predicted solution:
Fn(k+4,:)=Fn(k+3,:)+h*((55/24)*Fcn(t(k+3),Fn(k+3,1),Fn(k+3,2))-...
(59/24)*Fcn(t(k+2),Fn(k+2,1),Fn(k+2,2))+...
(37/24)*Fcn(t(k+1),Fn(k+1,1),Fn(k+1,2))-...
```

```
(3/8)*Fcn(t(k),Fn(k,1),Fn(k,2)));
% Corrected solution by ADAMS-Moulton 5-step method from here:
Fn(k+4,:)=Fn(k+3,:)+h*((251/720)*Fcn(t(k+4),Fn(k+4,1),Fn(k+4,2))+...
(646/720)*Fcn(t(k+3),Fn(k+3,1),Fn(k+3,2))-...
(264/720)*Fcn(t(k+2),Fn(k+2,1),Fn(k+2,2))+...
(106/720)*Fcn(t(k+1),Fn(k+1,1),Fn(k+1,2))-...
(19/720)*Fcn(t(k),Fn(k,1),Fn(k,2)));
end
plot(t, Fn(:,1), 'g--', 'linewidth', 2), grid on
title('Adams-Moulton method simulation of: $$ \frac{d^2y}{dt^2} = sin(t)-\
frac{dy}{dt} $$', 'interpreter', 'latex')
xlabel '\it t', ylabel('\it Solution, y(t)')
% Analytical solution of the problem:
Y=dsolve('D2u=-Du+sin(T)', 'u(0)=1, Du(0)=2','T');
Y=double(subs(Y,'T',t));
plot(t, Y, 'r-')
legend('1-step','2-step','3-step','4-step','5-step', 'DOLVE', 'location',
'southeast')
hold off
```

From the plot of the simulation results (shown in Figure 3-11) with Adams-Moulton methods of different orders or steps, it can be concluded that this method performs well for the given problem and its different order solvers produce well-converged results with the analytical solution of the problem computed with dsolve(). We can implement this method and the previous script for any second- or higher-order explicitly defined IVP of ODEs in a similar way with several changes in the script context according to a given problem. The only changes required to make in the script are the initial conditions and simulation time, and the final important change is an anonymous function (@) formulation that needs to be redefined according to a given problem formulation. For instance, we can easily edit the previous script to compute numerical solutions of another second-order ODE.

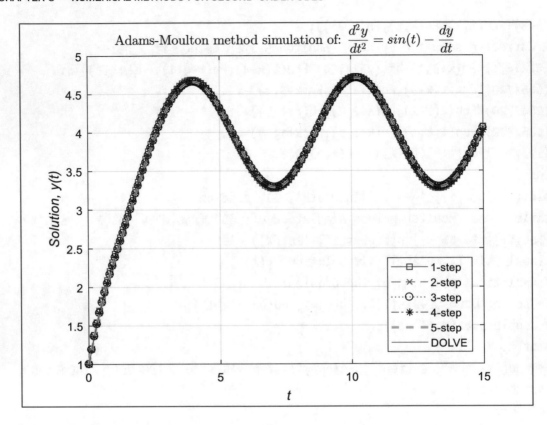

Figure 3-11. *Simulation of the example:* $\ddot{y}+\dot{y}=sin(t)$, $y(0)=1$, $\dot{y}(0)=2$

Example 12

Here's the example problem: $\ddot{y}+\dot{y}=sin(yt)$, $y(0)=1$, $\dot{y}(0)=2$.

By applying the one-, two-, three-, four-, and five-step Adams-Moulton methods, we can write a new script called Adams_Moulton_2ndODE_ex12.m by editing the previous script, Adams_Moulton_2ndODE_ex1.m, to solve the given nonhomogeneous ODE.

```
% Adams_Moulton_2ndODE_ex12.m
% EXAMPLE 12. y"+y'=sin(yt) with ICs: [1, 2]
% ADAMS-Moulton 1st-order method
% Part 1
clearvars; close all
Fn(1,:)=[1, 2];      % ICs: u0 at t0
tend=13;             % End of simulations
```

```
h=0.1;                  % Step size
t=0:h:tend;             % Simulation space
steps=length(t);   % # of steps
 % NB: y"+y'=sin(yt) =>DDy = f(t,y,dy)=sin(yt)-dy defined via          %
Anonymous Function (@):
Fcn =@(t, u1, u2)([u2, -u2+sin(u1*t)]);
for ii=1:steps-1
% 1-step: EULER's forward method is used here
Fn(ii,:)=Fn(ii,:)+h*Fcn(t(ii), Fn(ii, 1),Fn(ii,2));
% ADAMS-Moulton method starts from here
Fn(ii+1,:)=Fn(ii,:)+h*Fcn(t(ii+1),Fn(ii, 1),Fn(ii,2));
end
subplot(211)
plot(t, Fn(:,1), 'bs-'), grid on; hold on
```

%% ADAMS-Moulton 2-order Method

```
% Part 2
clearvars Fn
Fn(1,:)=[1, 2];
for ii=1:steps-1
% Predicted by EULER's forward method:
Fn(ii+1,:)=Fn(ii,:)+h*Fcn(t(ii), Fn(ii, 1),Fn(ii,2));
% Corrected by ADAMS-Moulton method
Fn(ii+1,:)=Fn(ii,:)+(h/2)*(Fcn(t(ii+1),Fn(ii+1,1),Fn(ii+1,2))+ ...
Fcn(t(ii),Fn(ii,1),Fn(ii,2)));
end
plot(t, Fn(:,1), 'm--p', 'markersize', 9), grid on
```

```
%% % ADAMS-Moulton 3rd-order Method
% Part 3
clearvars Fn
Fn(1,:)=[1, 2];
for ii=1:steps-2
% 1st: Predicted value by EULER's forward method
Fn(ii+1,:)=Fn(ii,:)+h*Fcn(t(ii), Fn(ii, 1),Fn(ii,2));
% 2nd: Corrected value by trapezoidal rule
```

```
Fn(ii+1,:)=Fn(ii,:)+(h/2)*(Fcn(t(ii+1),Fn(ii+1,1),Fn(ii+1,2))+...
Fcn(t(ii),Fn(ii,1),Fn(ii,2)));
% Predicted solution:
Fn(ii+2,:)=Fn(ii+1,:)+(3*h/2)*Fcn(t(ii+1),Fn(ii+1,1),Fn(ii+1,2))-...
(h/2)*Fcn(t(ii),Fn(ii,1),Fn(ii,2));
% ADAMS-Moulton method: 3-step starts from here
Fn(ii+2,:)=Fn(ii+1,:)+h*((5/12)*Fcn(t(ii+2),Fn(ii+2,1),Fn(ii+2,2))+...
(2/3)*Fcn(t(ii+1),Fn(ii+1,1),Fn(ii+1,2))-...
(1/12)*Fcn(t(ii),Fn(ii,1),Fn(ii,2)));
end
plot(t, Fn(:,1), 'k-', 'linewidth', 2.5), grid on
legend('1-step', '2-step','3-step'), hold off
title('Adams-Moulton method simulation: $$ \frac{d^2y}{dt^2} = sin(yt)-\
frac{dy}{dt} $$', 'interpreter', 'latex')
xlim([0, 13]), xlabel( '\it t'), ylabel('\it Solution, y(t)')

%% % ADAMS-Moulton 4-order Method
% Part 4
Clearvars Fn
Fn(1,:)=[1, 2];
for ii=1:steps-3
% 1st: Predicted value by EULER's forward method
Fn(ii+1,:)=Fn(ii,:)+h*Fcn(t(ii), Fn(ii, 1),Fn(ii,2));
% 2nd: Corrected value by trapezoidal rule
Fn(ii+1,:)=Fn(ii,:)+(h/2)*(Fcn(t(ii+1),Fn(ii+1,1), Fn(ii+1,2))+...
Fcn(t(ii),Fn(ii,1),Fn(ii,2)));
% 3rd: Predicted solution:
Fn(ii+2,:)=Fn(ii+1,:)+(3*h/2)*Fcn(t(ii+1),Fn(ii+1,1),Fn(ii+1,2))-...
(h/2)*Fcn(t(ii),Fn(ii,1),Fn(ii,2));
% 3rd: Corrected ADAMS-Moulton method:
Fn(ii+2,:)=Fn(ii+1,:)+h*((5/12)*Fcn(t(ii+2),Fn(ii+2,1),Fn(ii+2,2))...
+(2/3)*Fcn(t(ii+1),Fn(ii+1,1),Fn(ii+1,2))-...
(1/12)*Fcn(t(ii),Fn(ii,1),Fn(ii,2)));
% Predicted solution:
Fn(ii+3,:)=Fn(ii+2,:)+h*((23/12)*Fcn(t(ii+2),Fn(ii+2,1),Fn(ii+2,2))...
-(4/3)*Fcn(t(ii+1),Fn(ii+1,1),Fn(ii+1,2))+...
```

```
(5/12)*Fcn(t(ii),Fn(ii,1),Fn(ii,2)));
% Corrected solution by ADAMS-Moulton method 4-step from here:
Fn(ii+3,:)=Fn(ii+2,:)+h*((3/8)*Fcn(t(ii+3),Fn(ii+3,1),Fn(ii+3,2))+...
(19/24)*Fcn(t(ii+2),Fn(ii+2,1),Fn(ii+2,2))-...
(5/24)*Fcn(t(ii+1),Fn(ii+1,1),Fn(ii+1,2))+...
(1/24)*Fcn(t(ii),Fn(ii,1),Fn(ii,2)));
end
subplot(212)
plot(t, Fn(:,1), 'k:o', 'markersize',9, 'markerfacecolor', 'y'), hold on

%% % ADAMS-Moulton 5-order Method
% Part 5
clearvars Fn
Fn(1,:)=[1, 2];
for ii=1:steps-4
% 1st: Predicted value by EULER's forward method
Fn(ii+1,:)=Fn(ii,:)+h*Fun(t(ii), Fn(ii, 1),Fn(ii,2));
% 2nd: Corrected value by trapezoidal rule
Fn(ii+1,:)=Fn(ii,:)+(h/2)*(Fun(t(ii+1),Fn(ii+1,1),Fn(ii+1,2))+...
Fun(t(ii),Fn(ii,1),Fn(ii,2)));
% 3rd: Predicted solution:
Fn(ii+2,:)=Fn(ii+1,:)+(3*h/2)*Fun(t(ii+1),Fn(ii+1,1),Fn(ii+1,2))-...
(h/2)*Fun(t(ii),Fn(ii,1),Fn(ii,2));
% 3rd: Corrected ADAMS-Moulton method:
Fn(ii+2,:)=Fn(ii+1,:)+h*((5/12)*Fun(t(ii+2),Fn(ii+2,1),Fn(ii+2,2))+...
(2/3)*Fun(t(ii+1),Fn(ii+1,1),Fn(ii+1,2))-...
(1/12)*Fun(t(ii),Fn(ii,1),Fn(ii,2)));
% Predicted solution:
Fn(ii+3,:)=Fn(ii+2,:)+h*((23/12)*Fun(t(ii+2),Fn(ii+2,1),Fn(ii+2,2))...
-(4/3)*Fun(t(ii+1),Fn(ii+1,1),Fn(ii+1,2))+...
(5/12)*Fun(t(ii),Fn(ii,1),Fn(ii,2)));
% Corrected solution by ADAMS-Moulton method 4-step from here:
Fn(ii+3,:)=Fn(ii+2,:)+h*((3/8)*Fun(t(ii+3),Fn(ii+3,1),Fn(ii+3,2))+...
(19/24)*Fun(t(ii+2),Fn(ii+2,1),Fn(ii+2,2))-...
(5/24)*Fun(t(ii+1),Fn(ii+1,1),Fn(ii+1,2))+...
(1/24)*Fun(t(ii),Fn(ii,1),Fn(ii,2)));
```

```
% Predicted solution:
Fn(ii+4,:)=Fn(ii+3,:)+h*((55/24)*Fun(t(ii+3),Fn(ii+3,1),Fn(ii+3,2)) ...
-(59/24)*Fun(t(ii+2),Fn(ii+2,1),Fn(ii+2,2))+...
(37/24)*Fun(t(ii+1),Fn(ii+1,1),Fn(ii+1,2))-...
(3/8)*Fun(t(ii),Fn(ii,1),Fn(ii,2)));
% Corrected solution by ADAMS-Moulton 5-step method from here:
Fn(ii+4,:)=Fn(ii+3,:)+ ...        h*((251/720)*Fun(t(ii+4),Fn(ii+4,1),
Fn(ii+4,2))+...
(646/720)*Fun(t(ii+3),Fn(ii+3,1),Fn(ii+3,2))-...
(264/720)*Fun(t(ii+2),Fn(ii+2,1),Fn(ii+2,2))+...
(106/720)*Fun(t(ii+1),Fn(ii+1,1),Fn(ii+1,2))-...
(19/720)*Fun(t(ii),Fn(ii,1),Fn(ii,2)));
end
plot(t, Fn(:,1), 'r-', 'linewidth', 2.5), grid on
legend( '4-step','5-step')
xlabel( '\it t'), ylabel('\it Solution, y(t)');
hold off; shg
xlim([0, 13])
```

From the plotted results (shown in Figure 3-12), we can conclude that the solutions found via Adams-Moulton's higher-order solvers (two-step, three-step, four-step, and five-step solver methods) converge closely. They are much more accurate than the one-step solver. However, the computation time costs are substantially higher with higher order, similar to the Adams-Bashforth and Runge-Kutta methods.

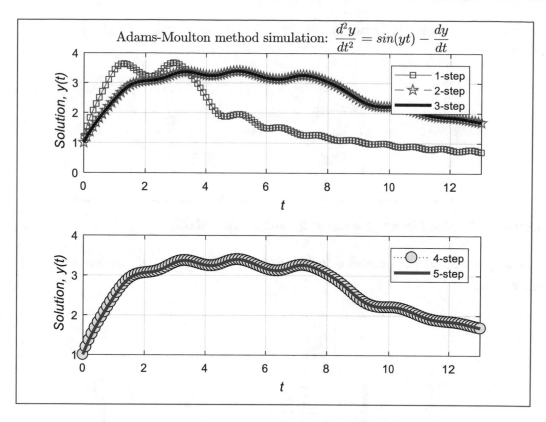

Figure 3-12. *Simulation of the example:* $\ddot{y} + \dot{y} = sin(ty)$, $y(0) = 1, \dot{y}(0) = 2$

Simulink Modeling

Solving second- or higher-order ODEs with Simulink modeling is very much like solving first-order ODEs except for needing one or more additional [Integrator] blocks to obtain a solution from a second- or higher-order derivative variable. Let's look at a few examples to demonstrate Simulink modeling to solve second-order ODEs.

Example 13

Here's the example problem: $\dfrac{1}{2}\ddot{u} + \dfrac{2}{5}\dot{u} + u = t$, $u(0) = 1, \dot{u}(0) = 2$.

We first rewrite the given second-order ODE before starting to model it, as shown here:

$$\ddot{u} = 2t - 0.8\dot{u} - 2u$$

Note that to obtain $u(t)$ from $\ddot{u}$, that must be integrated twice, as shown in Figure 3-13.

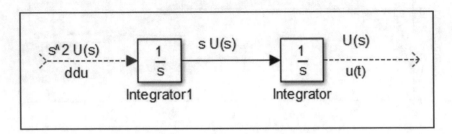

Figure 3-13. *Double integration with [Integrator] blocks*

The given problem is modeled in `Sim_Model_2ndODE_ex13.mdl`, which is shown in Figure 3-14.

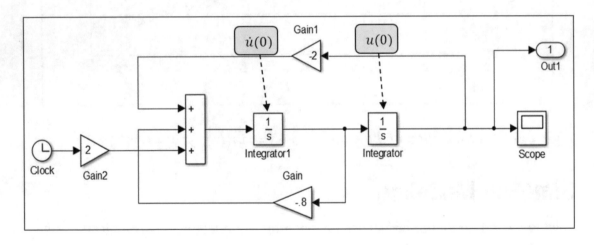

Figure 3-14. *Simulink model of the problem:* $\frac{1}{2}\ddot{u}+\frac{2}{5}\dot{u}+u=t$,
$u(0)=1$ *and* $\dot{u}(0)=2$

For the Simulink model shown in Figure 3-14, the [Integrator1] block has an internal initial condition value of 2.0, and the other one has an internal initial condition value of 1.0. By executing the model (Figure 3-14), the simulation results shown in the [Scope] block are obtained.

Simulation results of $\frac{1}{2}\ddot{u}+\frac{2}{5}\dot{u}+u=t$, $u(0)=1$, and $\dot{u}(0)=2$ are displayed in the [Scope] block, as shown in Figure 3-15. Note that in the [Scope] block shown, we have made some adjustments by setting its background color, plotting the data points from parameters of the block (a marker and line type), and setting the color of the plotted data points, which are similar to the plot tools of MATLAB.

Note   The options in the [Scope] block parameters that allow you to change the background color, change the plotted data's line type and marker type, change the axis color, and add legends are available starting in MATLAB 2012/Simulink 8.0.

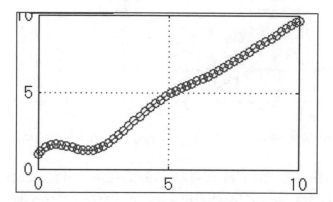

Figure 3-15. *Simulation results shown in the [Scope] block of the Simulink model from Figure 3-14*

Example 14

Here's the example problem: $\ddot{u} - K_2\dot{u} - K_1 u = 0$, $u(0) = 2, \dot{u}(0) = 0$.

The modeling procedures of this problem are similar to the previous problem except for the two gains, K_1 and K_2, that can be specified in the command window or can be initiated by selecting File ➤ Model Properties ➤ Callbacks ➤ InitFcn from a model's window menu.

The Simulink model, called `Sim_Model_2ndODE_ex14.mdl`, of the exercise is shown in Figure 3-16.

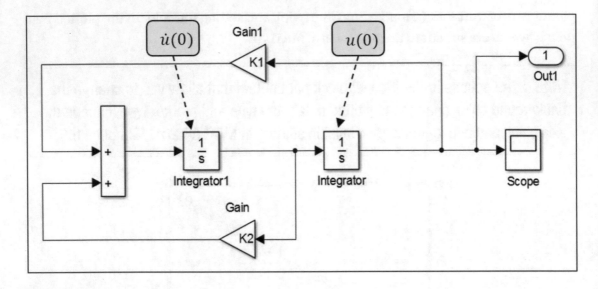

Figure 3-16. *Simulink model of* $\ddot{u} - K_2\dot{u} - K_1u = 0$, $u(0) = 2$, *and* $\dot{u}(0) = 0$

It is possible to have an integrator block with the external ICs. This can be set up by double-clicking the [Integrator] block and selecting the initial condition source from the drop-down menu, as shown in Figure 3-17.

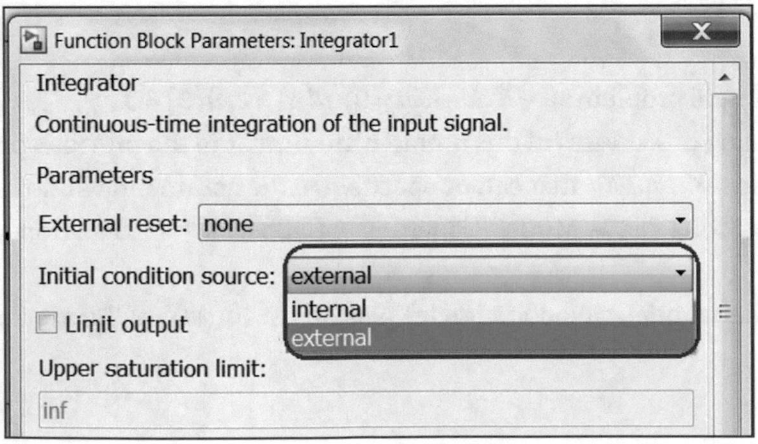

Figure 3-17. *[Integrator] block parameter settings*

After confirming by pressing [Apply] and [OK], we get the following updated model block:

To this changed [Integrator] block, now we add a [Constant] block as a source value of an initial condition, and our updated model (Sim_Model_2ndODE_ex2.mdl) will look like Figure 3-18.

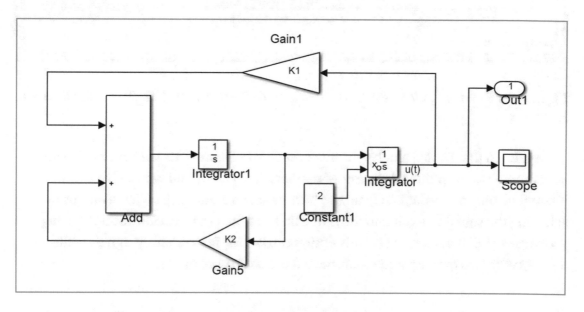

Figure 3-18. *Alternative Simulink model of* $\ddot{u} - K_2\dot{u} - K_1u = 0$, $u(0) = 2$ *and* $\dot{u}(0) = 0$

By initiating the values of gains ($K_1 = -6$, $K_2 = -9$) first and then simulating our created model, we get the outputs displayed in Figure 3-19 in the scope block.

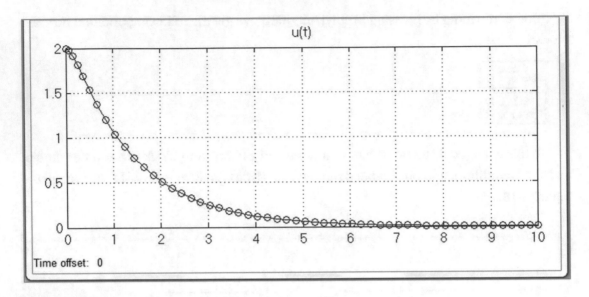

Figure 3-19. *Simulation results of $\ddot{u} - K_2\dot{u} - K_1u = 0$, $u(0) = 2, \dot{u}(0) = 0$ for $K_1 = -6$, $K_2 = -9$*

Again, in the [Scope] block parameters shown in Figure 3-19, we have made some adjustments of the plotted data points' markers, line colors, and types. Also in the [Scope] display of the output signal, we have assigned a title with $u(t)$. We do this by selecting the signal line going to the scope first and then right-clicking and choosing Properties ➤ Signal name: $u(t)$. This option can be helpful to identify signals while dealing with two or more signals connected to one [Scope] display.

Example 15

Here's the example problem: $\ddot{u} + \sin t = 0$, $u(0) = 0, \dot{u}(0) = 0$. The given problem can be rewritten as follows: $\ddot{u}(t) = -\sin t$.

This problem is relatively simple, and its numerical solutions can be computed from the model `Sim_Model_2ndODE_ex15.mdl` that is similar to the previous problem's simulation model with double integration.

In our developed Simulink model (Figure 3-20), model version A and model version B differ in input signal blocks. This is a good example of how easily we can substitute one type of blocks with another type to simplify our model. Both of the model versions produce identical results that can be observed from the results in [Scope2] shown in Figure 3-21.

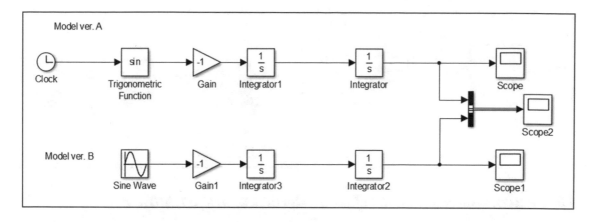

Figure 3-20. *Two versions of a Simulink model of* $\ddot{u} + sint = 0$, $u(0) = 0$, $\dot{u}(0) = 0$

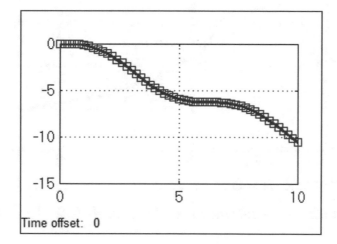

Figure 3-21. *Scope2 displays the simulation results of the two models shown in Figure 3-20*

Example 16

Here is the example problem: $\ddot{u} + \dot{u} - \sin(tu) = 0$, $u(0) = 1$, $\dot{u}(0) = 2$.

First, the given second-order ODE will be rewritten in the following form: $\ddot{u} = \sin(tu) - \dot{u}$. Note that this problem is modeled via a closed-loop feedback system (because in this case we have $\ddot{u} = \sin(tu) - \dot{u}$) unlike the previous case (we had only $\ddot{u} = -\sin(t)$) that was an open-loop system. Figure 3-22 shows the complete Simulink model, Sim_Model_2ndODE_ex16.mdl. Figure 3-23 shows the simulation results of the built Simulink model.

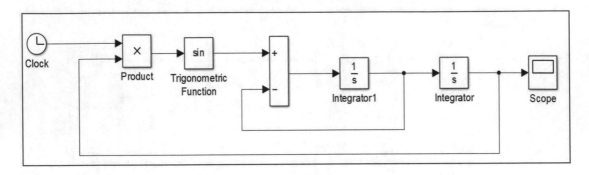

Figure 3-22. *Simulink model of* $\ddot{u} + \dot{u} - \sin(tu) = 0$, $u(0) = 1, \dot{u}(0) = 2$

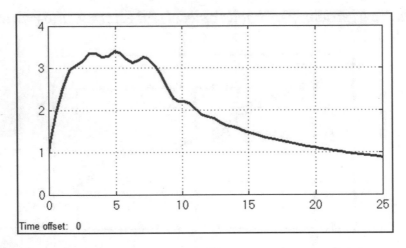

Figure 3-23. *Simulation results displayed in [Scope] block of the Simulink model from Figure 3-22*

From the plot shown in the [Scope] window, we can see that the result does not seem to be quite smooth, which is due to the inappropriately chosen step size and error tolerances of the model solver. All tunings and adjustments can be done by adjusting the configuration parameters accordingly, which can be accessed by clicking the menu panel Model Configuration Parameters [icon] (for MATLAB R2012b or newer versions); then select Solver or from the menu select Tools ➤ Model Explorer ➤ Configuration ➤ Solver.

Nonzero Starting Initial Conditions

Not all IVPs start at zero (time) initial conditions. Let's solve the IVP that starts at a nonzero initial point.

Example 17

Here's the example problem: $\ddot{x}+1=0$ with ICs: $x\left(\dfrac{\pi}{3}\right)=2$; $\dot{x}\left(\dfrac{\pi}{3}\right)=-4$.

Here is a Simulink model, Sim_Model_2ndODE_ex17.mdl, that was created in two different versions, one of which (Ver 1) is implemented with the [Second-Order Integrator] and the other (Ver 2) is implemented with two first-order [Integrator] blocks. Both versions of the model are equivalent, and two initial conditions are predefined as internal. The built Simulink simulation model is shown in Figure 3-24.

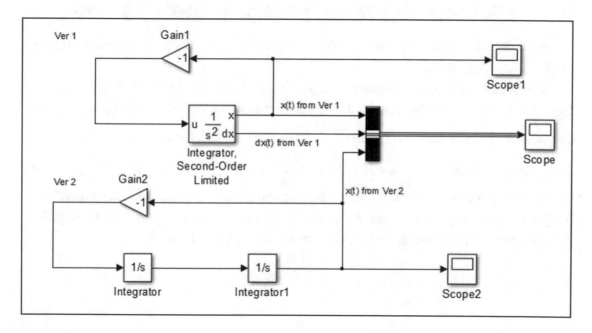

Figure 3-24. *Simulink model of $\ddot{x}+1=0$, $x\left(\dfrac{\pi}{3}\right)=2$; $\dot{x}\left(\dfrac{\pi}{3}\right)=-4$*

In this given example, IVP starts not at $t=0$ but at $t=\dfrac{\pi}{3}$. Thus, the simulation has to start at $t=\dfrac{\pi}{3}$, which can be easily implemented in the Simulink model parameters. To do that, we make changes in the configuration parameters. Select Solver ➤ Start Time and insert **pi/3** instead of 0, as highlighted in Figure 3-25.

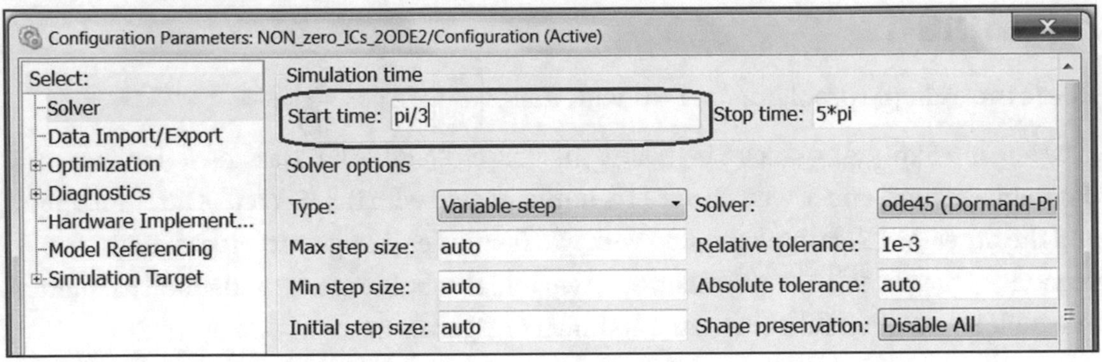

Figure 3-25. *Configuration of parameter settings to set up simulation time*

Note that the [Second-Order (double) Integrator] block was made available starting in MATLAB 2011/Simulink.

By simulating the model shown in Figure 3-24, we obtain the results (shown in Figure 3-27) from both versions of the model in the [Scope] window. The [Scope] block contains output signals of $x(t)$ and $\dfrac{dx(t)}{dt}$ from Ver 1 and $x(t)$ from Ver 2. We have assigned signal names for signals $x(t)$ and $\dfrac{dx(t)}{dt}$ from Ver 1 and $x(t)$ from Ver 2 by using the right-click options. Click the signal line, and then right-click and select Properties ➤ Signal name. Type in $x(t)$ and hit Apply and OK, as shown in Figure 3-26. Finally, the simulation results shown in Figure 3-27 contain all assigned legends.

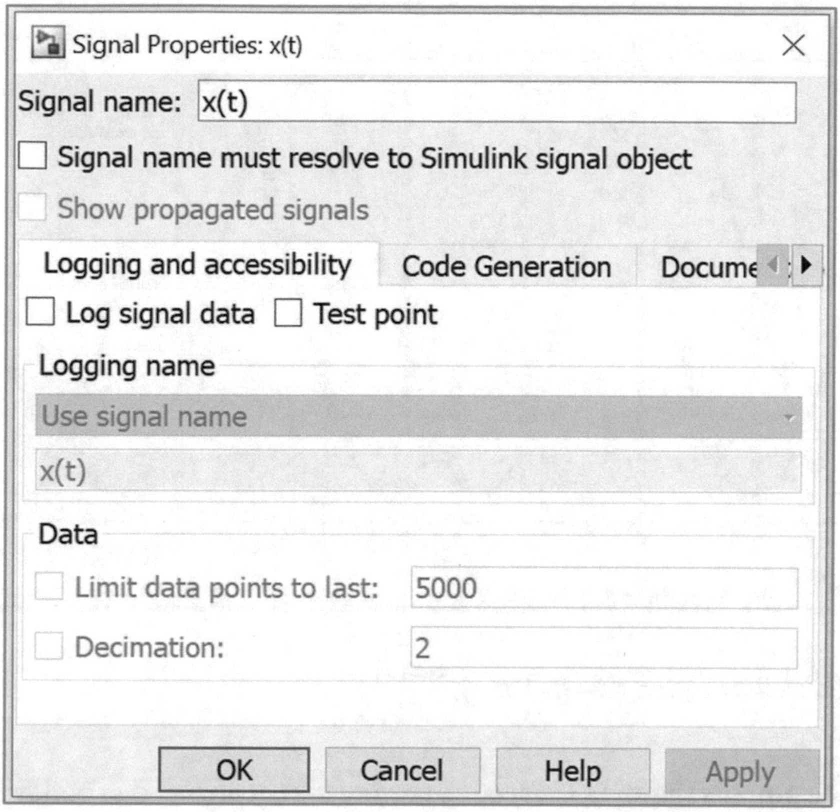

Figure 3-26. *Set up the signal name using the right-click options*

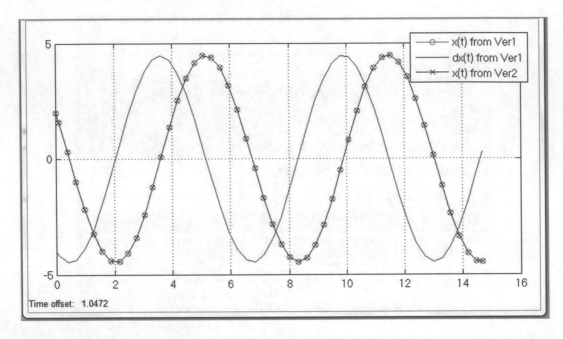

Figure 3-27. *Simulation results* $\left(x(t), \dfrac{dx(t)}{dt} \right)$

From the outputs shown in the scope in Figure 3-27, it is clear that both versions of the model give the same results.

To summarize all the examples of Simulink modeling, it is clear that programming or modeling ODEs in Simulink is relatively simple from a learning standpoint.

ODEx Solvers

We have already covered MATLAB's built-in solvers for first-order ODEs. In this section, we will employ built-in solvers and tools to solve second-order and higher-order ODEs. In this regard, all of the built-in tools that we have employed for first-order ODEs are valid and used with the same command syntaxes. When solving second or higher-order ODEs, we need to rewrite a given problem as a system of first-order ODEs.

Example 18

Here's the example problem: $\dfrac{1}{2}\ddot{u} + \dfrac{2}{5}\dot{u} + u = 2t,\ u(0) = 1,\ \dot{u}(0) = 2$.

Before writing a script of commands for MATLAB's built-in ODE solvers, we need to rewrite the given second-order ODE as a system of two first-order ODEs by introducing new variables.

$\ddot{u} = 4t - \dfrac{4}{5}\dot{u} - 2u$ is rewritten as follows: $\begin{cases} \dot{u}_1 = u_2 \\ \dot{u}_2 = 4t - \dfrac{4}{5}u_2 - 2u_1 \end{cases}$

Note that $u_1 = u$ and $\dot{u}_2 = \ddot{u}$.

The previously written system of first-order ODEs can be expressed by using matlabFunction, an anonymous function (@), a function file, and an inline function (which will be removed in future MATLAB releases) in scripts.

Note   ode45 is a recommended solver to try when solving the IVPs if the given problem is not stiff or implicitly defined.

The script (ODEx_solvers_EX18.m) embeds command syntaxes of the ODE solvers, such as ode45, ode23, and ode113, to compute numerical solutions of the given problem. The computed numerical solutions are shown in Figure 3-28.

```
% ODEx_solvers_EX18.m
clearvars; close all
t0=0;                % Start of simulations
tend=5;              % End of simulations
t=[t0, tend];
u(1,:)=[1; 2];       % Initial Conditions
% ode45 - RUNGGE-KUTTA 4/5 Order
Fun1 = inline('[u(2); 4*t-2*u(1)-u(2).*(4/5)]','t', 'u');
[T1, U1]=ode45(Fun1, t, u, []);
plot(T1, U1(:,1), 'rp', 'markersize', 9); grid on; hold on
% ode23 - RUNGGE-KUTTA 2/3 Order
Fun2=@(t, u)([u(2); 4*t-u(1).*2-u(2).*(4/5)]);
[T2, U2]=ode23(Fun2, t, u);
plot(T2, U2(:,1), 'b:o', 'markersize', 9)
% ode113 - ADAMS Higher Order
[T3, U3]=ode113(@Fun3, t, u);
```

```
plot(T3, U3(:,1), 'k-', 'linewidth', 2)
legend('ode45', 'ode23', 'ode113')
title('Simulation of: (1/2)ddy+(2/5)dy+y=2t')
xlabel('Time, t'), ylabel('Solution, u(t)'), axis tight
```

The function m-file, Fun3.m, recalled by the Adams method–based ode113 solver, is as follows:

```
function f = Fun3(t,u)
% Fun3.m is a function file.
% This function file is called by a solver ode113
f = [u(2); t.*4.0-u(1).*2.0-u(2).*(4.0./5.0)];
end
```

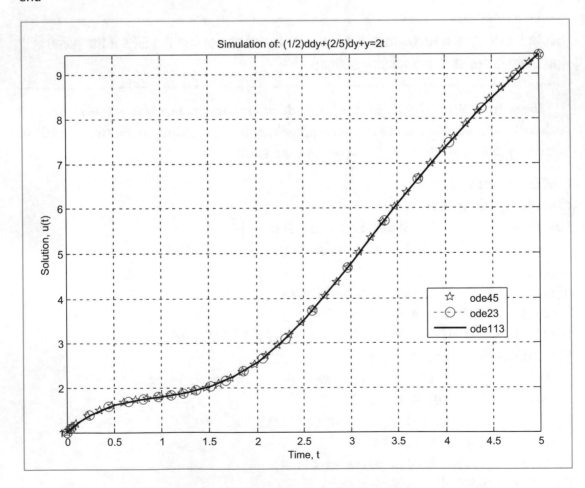

Figure 3-28. *Simulation of the example:* $\dfrac{1}{2}\ddot{u}+\dfrac{2}{5}\dot{u}+u=2t,\ u(0)=1,\ \dot{u}(0)=2$

Example 19

Here's the example problem: $\ddot{u} + 6\dot{u} + 9u = 0$, $u(0) = 2$, $\dot{u}(0) = 0$.

In this case, again like with the previous problem, we can rewrite the given problem via two first-order ODEs by introducing new variables.

$\ddot{u} = -6\dot{u} - 9u$ is rewritten as follows: $\begin{cases} \dot{u}_1 = u_2 \\ \dot{u}_2 = -6u_2 - 9u_1 \end{cases}$

Note that $u_1 = u$ and $\dot{u}_2 = \ddot{u}_1$.

The following is the script (ODEx_solvers_EX19.m) that embeds the Runge-Kutta method–based ode45, ode15s (for stiff problems), and ode113 (higher-order Adams) MATLAB solvers to compute numerical solutions of the problem:

```
% ODEx_solvers_EX19.m
clearvars; close all
t0=0; tend=5;   t=[t0, tend]; u(1,:)=[2 0];
% ode45 - 4/5 order RUNGGE-KUTTA
Fun1 = inline('[u(2); -6*u(1)-9*u(2)]','t', 'u');
[T1, U1]=ode45(Fun1, t, u, []);
plot(T1, U1(:,1), 'r*', 'markersize', 9), grid on; hold on
% ode15s - Stiff problem solver
Fun2=@(t, u)([u(2); -6*u(1)-9*u(2)]);
[T2, U2]=ode15s(Fun2, t, u,[]); plot(T2, U2(:,1), 'bo', 'markersize', 9)
% ode113 - ADAMS Higher Order
[T3, U3]=ode113(Fun2, t, u,[]);
plot(T3, U3(:,1), 'k-.', 'linewidth', 2)
legend('ode45', 'ode15s', 'ode113',0)
title('Simulation: ddy+9dy+6y=0. ode45, ode15s, ode113')
xlabel('Time, t'); ylabel('Solution, u(t)')
```

The simulation results (shown in Figure 3-29) show that there is a close convergence of solutions computed by ode45, ode15s, and ode113 solvers.

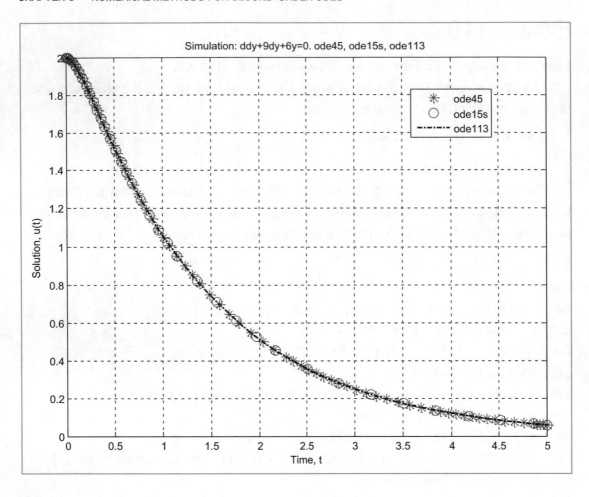

Figure 3-29. *Comparative simulation of* $\ddot{u}+6\dot{u}+9u=0,\ \ u(0)=2,\ \dot{u}(0)=0$

Example 20

Here's the example problem: $t\ddot{y}+2\dot{y}+4y=4,\ \ y(1)=1,\ \dot{y}(1)=1.$

Note that in the given problem, the initial condition is at $t=1$.

We start in this case again in the same way by rewriting the given second-order problem via two first-order ODEs by introducing new variables.

$$\ddot{y}=\frac{4-2\dot{y}-4y}{t}\quad \text{is rewritten as follows:} \quad \begin{cases} \dot{y}_1 = y_2 \\ \dot{y}_2 = \dfrac{4-2y_2-4y_1}{t} \end{cases}$$

Note that $y_1 = y$ and $\dot{y}_2 = \ddot{y}_1$.

The following script, called ODEx_solvers_EX20.m, embeds the ode45 solver and executes the Simulink model called ICs_non_Zero_2ODE.mdl to compute numerical solutions of the given problem:

```
% ODEx_solvers_EX20.m
% Given:  t*y"+2y'+4y=4 with ICs: y(1)=1, y'(1)=1
% NB: simulation start point is not "0" but 1
t=[1,50];  % NB: Simulation is to start at 1.
y_s=inline('[y(2); (1/t)*(4-2*y(2)-4*y(1))]', 't', 'y');
ICs=[1, 1]; options=odeset('reltol', 1e-6);
[time, y_sol]=ode45(y_s, t, ICs, options);
plot(time, y_sol(:,1),'r', 'linewidth', 2), grid on
title('Simulation of t*ddy+2dy+4y=4 with ICs y(1)=1, dy(1)=1')
xlabel('Time, t'), ylabel('Solution, y(t)'), ylim([.8 1.35]); hold on
% Simulation via SIMULINK: ode23
opts=simset('reltol',1e-6,'solver', 'ode23');
K=4; sim('ICs_non_Zero_2ODE_EX3', [1, 50], opts);
plot(SIM_outs(:,1), SIM_outs(:,2), 'bo-' ,'linewidth', 1)
legend('ode45', 'Simulink ode23', 0); hold off
```

Here is a Simulink model (ICs_non_Zero_2ODE_EX3.mdl) created in two different versions shown in Figure 3-30. Both of them output the same results. The latter model (Alternative model) is developed by employing the function block $\boxed{\; f(u) \;}$

Fcn

taken by choosing Simulink Library ➤ User Defined Functions. The initial conditions in [Integrator] blocks implemented as internal ICs and simulation time are set to start at $t = 1$ *sec*. The simulation results are shown in Figure 3-31.

Note   For Simulink models with a nonzero starting time of IVPs, the simulation has to start at a given initial time (value). For example, for $y\left(\dfrac{\pi}{2}\right) = \dfrac{1}{3}$, simulation has to start at $t = \dfrac{\pi}{2}$.

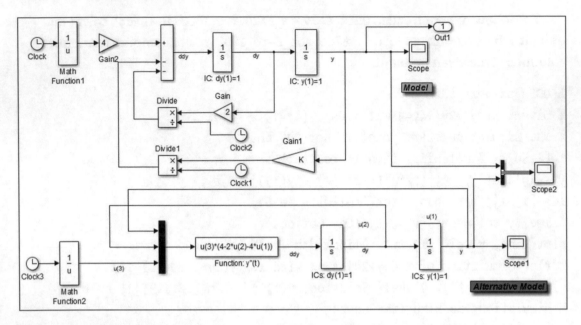

Figure 3-30. *Simulink model (`ICs_non_Zero_2ODE_EX3.mdl`) of Example 20*

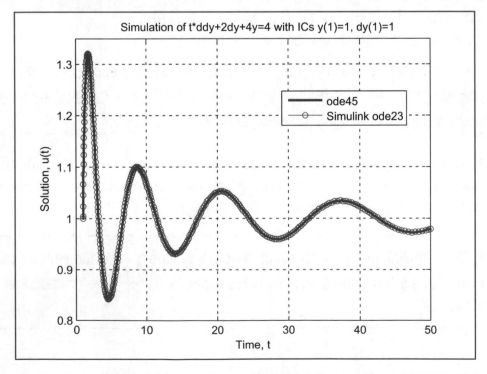

Figure 3-31. *Simulation of the problem using the `ode45` solver of MATLAB and the `ode23` solver of Simulink*

Example 21

Here's the example problem: $\ddot{y} - 2\dot{y} + 5y = 0$, $y\left(\dfrac{\pi}{2}\right) = 0$, $\dot{y}\left(\dfrac{\pi}{2}\right) = 1$.

Note that in the given problem, the initial conditions are at $t = \dfrac{\pi}{2}$.

We start in this example again in the same way by expressing the given second-order problem by two first-order ODEs by introducing new variables.

$\ddot{y} = 2\dot{y} - 5y$ is expressed with this: $\begin{cases} \dot{y}_1 = y_2 \\ \dot{y}_2 = 2y_2 - 5y_1 \end{cases}$

Note that $y_1 = y$ and $\dot{y}_2 = \ddot{y}$.

The script ODEx_solvers_EX21.m embeds ode23 of Simulink, ode45 of MATLAB, the 4th-order Runge-Kutta method implemented via a script to compute numerical solutions, and dsolve() to compute analytical solution values of the problem. The simulink models (ICs_non_Zero_2ODE_EX4.mdl) of the example are shown in Figure 3-32.

```
% ODEx_solvers_EX21.m
% Given: y"-2y'+5y=0, y(pi/2)=0, y'(pi/2)=2
clearvars; close all
%% SIMULINK Solution
%!!! NB: Computation starts at pi/2
opts=simset('reltol',1e-5,'solver', 'ode23');
[time, Yout]=sim('ICs_non_Zero_2ODE_EX4', [pi/2, 2*pi], opts);
plot(time, Yout(:,1), 'm--o', 'markersize', 9), hold on
%% RUNGE-KUTTA (ode45 solver based) Solution
% NB: Computation starts at pi/2 that is a set point of the ICs.
ICs=[0; 2];    %ICs
ts=pi/2;       %!!! NB: Computation starts at pi/2
tend=2*pi;
U(1,:)=[0, 2];
[time,u_sol]=ode45(@(t,u)([u(2); 2*u(2)-5*u(1)]),[ts,tend], ICs);
plot(time, u_sol(:,1), 'b>', 'markersize', 9), hold on
%% RUNGE-KUTTA (script based) Solution
% NB: Computation starts at pi/2 that is a set point of the ICs.
U(1,:)=[0, 2];    %ICs
ts=pi/2;          %!!! NB: Computation starts at pi/2
h=pi/20; tend=2*pi; t=ts:h:tend; steps=length(t);
FF=@(t, u1, u2)([u2, 2*u2-5*u1]);
for ii=1:steps-1
```

```
k1=FF(t(ii), U(ii,1), U(ii,2));
k2=FF(t(ii)+h/2,U(ii,1)+h*k1(:,1)/2,U(ii,2)+h*k1(:,2)/2);
k3=FF(t(ii)+h/2,U(ii,1)+h*k2(:,1)/2,U(ii,2)+h*k2(:,2)/2);
k4=FF(t(ii)+h,U(ii,1)+h*k3(:,1),U(ii,2)+h*k3(:,2));
U(ii+1,:)=U(ii,:)+h*(k1+2*k2+2*k3+k4)/6;
end
plot(t(:)',U(:,1),'k:p','markersize', 13, 'markerfacecolor', 'c')
%% Analytical Solution: via dsolve
y=dsolve('D2y=2*Dy-5*y', 'y(pi/2)=0', 'Dy(pi/2)=2', 'x');
display('y(t) is'); pretty(y); y=vectorize(y); x=t; Yt=eval(y);
plot(t, Yt, 'g-', 'linewidth', 2.5), grid on; hold on
legend('Simulink(ode23 ~var)','ODE45','RK 4th order script','Analytical
Solution')
title('ddy-2dy+5y=0, y(\pi/2)=0, dy(\pi/2)=2')
xlabel('t, time'), ylabel('y(t), solution'), hold off; shg
```

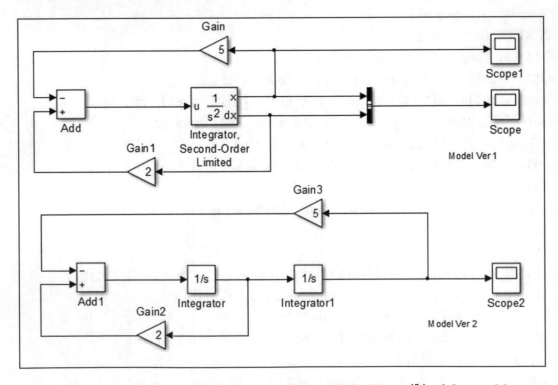

Figure 3-32. *Simulink models (ICs_non_Zero_2ODE_EX4.mdl) of the problem*

After executing the previously shown script (ODEx_solversEX4.m), which also executes the Simulink model (ICs_non_Zero_2ODE_EX21.mdl), the plot (shown in Figure 3-33) displaying all the results is obtained.

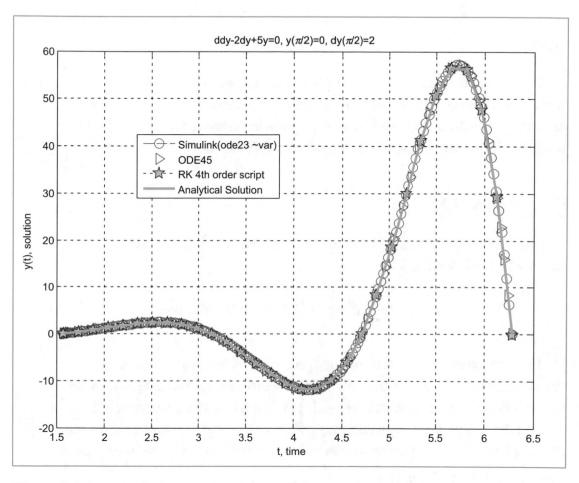

Figure 3-33. *Simulation results of the problem*

Moreover, the following symbolic expression, an analytical solution of the problem, is obtained in the command window:

y(t) is

```
                /   pi \
  - sin(2 x) exp| - -- | exp(x)
                \   2  /
```

From the plot shown in Figure 3-33, it is clear that the Simulink model with the ode23 with a variable step solver and 4th-order Runge-Kutta method computes well-converged numerical solutions with the analytical solution of the given second-order ODE.

diff( )

diff() is a built-in function of MATLAB that computes a difference and an approximate derivative. It works similar to the Taylor series method to evaluate approximate derivatives of a function to compute differences in computation evolutions. Let's consider the following example and apply the diff function.

Example 22

The Mars lander's velocity and acceleration equation of motion is given when it is landing by the following:

$$\ddot{y} + \left(\frac{k}{m}\right)\dot{y}^2 = g, \ y(0) = 20, \ \dot{y}(0) = 67.0556$$

In this problem, unlike the previous problems, we are not interested in defining the lander's displacement $y(t)$; instead, we need to evaluate its velocity ($\dot{y}(t) = v(t)$) and acceleration ($\ddot{y}(t) = a(t)$) when it is landing, where m is the mass of the Mars lander, k is a damping coefficient, and g is gravity (acceleration due to gravity) in Mars. For this numerical simulation, we use the following values:

$m = 150 [kg], k = 1.2 [N \ m \ s], g = 3.885 \left[\dfrac{m}{s^2}\right].$

We start in this case again by rewriting the given second-order ODE as two first-order ODEs by introducing new variables.

$\ddot{y} + \left(\dfrac{k}{m}\right)\dot{y}^2 = g$ is expressed with this: $\begin{cases} y_2 = \dot{y}_1 \\ \dot{y}_2 = g - \left(\dfrac{k}{m}\right)y_2^2 \end{cases}$

Note that $y_1 = y$ and $\dot{y}_2 = \ddot{y}_1$.

Here is the script (MARS_lander.m) that embeds the ode45 solver and diff:

```
%%  MARS Lander Vehicle velocity and acceleration while landing
%    MARS_lander.m
t=0:.05:6; u0=[20, 67.0556]; g=3.885; k=1.2; m=150;
%F=inline('[u(2); g-(k/m)*(u(2).^2)]', 't', 'u');
Flander=@(t, u)([u(2); g-(k/m)*(u(2).^2)]);
[time, Uout]=ode45(Flander, t, u0, []);
% Acceleration is a derivative of velocity!
acc = diff(Uout(:,2))/(time(2)-time(1));
plot(t,Uout(:,2),'ro',t(2:end),acc,'bs'); hold on
% Time derivative:
Ttt=[diff(t);diff(t)];
% Velocity:
Utt=diff(Uout);
% Acceleration:
UT=Utt./Ttt';
plot(t(2:end),UT(:,1),'b-',t(2:end),UT(:,2), ...
'm--','linewidth',2.5)
legend({'Velocity1 [m/s]';'Acceleration1 [m^2/s]';...
'Velocity2'; 'Acceleration2'},0); grid minor
title('Mars lander"s velocity and acceleration during landing')
xlabel('time'), ylabel('v, a'), shg
```

From the plot shown in Figure 3-34, we can see that both methods such as ode45 and diff for the given problem (the Mars lander's equation of motion while landing) have produced well-converged solutions.

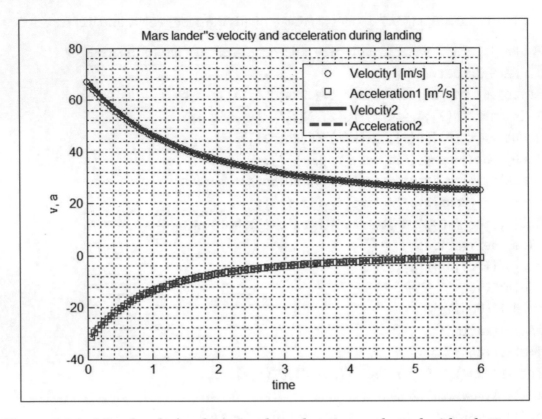

Figure 3-34. *Mars lander's velocity and acceleration evaluated with ode45 and diff*

It is easy to employ `diff()` when there is a need to compute the symbolic derivatives of complex functions. Here's an example:

```
>> syms x y t
>> F=[sin(10*t); x^2+y^3+2*x*y^2; t^2*sinc(2*x/y)];
dF=[diff(F(1),t); diff(F(2), x, y); diff(F(3), x, y)]
dF =

                                          10*cos(10*t)
                                          3*y^2 + 4*x*y
 (t^2*sin((2*pi*x)/y))/(2*pi*x) - (t^2*cos((2*pi*x)/y))/y
>> pretty(dF)
```

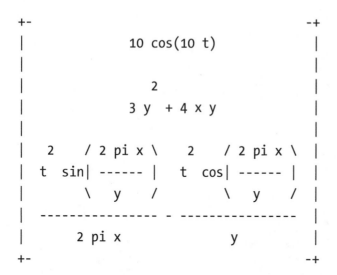

Via a few simulation examples of second-order ODES, we have demonstrated how to implement the Euler, Runge-Kutta, Runge-Kutta-Gill, Runge-Kutta-Fehlberg, Adams-Bashforth, Milne, Adams-Moulton, and Taylor Series methods to solve ODEs. In addition, we have used MATLAB's built-in ODE solvers and tools, such as ode23, ode23tb, ode45, ode113, odeset, and diff(). Moreover, we have demonstrated the Simulink modeling techniques with solvers, such as ode2, ode3, ode23, ode45, and tools, such as simset and sim. All of these methods and tools are applicable for all explicitly defined ODEs and systems of ODEs. The efficiency and accuracy of these methods depend on the given ODE equation type, such as linear or nonlinear, homogeneous or nonhomogeneous, and stiff or nonstiff. In addition, the accuracy and efficiency of employed numerical methods depend on the user's chosen/set error tolerances and step sizes.

CHAPTER 4

Stiff ODEs

In this chapter, we'll explore a few different ways to solve stiff ODEs. We will cover again how to employ the Euler, Runge-Kutta, and predictor-corrector methods. We will also use the Adams-Moulton methods and MATLAB's built-in solvers, namely, `ode15s`, `ode23s`, `ode23tb`, `ode45`, and `ode113`, and then we'll compare all of them to find the most efficient one in terms of accuracy and computation time. In many MATLAB references for very stiff ODEs, the solver `ode15s` is recommended. We'll see if it's the most efficient.

Example 1

Given a moderately stiff ODE, here's the example problem: $\dfrac{dy}{dt} + 100y = 0,\ y(0) = 1.$

For a comparative analysis of ODE solvers and methods, we will first compute the numerical solutions of the given exercise by using the forward Euler method with a (time) step size of 0.015 and then repeat the simulation with a step size of 0.0075, which is two times smaller than the first one. Then, we evaluate an analytical solution using `dsolve()`, and again we find the numerical solutions with the 4th-/5th-order Runge-Kutta method with a step size of 0.0075. All these steps are implemented in the following script, called `STIFF_1ODE_EX1.m`:

```
%% STIFF_1ODE_EX1.m
% Analytic solution of the problem is y(t)=exp(-100t).
% EULER-Forward Method:
close all; clearvars; clc
y0=1;                   % Initial Condition
dt1=1.5e-2;             % Step size 1
dt2=0.75e-2;            % Step size 2
tstart=0;
```

© Sulaymon L. Eshkabilov 2020
S. L. Eshkabilov, *Practical MATLAB Modeling with Simulink*, https://doi.org/10.1007/978-1-4842-5799-9_4

```
tend=.25;
t1=tstart:dt1:tend;     % Simulation time space for 1st try
t2=tstart:dt2:tend;     % Simulation time space for 2nd try
steps1=length(t1);      % 1st Iteration cycle number
f=@(y)(-100*y);
y=[y0, zeros(1,steps1-1)]; h=dt1;
for ii=1:steps1-1
    y(ii+1)=y(ii)+f(y(ii))*h;
end
figure, plot(t1, y, 'bo-'); hold on
steps2=length(t2);       % 2nd Iteration cycle number
y=[y(1), zeros(1,steps2-1)]; h=dt2;
for ii=1:steps2-1
    y(ii+1)=y(ii)+f(y(ii))*h;
end
plot(t2, y, 'rd-'); Y=dsolve('Dy=-100*y', 'y(0)=1', 't');
YY=vectorize(Y); t=t1;
YY=eval(YY);
plot(t1, YY, 'k-', 'linewidth',1.5)
title('\it Solutions of Stiff Problem: $$ \frac{dy}{dt}=-100*y, y_0=1 $$',
'interpreter', 'latex')
xlabel('t'); ylabel( 'y(t)'); xlim([0, .3])
grid on
%% RUNGE-KUTTA Method of 4th Order via SCRIPT
u(1)=1;      % IC: u(0)=1 starting at the index of 1
u=[u(1), zeros(1,steps2-1)];
for ii=1:steps2-1
k1=f(u(ii)); k2=f(u(ii)+h*k1(:,1)/2); k3=f(u(ii)+h*k2(:,1)/2);
k4=f(u(ii)+h*k3(:,1)); u(ii+1)=u(ii)+h*(k1+2*k2+2*k3+k4)/6;
end
plot(t2, u, 'mp--', 'linewidth',1.0)
legend('EULER-forward: h=0.015', 'EULER-forward: h=0.0075',...
'dsolve (solution)', 'Runge-Kutta: h=0.0075')
grid on
axis tight
hold off; shg
```

From the simulation results (shown in Figure 4-1), we can conclude that if we take a bigger step size than required, we will observe considerable fluctuations and unstable solutions when using the forward Euler method for stiff problems. Thus, there is a concern of what step size to choose while employing the Euler and Runge-Kutta methods, which are predictor types of ODE solvers.

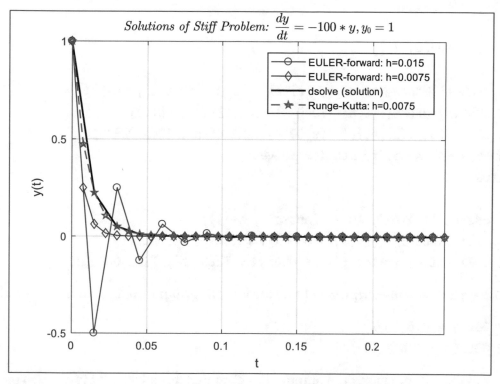

Figure 4-1. *Simulation results of a stiff first-order ODE:* $\dfrac{dy}{dt}+100y=0$ *with* $y(0)=1$

Example 2

This is a much stiffer problem: $\dfrac{dy}{dt}+1000y=0,\ y(0)=1$. In this problem simulation, we employ MATLAB's built-in ODE solvers, namely, ode45 and ode15s, and compare their efficiency level in terms of computation time (STIFF_1ODE_EX2.m).

```
%% STIFF_1ODE_EX2.m
close all; clearvars
%{
NOTE: The solution of this problem decays to zero. If a solver
```

produces a negative approximate solution, it begins to track the
solution of the ODE through this value, the solution goes off to
minus infinity, and the computation fails. Thus, using the NonNegative
property of ODESET prevents this from happening.
Nonnegativity is available for ode45, ode113 and not available for ode15i,
ode15s, ode23s, ode23t, ode23tb.
%}

```
% x1=1; time=[0, 15];
tic;
opts=odeset('NonNegative',1,'reltol', 1e-3, 'abstol', 1e-5);
[t1, xODE45]=ode45(@(t,x)(-1000*x), [0, 15], 1, opts);
Time_ODE45=toc; fprintf('Time_ODE45 = %2.6f\n', Time_ODE45)
%% ODE15s MATLAB Stiff problem Solver
clearvars
tic;
opts=odeset('reltol', 1e-3, 'abstol', 1e-5);
[t, xODE15s]=ode15s(@(t,x)(-1000*x), [0, 15], 1, opts);
Time_ODE15s=toc; fprintf('Time_ODE15s = %2.6f\n', Time_ODE15s)
```

After executing the script, we obtain the following outputs in the command window:

```
Time_ODE45 = 0.632997
Time_ODE15s = 0.013348
```

From the computation time outputs, it is clear that the solver ode15s is much more efficient in terms of computation time. If we plot the simulation results from the two solvers, we will obtain a very close convergence of results with an analytical solution of the problem. For moderately stiff problems, ode45 can perform relatively well. Note that not all ODEx solvers are capable of evaluating correct numerical solutions of very stiff IVPs with their default settings. Let's consider the next first-order stiff ODE exercise.

Example 3

Here's the example problem: $\dot{y} - \cos(1001t) = 0$ with $y(0) = 1.001$. This has an analytical solution, as shown here: $y(t) = \dfrac{\sin(1001t)}{1001} + 1.001$.

We compute the numerical solutions of the given first-order ODE by using built-in solvers, such as ode23, ode23s, ode45, ode113, and ode15s, as well as the forward Euler method (STIFF_1ODE_EX3.m).

```
% STIFF_1ODE_EX3.m
clearvars; close all
% Numerical solutions with ODEx:
Fun=@(t,y)(cos(1001*t)); Time=[0,0.75]; y0=1.001;
[t1, y1]=ode23s(Fun, Time, y0);
[t2, y2]=ode45(Fun, Time, y0);
[t3, y3]=ode113(Fun, Time, y0);
[t4, y4]=ode15s(Fun, Time, y0);
subplot(211)
plot(t1, y1, 'rs--', 'markersize', 9, 'markerfacecolor', 'c')
hold on
plot(t2, y2, 'bh-.', 'markersize', 9, 'markerfacecolor', 'y')
plot(t3, y3, 'm-', 'linewidth', 1.5)
legend('ode23s','ode45','ode113')
title('\it Solutions of Stiff Problem: $$ \frac{dy}{dt}=cos(1001*t,
y_0=1.001 $$', 'interpreter', 'latex')
xlabel('\it t'); ylabel( '\it y(t)'); axis tight; grid on
subplot(212)
plot( t4, y4, 'b--o', 'markersize', 9, 'markerfacecolor', 'y'), hold on
% Euler forward method:
y(1)=y0; h=1e-2; t=0:h:0.75; steps=length(t);
f=@(t, y)(cos(1001*t)); y=[y(1), zeros(1, length(t)-1)];
for ii=1:steps-1
    F=f(t(ii), y(ii));
    y(ii+1)=y(ii)+h*F;
end
plot(t, y, 'm', 'linewidth', 1.5)
% Analytical Solution:
clearvars t y
syms y(t)
Dy=diff(y); y=dsolve(Dy==cos(1001*t), y(0)==1.001); y=vectorize(y);
t=0:.01:0.75; y=eval(y); plot(t, y, 'k--', 'linewidth', 1.5),
```

```
hold on;
legend( 'ode15s', 'Euler forward','Analytical Sol')
xlabel('\it t'); ylabel( '\it y(t)'); shg
axis([0 .75 .95 1.33]), grid on
```

In this exercise, it is clear for the given first-order stiff ODE that the solvers ode23, ode23s, ode45, and ode15s (even though recommended) give wrong or false solutions (shown in Figure 4-2). On the contrary, the higher-order solver ode113 and crude approximate method of the forward Euler method result in adequately correct solutions that are well converged with the analytic solution results. Also, it is worth pointing out that when computing numerical solutions of stiff ODEs with built-in ODEx solvers with their default settings, such as relative and absolute error tolerances, it is not always viable to obtain accurate numerical solutions, even if we select the recommended solver ode15s.

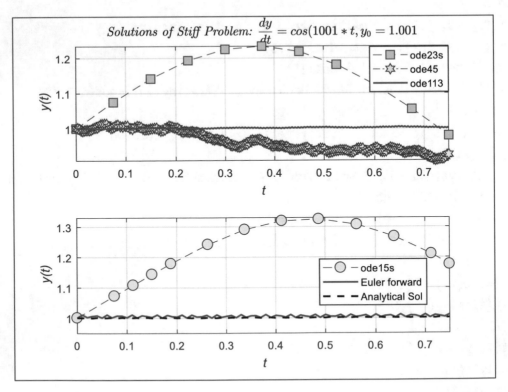

Figure 4-2. *Simulation of first-order stiff ODE: $\dot{y} - cos(1001t) = 0$ with $y(0) = 1.001$*

Note   Most built-in ODEx solvers including ode15s are not capable of computing accurate numerical solutions of very stiff IVPs with their default settings. Instead, the settings must be tuned accordingly. The higher-order solver ode113 may be exception when solving some very stiff ODEs.

Let's consider another moderately stiff second-order homogeneous ODE exercise with several built-in ODEx solvers.

Example 4

Here's the example problem: $\ddot{y} + 100\dot{y} + 10.9y = 0$, $y(0) = 1$, $\dot{y}(0) = 0$.

First, we rewrite the given second-order ODE as a system of two first-order ODEs, as shown here:

$$\ddot{y} = -100\dot{y} - 10.9y \implies \begin{cases} x_2 = \dot{x}_1 \\ \dot{x}_2 = -100x_2 - 10.9x_1 \end{cases}$$

The system is coded in a function file called A1_Stiff.m.

```
function Dxdt=A2_Stiff(T, X)
%%% Note: Given Equation is 2nd Order ODE.
%       ddy+100*dy+10.9*y(t)=0
%       The Function File Is Saved Under The Name: A2_Stiff.M
Dxdt=[X(2);
    -100*X(2)-10.9*X(1)];
end
```

The function file, A2_stiff.m, is recalled from the script STIFF_2ODE_EX4.m to simulate the given second-order ODE with a few built-in ODEx solvers.

```
% STIFF_2ODE_EX4.m
ICs=[1 0]; ts=[0, 100];
OPTs=odeset('RelTol', 1e-6, 'AbsTol', 1e-8);
tic;
[t, x_ODE45]=ode45(@A2_Stiff,  ts, ICs, OPTs);
t_ODE45=toc;
```

```
fprintf('Time_ode45 = %2.6f \n', t_ODE45)
clearvars
ICs=[1 0]; ts=[0, 100];
OPTs=odeset('RelTol', 1e-6, 'AbsTol', 1e-8);
tic;
[t1, x_ODE23s]=ode23s(@A2_Stiff, ts, ICs, OPTs);
t_ODE23s=toc;
fprintf('Time_ode23 = %2.6f \n', t_ODE23s)
clearvars
ICs=[1 0]; ts=[0, 100];
OPTs=odeset('RelTol', 1e-6, 'AbsTol', 1e-8);
tic;
[t2, x_ODE113]=ode113(@A2_Stiff, ts, ICs, OPTs);
t_ODE113=toc;
fprintf('Time_ode113 = %2.6f \n', t_ODE113)
clearvars
ICs=[1 0]; ts=[0, 100];
OPTs=odeset('RelTol', 1e-6, 'AbsTol', 1e-8);
tic;
[t3, Y_ODE23tb]=ode23tb(@A2_Stiff, ts, ICs, OPTs);
t_ODE23tb=toc;
fprintf('Time_ode23tb = %2.6f \n', t_ODE23tb)
clearvars
ICs=[1 0]; ts=[0, 100];
OPTs=odeset('RelTol', 1e-6, 'AbsTol', 1e-8);
tic;
[t4, x_ODE15s]=ode15s(@A2_Stiff,  ts, ICs, OPTs);
t_ODE15s=toc;
fprintf('Time_ode15s = %2.6f \n', t_ODE15s)
```

After executing the script on a laptop computer (with Windows 10, Intel Core i7 chip, 2.5 GHz, and 8GB of RAM), the following results are obtained. It is clear that ode15s outperforms all other ODE solvers considerably in terms of computation (elapsed) time efficiency.

```
Time_ode45 = 0.232020
Time_ode23 = 0.098099
```

```
Time_ode113 = 0.480924
Time_ode23tb = 0.084671
Time_ode15s = 0.028107
```

This demonstrates the efficiency of the solvers available in MATLAB/Simulink for stiff ODE problems.

Jacobian Matrix

Using the Jacobian matrix to solve stiff ODEs with built-in ODEx solvers may speed up a computation process. The Jacobian matrix properties are relevant only to the stiff problem solvers, meaning ode15s, ode23s, ode23t, ode23tb, and ode15i, and they can play an important role not only in computation but also with reliability issues. Note that there are different properties of the Jacobian matrix for ode15i.

The Jacobian matrix is a matrix of partial derivatives of the given function that defines the differential equations. For example, given a vector h:

$$h(x,y,z)=\begin{cases} x^2+y^2+z^2 \\ x^2y+y^2z-2yz \\ 3xyz+2xy \end{cases}$$

the Jacobian matrix of the vector $h(x, y, z)$ with respect to a vector $g(x,y,z)=\begin{cases} x \\ y \\ z \end{cases}$ can be

computed in the following way via the Symbolic Math Toolbox tools/commands:

```
>> syms x y z
>> h=[x^2+y^2+z^2; x^2+y^2*z-2*y*z; 3*x*y*z-2*x*y];
>> g=[x, y, z];
>> Jac_h=jacobian(h, g)

Jac_h =

[           2*x,           2*y,        2*z]
[           2*x, 2*y*z - 2*z, y^2 - 2*y]
[ 3*y*z - 2*y, 3*x*z - 2*x,      3*x*y]
```

Now, we can verify the importance of the Jacobian matrix in solving stiff problems by using their Jacobian matrices. Note that we'll look at the second-order ODE (the Chebyshev differential equation) next to explain the Jacobian matrices.

Example 5

The Chebyshev differential equation is defined as follows:

$$\left(1-t^2\right)\ddot{y}-t\dot{y}+n^2y=0 \text{ with } |t|<1 \text{ and } y(0)=1, \ \dot{y}(0)=1$$

Let's compute numerical values of the Chebyshev differential equation solution for $n = 0, 1, 2, …, 9$.

In the first step, we need to rewrite the Chebyshev equation as two first-order differential equations by introducing new variables.

$$\ddot{y}=\frac{t\dot{y}-n^2y}{\left(1-t^2\right)} \ => \ \begin{cases} y_2 = \ddot{y}_1 \\ \dot{y}_2 = \dfrac{ty_2 - n^2y_1}{1-t^2} \end{cases}$$

In the second step, we find the Jacobian matrix of the system of differential equations depicted earlier.

```
>> syms y1 y2 n t
>> C1=[y2; (t*y2-n^2*y1)/(1-t^2)];
>> G=[y1, y2];
>> JC1=jacobian(C1, G)

JC1 =

[                0,              1]
[ n^2/(t^2 - 1), -t/(t^2 - 1)]

>> pretty(JC1)
```

```
+-                   -+
|    0,         1    |
|                    |
|    2               |
|    n          t    |
| ------, - ------   |
|    2          2    |
|  t - 1      t - 1  |
+-                   -+
```

Based on the evaluated Jacobian matrix of the Chebyshev differential equation and its initial conditions, we write the next script, named ChebyshevDemo.m:

```
function ChebyshevDemo
% HELP: ChebyshevDemo.m is a script to simulate
%       numerical solutions of Chebyshev Equation:
%       (1-t^2)*ddy-t*dy+n^2*y=0
%       for n=0,1,2,3,4,5,6, ...
%       t is defined to be |t|<1
%       Initial conditions: y(0)=0, dy(0)=1
%       Colorit (color type): b-blue; g-green; r-red; m-magenta;
%       c-cyan; y-yellow; k-black.
%       Lineit (line type): - solid line; : colon; -- dashed line;
%       Markit (marker type): o-circle; d-diamond; x-cross;
%       s-square; h-hexagon; + plus sign; * asterisk; etc.
figure;
y0=0; dy0=1;   % Initial Conditions
t=[0, 0.999];  % Simulation time space
Labelit = {};
Colorit = 'bgrcymkgrckmbgrygr';
Lineit  = '--:-:--:-:--:----:----:--';
Markit  = 'odxsh+*^v<p>.xsh+od+*^v';
options = odeset('RelTol',1e-4,'AbsTol',1e-5,'Jacobian',@JC1);
for ii = 1:10
  [t, Y]    = ode15s(@D2Y, t,[y0, dy0], options, ii-1);
  Stylo     = [Colorit(ii) Lineit(ii) Markit(ii)];
```

```
  Labelit{ii} = ['n= ' num2str(ii-1)]; plot(t, Y(:,1),Stylo), hold on
end
xlabel('\it t'), ylabel('\it Solution, y(t)')
legend(Labelit{:}, 'location', 'northwest'), grid on
title('\it Chebyshev Equation Solutions for n=0,1...9'), hold off
function ddy=D2Y(t, y, n)
% Chebyshev differential equation
        ddy=[y(2);
            (t.*y(2)-n.^2*y(1))./(1-t.^2)];
function J=JC1(t, y, n)
% Jacobian matrix is computed
            J=[0, 1;
               n^2/(t^2-1), -t/(t^2-1)];
```

By executing the script ChebyshevDemo.m, we obtain the plot shown in Figure 4-3 containing 10 plots of the Chebyshev differential equations for different values of $n = 0, 1, ... 9$.

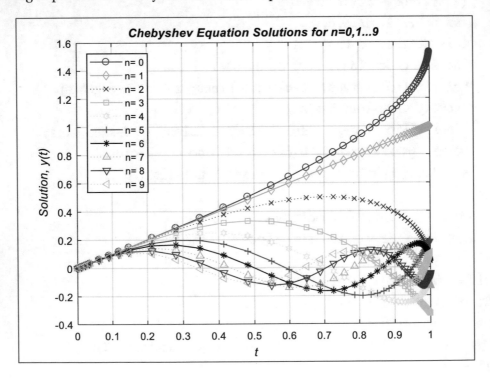

Figure 4-3. *Simulation of the Chebyshev differential equation:* $(1-t^2)\ddot{y}-t\dot{y}+n^2y=0$ *with* $|t| < 1$ *and* $y(0)=1, \dot{y}(0)=1$ *for* $n = 0, 1, 2, ... , 9$

Note   The Jacobian matrix is an essential component of ODEx solvers. Thus, if the Jacobian matrix is provided, the computation efficiency of solvers will be improved, especially when the solution space is large.

Example 6

This is another good example for relatively stiff differential equations. The given problem formulates an electric rectifier circuit voltage: $\ddot{u} + 100\,\dot{u} + 10^4 u = 10^4 \left| \sin(333\,t) \right|$, $u(0) = 0, \dot{u}(0) = 0$.

First, the problem is rewritten as two first-order differential equations, as we have done in the previous examples.

$$\ddot{u} = 10^4 \left| \sin(333\,t) \right| - 100\ddot{u} - 10^4 u \;=>\; \begin{cases} u_2 = \dot{u}_1 \\ \dot{u}_2 = 10^4 \left| \sin(333\,t) \right| - 100 u_2 - 10^4 u_1 \end{cases}$$

The simulation model of this second-order stiff ODE is implemented via the script STIFF_2ODE_EX6.m and the Simulink model EL_Circuit.mdl in two versions.

```
% Stiff_2ODE_EX6.m
% EXAMPLE: Second order ODEs ~ Rectifier Circuit Voltage
clearvars; close all
u(1,:)=[0, 0];          % Initial Conditions
h=1e-4;                 % Time step
t=0:h:.1;               % Simulation time space
steps=length(t);        % Number of steps
f=@(t, u1,u2)([u2, 1e4*abs(sin(333*t))-1e4*u1-100*u2]);
tic;
for ii=1:steps-1
    F=f(t(ii), u(ii,1), u(ii,2));
    u(ii+1,:)=u(ii,:)+h*F;
end
t_EF=toc;
plot(t, abs(sin(333*t)),'r', t, u(:,1), 'bd')
axis([0, .1 0, 1.2]),
xlabel( '\it t')
```

```
ylabel('\it Output voltage, u(t)')
title('Simulation of: $$ \frac{d^2u}{dt^2}+100*\frac{du}
{dt}+10^4*u=10^4*|sin(333*t)| $$', ...
    'Interpreter', 'latex')
hold on
fprintf('Sim. time with Euler forward is:  %3.5f\n', t_EF)
% ODE23s
clearvars
t=[0,.1];ICs=[0, 0];
FFm=@(tm, um)([um(2); 1e4*abs(sin(333*tm))-1e4*um(1)-100*um(2)]);
options=odeset('reltol', 1e-4);
tm=t;
tic
[Tm, Um]=ode23s(FFm, tm', ICs, options);
t_ode23s=toc;
fprintf('Sim. time with ode23s:          %3.5f\n', t_ode23s)
plot(Tm,Um(:,1),'k--+'),
% via SIMULINK
% Simulink model: EL_circuit.mdl is created in two versions
% Relative Tolerance is set at: 1e-4
clearvars
 options=simset('reltol', 1e-4);
 tic
 [tout, yout]=sim('EL_Circuit', [0, .1], options);
 t_SIM=toc;
 plot(tout, yout(:,1), 'm-', 'linewidth', 1.5)
 legend('Input: |sin(333*t)|','EULER forward script', ...
'ode23s', 'Simulink: ode15s','location', 'northeast'), shg
fprintf('Sim. time with Simulink ode15s: %3.5f\n', t_SIM)
```

By executing the previous script (STIFF_2ODE_EX6.m) that recalls the Simulink model EL_Circuit.mdl shown in Figure 4-4, we obtain the simulation results, as shown in Figure 4-5, namely, computation time efficiency and numerical results.

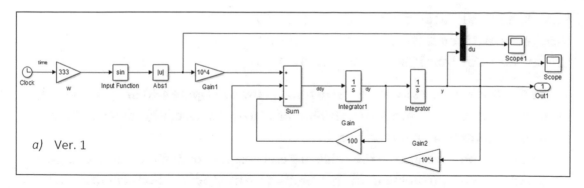

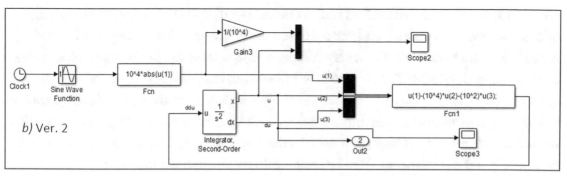

Figure 4-4. *Two Simulink model versions of EL_Circuit.mdl*

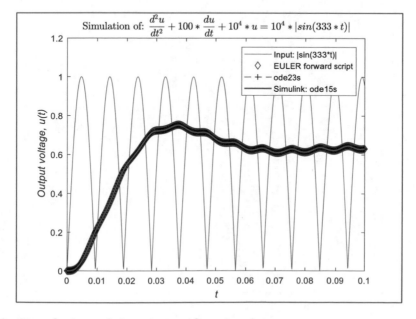

Figure 4-5. *Simulation of electric rectifier circuit*

```
Sim. time with Euler forward is:    0.03180
Sim. time with ode23s is:           0.63395
Sim. time with Simulink ode15s is: 0.27289
```

In this exercise, the forward Euler method (shown in Figure 4-5) has considerably outperformed the other two MATLAB/Simulink ODE solvers, ode23s and ode15s, in terms of its computation time efficiency.

In this chapter, we studied the efficiency and accuracy of different numerical solvers via simulations of stiff problems in six exercises; we used ode15s (as recommended), ode23s (as recommended), ode23 (not recommended), ode113 (not recommended), ode45 (not recommended), and scripts based on the forward Euler, trapezoidal, and Adams-Moulton methods. For a very stiff problems, ode15s outperforms all other solvers as recommended by MATLAB's tutorials, but we need to tune its settings, such as the relative and absolute error tolerances, carefully and, if it is feasible, provide the Jacobian matrix of the problem (see Example 5) with odeset when we are dealing with very stiff problems. Moreover, in Example 6, we saw the case in which the forward Euler method can be applied as alternative or to have preliminary evaluations for some stiff problems due to its high computation time efficiency.

CHAPTER 5

Higher-Order and Coupled ODEs

This chapter is dedicated to solving higher-order and coupled ODEs numerically. In general, the procedures used to solve higher-order ODEs and coupled systems of ODEs are similar to the ones used for second-order ODEs except for the additional steps required for new variables while rewriting the given higher-order ODEs as a system of first-order ODEs.

Fourth-Order ODE Problem

Let's consider the following fourth-order nonhomogeneous ODE:

$$\frac{d^4 y}{dt^4} + 3\frac{d^3 y}{dt^3} - \sin(t)\frac{dy}{dt} + 8y = t^2, \ y(0) = 50, \ dy(0) = 40, \ d^2y(0) = 30, \ d^3y(0) = 20.$$

The given fourth-order ODE can be rewritten via a system of four first-order ODEs as follows:

$$\begin{cases} y_2 = \dot{y}_1 \\ y_3 = \dot{y}_2 \\ y_4 = \dot{y}_3 \\ \dot{y}_4 = t^2 - 3y_4 + \sin(t)y_2 - 8y_1 \end{cases}$$

This system is implemented via the script SOLVE_4order_ODE.m.

```
function SOLVE_4order_ODE
% HELP: EXE4orderODE.m with f4ode (nested function)
%       simulates the problem.
```

© Sulaymon L. Eshkabilov 2020
S. L. Eshkabilov, *Practical MATLAB Modeling with Simulink*, https://doi.org/10.1007/978-1-4842-5799-9_5

```
close all
time=[0, 2*pi]; y0=[50, 40, 30, 20];
[t, y]=ode45(@f4ode, time, y0, odeset('reltol', 1e-8));
figure('name', 'Simulate with ODE45'),
plot(t,y(:,1), 'b-', t,y(:,2), 'r--',t,y(:,3), ...
  'k:',t,y(:,4), 'm-.','linewidth', 1.5)
legend('\it y(t)', '\it dy','\it ddy','\it dddy',... 'location','best')
title('ODE45 Simulation of: $$ \frac{d^4y}{dt^4}+3*\frac{d^3y}{dt^3}-
sin(t)*\frac{dy}{dt}+8*y=t^2 $$', 'interpreter', 'latex')
grid on, ylim([-700 300]), xlim([0, 2*pi])
xlabel('\it t')
ylabel('\it Solutions: $$ \frac{d^3y}{dt^3}, \frac{d^2y}{dt^2}, \frac{dy}
{dt}, y(t) $$', 'interpreter', 'latex')
grid on
figure('name','Simulate with ODE23');
ode23(@f4ode, time, y0, odeset('reltol', 1e-8))
ylim([-700 300]), xlim([0, 2*pi])
legend('\it y(t)', '\it dy','\it ddy','\it dddy', 'location','best')
title('ODE23 Simulation of: $$ \frac{d^4y}{dt^4}+3*\frac{d^3y}{dt^3}-
sin(t)*\frac{dy}{dt}+8*y=t^2 $$', 'interpreter', 'latex')
xlabel('\it t')
ylabel('\it Solutions: $$ \frac{d^3y}{dt^3}, \frac{d^2y}{dt^2}, \frac{dy}
{dt}, y(t) $$', 'interpreter', 'latex')
grid on
function f=f4ode(t, y)
% The given 4th order ODE is: y""+3*y"'-sin(t)*y'+8*y=t^2
f=[y(2); y(3); y(4); t.^2-3.*y(4)+sin(t).*y(2)-8.*y(1)];
end
end
```

The simulation results with ode45 (shown in Figure 5-1) match well with the ones obtained with ode23 (plot figure not shown here). Note that in the script SOLVE_4order_ODE.m, a nested function (f4ode) is employed to express the given fourth-order ODE that can be substituted with the anonymous function. Here's an example:

```
f4ode=@(t, y)([y(2); y(3); y(4); t.^2-3.*y(4)+sin(t).*y(2)-8.*y(1)]);
```

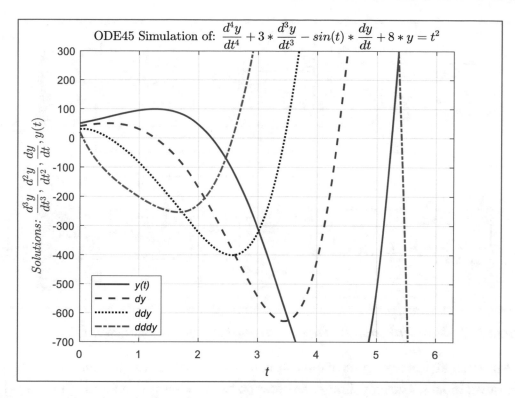

Figure 5-1. *Simulation of the fourth-order nonhomogeneous and nonlinear ODE problem*

Then the syntax of calling ode45 or ode23 would be as follows:

```
[t, y]=ode45(f4ode, time, y0, odeset('reltol', 1e-8));
```

Figure 5-2 shows two versions of a Simulink model (Fourth_order_ODE.slx) of the given fourth-order ODE problem.

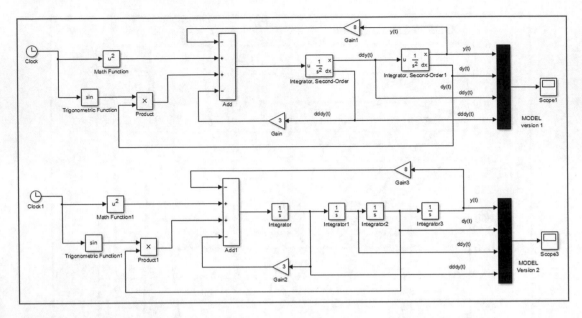

Figure 5-2. *Simulink model of the fourth-order nonhomogeneous ODE*

Note that all ICs are inputs of internal sources in the first and [Second-Order Integrator] blocks. The only difference between two models is that for the [Integrator] block types that are in the first model, two [Second-Order Integrator] blocks are used, and in the second one, four first-order [Integrator] blocks are employed. Note that in the simulation results shown in Figure 1-3, the output signal names are displayed in legends. That is, they are attained by naming each signal going to the [Mux] block as y(t), dy(t), ddy(t), dddy(t). You can rename signals is done by right-clicking and renaming the signals one by one. The simulation results (Figure 5-3) from the Simulink model Fourth_ order_ODE.slx are consistent with the ode45 solver outputs (shown in Figure 5-1) from the script SOLVE_4order_ODE.m.

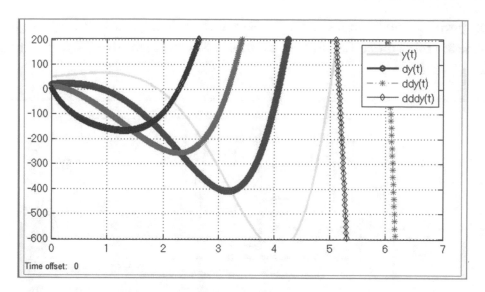

Figure 5-3. *Simulink model simulation results in [Scope]*

From our simulations of higher-order ODEs, we can conclude that computing the numerical solutions of higher-order ODEs is like with second-order ODEs, and the tools and functions used for simulations of higher-order ODEs are the same as what we have employed for first-order and second-order ODEs. The only difference we have in higher-order ODEs is that additional steps are required while rewriting the higher-order ODEs as first-order ones in scripts, and in Simulink modeling, additional [Integrator] blocks are needed.

Robertson Problem

The Robertson problem [1] describes chemical reactions and is used as a test problem for ODE solvers intended for stiff IVPs. It is a good example to use for coupled systems and also for stiff ODE problems. The Robertson problem is formulated by the following coupled first-order differential equations:

$$\begin{cases} \dot{y}_1 = 0.04\, y_1 + 10^4\, y_2 y_3 \\ \dot{y}_2 = 0.04\, y_1 - 10^4\, y_2 y_3 - 3*10^7\, y_2 \\ \dot{y}_3 = 3*10^7\, y_2 \end{cases}$$

with the initial conditions of $y_1(0) = 1$, $y_2(0) = 0$, $y_3(0) = 0$.

197

This problem can be implemented via Simulink modeling relatively easily. Figure 5-4 shows the Simulink model (`Robertson_problem.mdl`), and Figure 5-5 shows the simulation results.

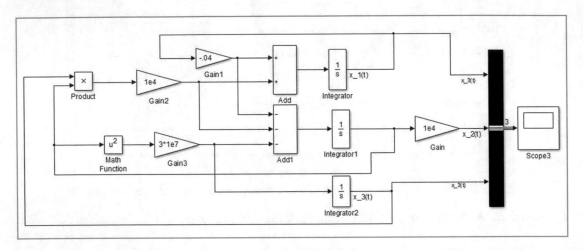

Figure 5-4. *Robertson problem model (Robertson_problem.mdl) in Simulink*

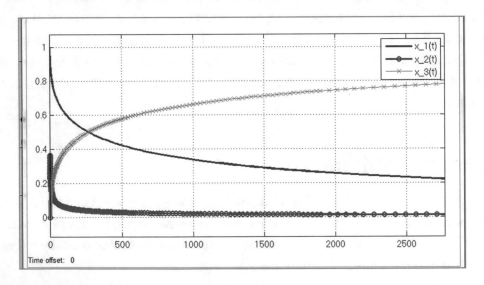

Figure 5-5. *The Robertson problem's Simulink model simulation results in a zoomed-in view*

The Robertson problem is a good demonstration to use for stiff ODEs. Its numerical solutions are computed with the Simulink modeling approach, and the computed solutions are found to be considerably stable. However, to assess the accuracy and computation time efficiency of the developed Simulink model, it is required that you develop other numerical simulation models using other methods, such as Milne, Adams-Moulton, or Euler's improved methods, or the built-in ODEx solvers such as ode15, ode23s, and ode113.

Akzo-Nobel Problem

The Akzo-Nobel problem originates from Akzo Nobel Central Research in Arnhem, the Netherlands. It describes a chemical process in which two species, FLB and ZHU, are mixed, while carbon dioxide is continuously added. A resulting species of importance is ZLA. The names of the chemical species are fictitious here due to their commercial values. This Akzo-Nobel problem is a version of the chemical Akzo-Nobel problem taken from ODE problems compiled by W. Lioen and H. de Swart [2]. The mathematical formulation of the problem has the following form:

$$M\frac{dy}{dt} = f(y), \quad y(0) = y_0, \quad y'(0) = y_0'$$

$$\text{with } y \in \Re^6, \quad 0 \le t \le 180$$

The matrix M is of rank 5 and is as follows:

$$M = \begin{pmatrix} 1 & 0 & 0 & 0 & 0 & 0 \\ 0 & 1 & 0 & 0 & 0 & 0 \\ 0 & 0 & 1 & 0 & 0 & 0 \\ 0 & 0 & 0 & 1 & 0 & 0 \\ 0 & 0 & 0 & 0 & 1 & 0 \\ 0 & 0 & 0 & 0 & 0 & 0 \end{pmatrix}$$

The function f is as follows:

$$f(y) = \begin{pmatrix} -2r_1 & +2r_2 & -r_3 & -r_4 & 0 & 0 \\ -\frac{1}{2}r_1 & 0 & 0 & -r_4 & -\frac{1}{2}r_5 & +F_{in} \\ r_1 & -r_2 & +r_3 & 0 & 0 & 0 \\ 0 & -r_2 & +r_3 & -2r_4 & 0 & 0 \\ 0 & r_2 & -r_3 & 0 & +r_5 & 0 \\ K_s y_1 y_4 - y_6 & 0 & 0 & 0 & 0 & 0 \end{pmatrix}$$

where r_i and F_{in} are auxiliary variables, as follows:

$$r_1 = k_1 \ y_1^4 \ y_2^{\frac{1}{2}} ; r_2 = k_2 y_3 y_4 ; r_3 = \frac{k_2}{K} y_1 y_5; \ r_4 = k_3 y_1 y_4^2 ; r_5 = k_4 y_6^2 y_2^{\frac{1}{2}} ; \ F_{in} = klA \left(\frac{p(CO_2)}{H} - y_2 \right)$$

The values of parameters k_1, k_2, k_3, k_4, K, klA, $p(CO_2)$, and H are as follows:

$$k_1 = 18.7, \quad k_4 = 0.42, \quad K_s = 115.83,$$
$$k_2 = 0.58, \quad K = 34.4, \quad p(CO_2) = 0.9,$$
$$k_3 = 0.09, \quad klA = 3.3, \quad H = 737.$$

The consistent initial condition vectors are as follows:

$$y_0 = \left(0.437, \ 0.00123, \ 0, \ 0.007, \ 0, \ K_s y_{0,1}, \ y_{0,4}\right)^T$$

Based on the previous differential equations and their initial conditions, we have developed a script called F_NOBEL.m with the Runge-Kutta method and the Simulink model AKZO_NOBEL.mdl - shown in Figure 5-6. The simulation results of the exercise are shown in Figure 5-7 and 5-8.

```
function F_ANOBEL
% HELP: Akzo-NOBEL Problem.
% Akzo-NOBEL implemented via Runge-Kutta Method in scripts
clearvars, close all
% ICs:
Ks=115.83; y(1,:)=[.437, .00123, 0, .007, 0, Ks*.437*.007];
% NOTE: VERY important to show the ICs with proper indexing,
% such as, y(1, :) or y(:,:)or :# of rows & :# of columns.
% Otherwise, anonymous function evaluation gives an error on
% matrix dimensions exceed ....
tstart=0;
hstep=.1;
tend=180;
t=hstart:hstep:hend;
for ii=1:length(t)-1
 k1=F_anobel(y(ii,1),y(ii,2),y(ii,3), y(ii,4), y(ii,5), y(ii,6));
  k2=F_anobel(y(ii,1)+hstep*k1(:,1)/2, y(ii,2)+hstep*k1(:,2)/2,...
         y(ii,3)+hstep*k1(:,3)/2, y(ii,4)+hstep*k1(:,4)/2,...
         y(ii,5)+hstep*k1(:,5)/2, y(ii,6)+hstep*k1(:,6)/2);
```

```
  k3=F_anobel(y(ii,1)+hstep*k2(:,1)/2,y(ii,2)+hstep*k2(:,2)/2,...
        y(ii,3)+hstep*k2(:,3)/2, y(ii,4)+hstep*k2(:,4)/2,...
        y(ii,5)+hstep*k2(:,5)/2, y(ii,6)+hstep*k2(:,6)/2);
  k4=F_anobel(y(ii,1)+hstep*k3(:,1), y(ii,2)+hstep*k3(:,2), ...
        y(ii,3)+hstep*k3(:,3)/2, y(ii,4)+hstep*k3(:,4)/2,...
        y(ii,5)+hstep*k3(:,5)/2, y(ii,6)+hstep*k3(:,6)/2);
 y(ii+1,:)=y(ii,:)+hstep*(k1+2*k2+2*k3+k4)/6;
end
plot(t, y(:,1), 'r', t, y(:,2), 'b--', t, y(:,3),'k:', t, y(:,4),...
'c-.', t, y(:,5),'m', t, y(:,6), 'g-.', 'linewidth', 1.5 )
axis tight, legend ('u1', 'u2','u3','u4','u5','u6')
title('The AKZO-NOBEL problem Simulation with R-K 4th order method')
xlabel('Simulation time cycle, t'),
ylabel('u_1(t), u_2(t), ...u_6(t)')
figure; subplot(211)
plot(t,y(:,1),'r',t,y(:,3),'k:', t, y(:,6),'g-.','linewidth', 1.5)
axis tight, legend ('u1', 'u3','u6')
title('The AKZO-NOBEL problem Simulation with R-K 4th order method')
xlabel('Simulation time cycle, t'),
ylabel('u_1(t), u_3(t), u_6(t)')
subplot '212'
plot(t, y(:,2),'r--',t, y(:,4),'b',t,y(:,5),'m:','linewidth', 1.5)
axis tight; legend('u2','u4','u5','best')
xlabel('Simulation time cycle, t')
ylabel('u_2(t), u_4(t), u_5(t)')
function F=F_anobel(y1,y2,y3,y4,y5,y6)
k1=18.7; k2=.58; k3=.09; k4=.42; K=34.4; klA=3.3;
Ks=115.83; pCO2=.9; H=737;
F=[-2*k1*y1^4*y2^.5+k2*y3*y4-(k2/K*y1*y5)-(k3*y1*y4^2),...
-0.5*(k1*y1^4*y2^.5)-(k3*y1*y4^2)- ... 0.5*(k4*y6^2*y2^.5)+(klA*(pCO2/
H-y2)),...
(k1*y1^4*y2^.5)-k2*y3*y4+(k2/K*y1*y5),...
-k2*y3*y4+(k2/K*y1*y5)-2*(k3*y1*y4^2),...
k2*y3*y4-(k2/K*y1*y5)+(k4*y6^2*y2^.5), Ks*y1*y4-y6];
```

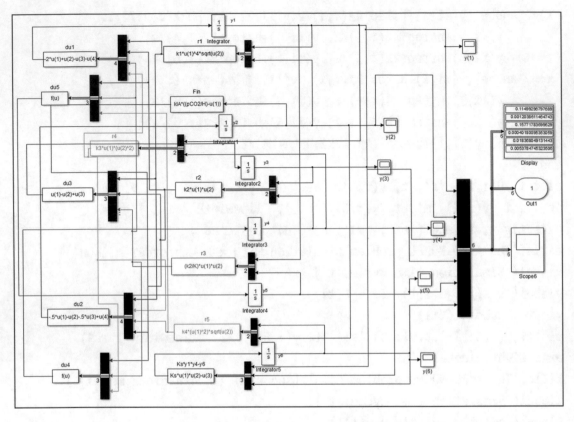

Figure 5-6. *Simulink model (AKZO_NOBEL.mdl) of the Akzo-Nobel problem*

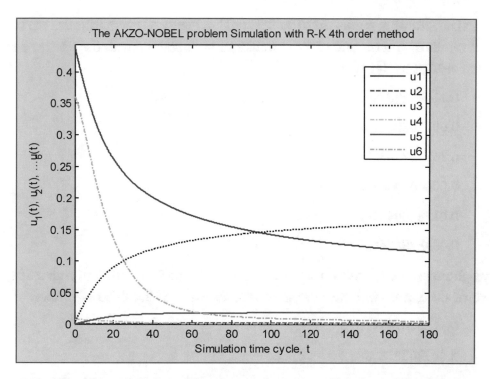

Figure 5-7. *The Akzo-Nobel problem's simulation results shown in one plot*

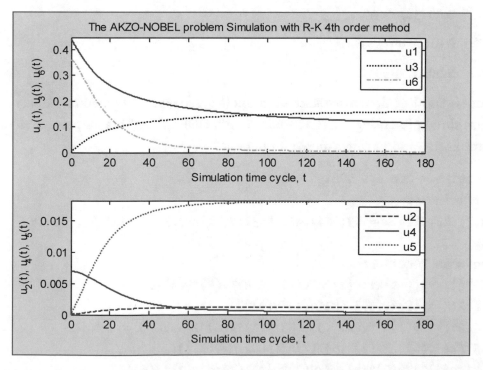

Figure 5-8. *The Akzo-Nobel problem's simulation results shown in subplots*

From the Simulink model (AKZO_NOBEL.mdl) in Figure 5-6 with the ode23tb solver, the final values of y_i at the end of the simulations are obtained with the relative error tolerance of 10^{-3} (by default).

> 0.1149929679...
>
> 0.0012038611...
>
> 0.15771718369...
>
> 0.00040180095...
>
> 0.01836804913...
>
> 0.0053784745...

By tightening the relative error tolerance to 10^{-10}, the final values of y_i after 180 seconds of simulation with the ode15s solver have been found to be as follows:

> 0.1150237271...
>
> 0.001203853338...
>
> 0.1577023801...
>
> 0.00040169047...
>
> 0.018356971283...
>
> 0.0053778600...

Here is another alternative solution script (AN_ode15s.m) of the Akzo-Nobel problem simulation using MATLAB's ode15s stiff ODE solver. This script is much more comprehensive and provides more output plots.

```
% AN_ode15s.m employs ODE15S solver
% Coefficients:
k1=18.7; k2=.58; k3=.09; k4=.42; K=34.4; k1A=3.3; Ks=115.83; pCO2=.9;
H=737;
% Anonymous Function:
F_NAN=@(t, y)([-2*k1*y(1)^4*y(2)^.5+k2*y(3)*y(4)-...
    ((k2/K)*y(1)*y(5))-(k3*y(1)*y(4)^2);
    -0.5*(k1*y(1)^4*y(2)^.5)-(k3*y(1)*y(4)^2)-...
    .5*(k4*y(6)^2*y(2)^.5)+(k1A*(pCO2/H-y(2)));
    (k1*y(1)^4*y(2)^.5)-k2*y(3)*y(4)+(k2/K*y(1)*y(5));
```

```
     -k2*y(3)*y(4)+(k2/K*y(1)*y(5))-2*(k3*y(1)*y(4)^2);
     k2*y(3)*y(4)-(k2/K*y(1)*y(5))+(k4*y(6)^2*y(2)^.5); ...
     Ks*y(1)*y(4)-y(6)]);
% ICs:
y0=[.437, .00123, 0, .007, 0, Ks*.437*.007];
hstart=0;
hstep=.1;
hend=180;
t=hstart:hstep:hend;
[tout, Yout]=ode15s(F_NAN, t, y0);
plot(tout, Yout, 'linewidth', 1.5),
legend('y_1', 'y_2','y_3','y_4','y_5','y_6'), grid on,
title('AKZO-NOBEL Problem simulation with ODE15s ')
xlabel('time'), ylabel('y_i(t)')
figure,
subplot(411), plot(tout, Yout(:,1), 'r-', 'linewidth', 1.5),
title('AKZO-NOBEL Problem: variable y(1)'), grid on
subplot(412), plot(tout, Yout(:,2), 'b-', 'linewidth', 1.5),
title('AKZO-NOBEL Problem: variable y(2)'), grid on
subplot(413), plot(tout, Yout(:,3), 'k-', 'linewidth', 1.5),
title('AKZO-NOBEL Problem: variable y(3)'), grid on
subplot(414), plot(tout, Yout(:,4), 'm-', 'linewidth', 1.5),
title('AKZO-NOBEL Problem: variable y(4)'), grid on
figure
subplot(211), plot(tout, Yout(:,5), 'b-', 'linewidth', 1.5),
title('AKZO-NOBEL Problem: variable y(5)'), grid on
subplot(212), plot(tout, Yout(:,6), 'm-', 'linewidth', 1.5),
title('AKZO-NOBEL Problem: variable y(6)'), grid on
figure, plot(tout, Yout(:,2), 'b-', 'linewidth', 1.5),
title('AKZO-NOBEL Problem: y(2) to view a change in [0,3]'),
grid on; xlim([0, 3])
```

Note that the plots of the computation results from the previous script with the solver ode15s are not shown here. They are left for you to obtain and compare with outputs (Figure 5-7, Figure 5-8) from the Simulink model (Figure 5-6). The three simulation models developed in this chapter, meaning the script based on the fourth-order

Runge-Kutta method, the Simulink model (with `ode23tb`, `ode15s`), and the `ode15s` solver simulation script of the Akzo-Nobel problem, have produced well-converged results, and there are relatively small discrepancies—less than 0.1 percent in final solutions. Thus, we can conclude that all of the approaches for this exercise are appropriate in terms of accuracy, just not for the computation time costs.

HIRES Problem

The High Irradiance RESponse (HIRES) problem originated from plant physiology taken from the test ODE problems compiled by W. Lioen and H. de Swart [2]. The HIRES problem is a test problem for stiff differential equations and is one of the most widely used test simulation problems for verifying the efficiency of numerical simulation methods. It is composed of eight variables and is formulated as follows:

$$
\begin{cases}
\dot{u}_1 = -1.7u_1 + 0.43u_2 + 8.32u_3 + 0.0007 \\
\dot{u}_2 = 1.71u_1 - 8.75u_2 \\
\dot{u}_3 = -10.03u_3 + 0.43u_4 + 0.35u_5 \\
\dot{u}_4 = 8.32u_2 + 1.71u_3 - 1.12u_4 \\
\dot{u}_5 = -1.75u_1 + 0.43u_6 + 0.43u_7 \\
\dot{u}_6 = -280u_6u_8 + 0.69u_4 + 1.71u_5 - 0.43\,u_6 + 0.69\,u_7 \\
\dot{u}_7 = 280u_6u_8 - 1.81u_7 \\
\dot{u}_8 = -280u_6u_8 + 1.81u_7
\end{cases}
$$

with the initial conditions of $u_1(0) = 1$, $u_2(0) = 0$, $u_3(0) = 0$, $u_4(0) = 0$, $u_5(0) = 0$, $u_6(0) = 0$, $u_7(0) = 0$, $u_8(0)=0.0057$. The implementation of the HIRES problem is rather straightforward since it is in the form of first-order ODEs. The developed simulation models (`HIRES_EM.m`, `HIRES_RK.m`) of the problem contain forward Euler and 4th-/5th-order Runge-Kutta methods implemented via scripts only. The simulation results are shown in Figure 5-9.

```
function HIRES_EM
%HELP. HIRES_EM.m solves the HIRES problem with EULER-Forward
% ICs:
u=[1 0 0 0 0 0 0 .0057];
tstart=0;
hstep=.01;
```

```
tend=321.8122;
% Need to integrate over [0, 321.8122],
% but for better visualization, we can limit with 5.8122
t=tstart:hstep:tend;
g=zeros(numel(t), 8);
for ii=1:length(t)-1
    g(ii,:)=Gee(u(ii,1), u(ii,2), u(ii,3), u(ii,4), u(ii,5), ...
    u(ii,6), u(ii,7), u(ii,8));
    u(ii+1, :)=u(ii,:)+g(ii,:)*hstep;
end
figure
plot(t, u(:,1), 'r', t, u(:,2), 'b--',t, u(:,3),'g:',...
t, u(:,4),'c-.',t, u(:,5),'m',...
t, u(:,6),'g-.',t, u(:,7), 'b:', t,u(:,8),'k', 'linewidth', 1.5 ), axis
tight
legend ('u_1', 'u_2','u_3','u_4','u_5','u_6','u_7','u_8')
xlim([0, 25]),
title('The HIRES problem solved with EULER forward')
xlabel('time'), ylabel('u_i(t)')
% NOTE: Interesting phenomena observed within [10, 18] of time
% interval. There are certain instabilities with u6, u7, u8.
function F=Gee(u1,u2,u3,u4,u5,u6,u7,u8)
F=[-1.7*u1+.43*u2+8.32*u3+.0007; 1.71*u1-8.75*u2;
    -10.03*u3+.43*u4+.035*u5; 8.32*u2+1.71*u3-1.12*u4;
   -1.75*u5+.43*u6+.43*u7;-280*u6*u8+.69*u4+1.71*u5-.43*u6+.69*u7;
    280*u6*u8-1.81*u7; -280*u6*u8+1.81*u7];
return
```

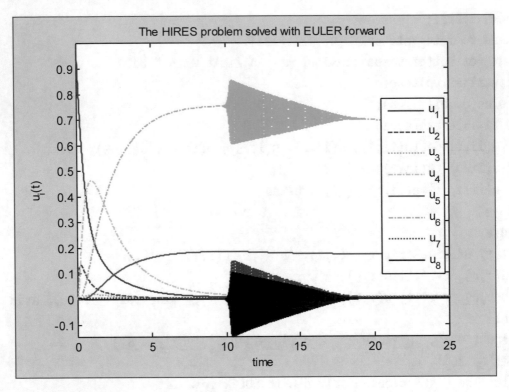

Figure 5-9. *The HIRES problem is solved with the forward Euler method*

```
function HIRES_RK
%HIRES_RK.m solves hires problem with Runge-Kutta method scripts
u=[1 0 0 0 0 0 0 .0057]; % Initial Conditions
tstart=0;
hstep=.01;
tend=321.8122;
% need to integrate over [0, 321.8122],
% but for better visualization, we can limit with 5.8122
t=tstart:hstep:tend;
for ii=1:length(t)-1
    k1=ff(u(ii,1), u(ii,2),u(ii,3), u(ii,4), u(ii,5), u(ii,6),...
        u(ii,7), u(ii,8));
    k2=ff(u(ii,1)+hstep*k1(:,1)/2,u(ii,2)+hstep*k1(:,2)/2,...
        u(ii,3)+hstep*k1(:,3)/2,u(ii,4)+hstep*k1(:,4)/2,...
        u(ii,5)+hstep*k1(:,5)/2,u(ii,6)+hstep*k1(:,6)/2, ...
        u(ii,7)+hstep*k1(:,7)/2,u(ii,8)+hstep*k1(:,8)/2);
```

```
    k3=ff(u(ii,1)+hstep*k2(:,1)/2,u(ii,2)+hstep*k2(:,2)/2, ...
        u(ii,3)+hstep*k2(:,3)/2,u(ii,4)+hstep*k2(:,4)/2,...
        u(ii,5)+hstep*k2(:,5)/2,u(ii,6)+hstep*k2(:,6)/2, ...
        u(ii,7)+hstep*k2(:,7)/2,u(ii,8)+hstep*k2(:,8)/2);
    k4=ff(u(ii,1)+hstep*k3(:,1),u(ii,2)+hstep*k3(:,2), ...
        u(ii,3)+hstep*k3(:,3)/2,u(ii,4)+hstep*k3(:,4)/2,...
        u(ii,5)+hstep*k3(:,5)/2,u(ii,6)+hstep*k3(:,6)/2, ...
        u(ii,7)+hstep*k3(:,7)/2,u(ii,8)+hstep*k3(:,8)/2);
    u(ii+1,:)=u(ii,:)+hstep*(k1+2*k2+2*k3+k4)/6;
end
plot(t, u(:,:), 'linewidth', 1.5), axis([0 321 0, 0.8])
title('the HIRES problem solved with Runge-Kutta method')
figure; subplot(311)
plot(t, u(:,1),'r:',t,u(:,4),'b-.',t,u(:,6),'m','linewidth',1.5 ),
ylabel('u_i(t)')
title('the HIRES problem solved with Runge-Kutta method')
legend ('u_1','u_4','u_6'), axis([0, 25 0 0.8])
subplot(312)
plot( t, u(:,7), 'b',t, u(:,8),'k-.','linewidth', 1.5)
legend ('u_7','u_8'), ylabel('u_i(t)'), xlim([0, 5]),
subplot(313)
plot(t, u(:,2),'b--',t,u(:,3),'k:',t,u(:,5),'m-','linewidth',1.5)
xlabel('t'), ylabel('u_i(t)'), legend ('u_2','u_3','u_5')
xlim([0, 15])
function f=ff(u1, u2, u3, u4, u5, u6, u7, u8)
f=[-1.7*u1+.43*u2+8.32*u3+.0007,1.71*u1-8.75*u2,...
-10.03*u3+.43*u4+.035*u5, 8.32*u2+1.71*u3-1.12*u4, ...
-1.75*u5+.43*u6+.43*u7,-280*u6*u8+.69*u4+1.71*u5-.43*u6+.69*u7,...
    280*u6*u8-1.81*u7, -280*u6*u8+1.81*u7];
end
```

There are some considerably instable regions in the numerical solutions (Figure 5-10) of variables u_6, u_7, u_8 computed with the forward Euler method.

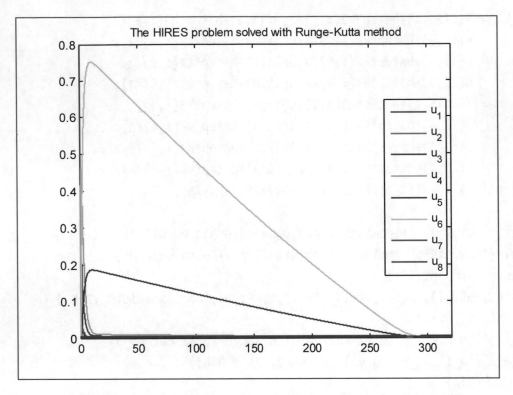

Figure 5-10. *The HIRES problem is solved with the 4/5 -order Runge-Kutta method*

From simulations of the HIRES problem with the forward Euler (Figure 5-9) and Runge-Kutta (Figure 5-10, Figure 5-11) methods, it is clear that for such higher-order stiff and coupled ODE problems, the forward Euler method may lead to some considerably instable results. In addition to these two methods, another simulation model in Simulink can be developed easily.

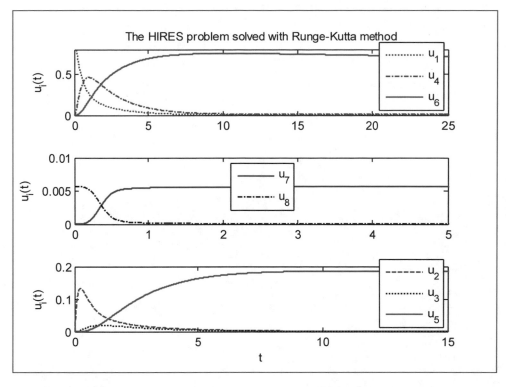

Figure 5-11. *The HIRES problem solutions found with the 4/5-order Runge-Kutta method shown in subplots*

From the simulations of higher-order and coupled ODEs, we can conclude that solving such ODEs or coupled systems numerically by using MATLAB's built-in ODE solvers, by using Simulink models, and by writing scripts based on the Euler or Runge-Kutta methods is similar to solving first- or second-order ODEs. However, we have to choose the solution methods carefully and adjust the settings of the built-in ODE solvers cautiously for the efficiency and accuracy of the simulation results.

Reference

[1] Edsberg, L., "Integration package for chemical kinetics," in *Stiff Differential Systems*, Willoughby, R. A., ed., Plenum Press, New York (1974), pp. 81–94.

[2] Lioen, W., de Swart, H., "Test for initial value problem solvers," CWI Report MAS-R9832, ISSN 1386-3703, Amsterdam (1998).

CHAPTER 6

Implicit ODEs

Modern theories and methods for solving ODEs have been implemented in a broad range of software applications and tools. The theory of numerical search solution is well understood, and there are robust methods for solving different types of problems. However, the theory of differential algebraic equations (DAEs) has not been studied to the same extent as ODEs. There were some extended studies carried out by Ascher and Petzold [1], but that was in the 1970s. Despite that there are many important applications of DAEs to model various systems, the methods to solve DAEs are limited to a certain extent because the problems used with DAEs are difficult to solve, and the theories behind DAEs are hard to understand. Therefore, there are a limited number of studies and literature devoted to DAEs. In this chapter, we will show some of the available tools of MATLAB/Simulink that can solve DAEs.

Let's first define implicit ODEs. A given problem can be expressed in the following general form:

$$F(t, y(t), y'(t)) = 0 \qquad (6\text{-}1)$$

where F is a continuous and sufficiently smooth function that is called the *first-order implicit differential equation*. If this equation can be solved for $y'(t)$, we will obtain one or several explicitly defined differential equations, which are the explicit ODEs that we have covered in previous chapters.

In this section, we will demonstrate via examples of several implicitly defined (or intentionally treated as implicit) ODEs how to obtain numerical solutions of implicit IVPs using MATLAB's built-in ODE solver ode15i. We will also compare the solutions obtained from ode15i to the ones obtained from the ode15s, ode45, and Simulink models. We will explain implicitly defined IVP-solving issues via specific examples and some standard procedures that are applicable to a broad range of IVP problems in order to obtain numerical solutions using the MATLAB tools.

© Sulaymon L. Eshkabilov 2020
S. L. Eshkabilov, *Practical MATLAB Modeling with Simulink*, https://doi.org/10.1007/978-1-4842-5799-9_6

In fact, one of the major difficulties in solving implicit ODEs is that the highest-order (differential) variable cannot be computed by allocating it on one side of the equation or by separating variables. Another difficulty is that all initial conditions aren't defined explicitly. In newer versions of MATLAB (starting with MATLAB version 7/8), there is a built-in ODE solver called `ode15i` and function called `decic()` to compute consistent initial values of a given implicit ODE.

Let's start by studying how to compute numerical solutions of a fully implicit set of first-order differential equations in the form of $F(t, y(t), y'(t)) = 0$ with a time interval of $[t_0, t_{end}]$ and the initial conditions at t_0. The consistent initial conditions $t_0, y_0(t_0)$, and $y'_0(t_0)$ are to satisfy the algebraic equation $F(t_0, y_0(t_0), y'_0(t_0)) = 0$, where F is a smooth function.

The general syntax of the solver `ode15i` is similar to the other MATLAB ODE solvers, including `ode23`, `ode45`, `ode113`, etc. But as mentioned, there is an additional tool/function used with `ode15i` for implicit ODEs, called `decic()`. This function defines a consistent initial value of $y'_0(t_0)$ for first-order ODEs, $y''_0(t_0)$ for second-order ODEs, and so forth.

Example 1

Here is the example problem: $t^2 \dot{y} + t \dot{y}^2 + 2\cos(t) = \dfrac{1}{y}$.

with the initial condition $y(0) = \dfrac{1}{2}$. The given ODE can be rewritten in the form of an implicit differential equation expressed in Equation (8.44), as shown here:

$$t^2 \dot{y} + t \dot{y}^2 + 2\cos(t) - \frac{1}{y} = 0$$

Subsequently, this can also be expressed as follows:

$$\frac{t^2 dy}{dt} + \frac{t dy^2}{dt} + 2\cos(t) - \frac{1}{y} = 0$$

$$F\left(t, y, \frac{dy}{dt}\right) = \frac{t^2 dy}{dt} + \frac{t dy^2}{dt} + 2\cos(t) - \frac{1}{y}$$

and as follows:

$$F\left(t, y, \frac{dy}{dt}\right) = 0$$

Note that the expression of $F(t, y, dy) = 0$ is the main difference in defining F in implicit ODEs from explicit ones. In this example, the F function, containing three arguments (which are $t, y,$ and $\frac{dy}{dt}$), can be expressed via an anonymous function (@) directly. Here is the complete solution, called IMPLICIT_1ODE_EX1.m:

```
% IMPLICIT_1ODE_EX1.m
%{
EXAMPLE 1. Problem: t^2*y'+t*(y')^2+2*cos(t)=1/y with ICs: y(0)=1/2;
Part 1.
The problem function is defined via function handle under name of Yin.
A new variable is introduced: dy=yp.
%}
clearvars; close all
Yin=@(t,y, yp)(t^2*yp+t*yp^2+2*cos(t)-1/y);
y0=1/2; yp0=0;  % Initial conditions
[t, yt]=ode15i(Yin, [0, 6*pi], y0, yp0);
plot(t, yt, 'bo-')
title('\it Simulation of: $$ t^2*\frac{dy}{dt}+t*(\frac{dy}
{dt})^2+2*cos(t)=y^{-1}, y_0=\frac{1}{2} $$', 'interpreter', 'latex')
grid on;
xlabel('\it t'),
ylabel( '\it Solution, y(t)'),
hold on

% Part 2. Another way of solving this problem is SIMULINK with
% simulation start time set at 1e-3 and end time set at 6*pi.
[t, Y_t]=sim('IMPLICIT_1ODE_EX1_sim', [1e-3 6*pi]);
plot(t, Y_t(:,1), 'rx--')
legend('ode15i (script)', 'Simulink ode15s', 'location', 'best')
```

It is also feasible to employ Simulink tools with the ode15s solver to model the given implicit ODE to simulate it. Note that the initial condition ($y(0) = \dfrac{1}{2}$) is inserted as an internal in the Integrator block. Figure 6-1 shows a complete Simulink model, called IMPLICIT_1ODE_EX1_sim.mdl.

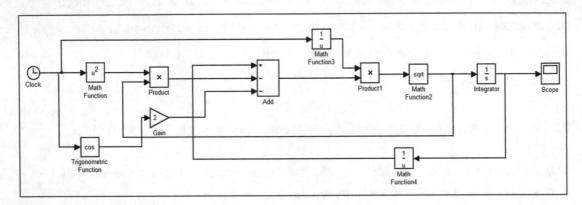

Figure 6-1. *The Simulink model - IMPLICIT_1ODE_EX1_sim.mdl is a solution model of the problem:* $t^2\dot{y} + t\dot{y}^2 + 2\cos(t) = \dfrac{1}{y}$.

By executing the previously shown parts 1 and 2 of the model (ode15i), called IMPLICIT_1ODE_EX1.m, and the Simulink model (ode15s), called IMPLICIT_1ODE_EX1_sim.mdl, we get the numerical results of the problem, as displayed in Figure 6-2.

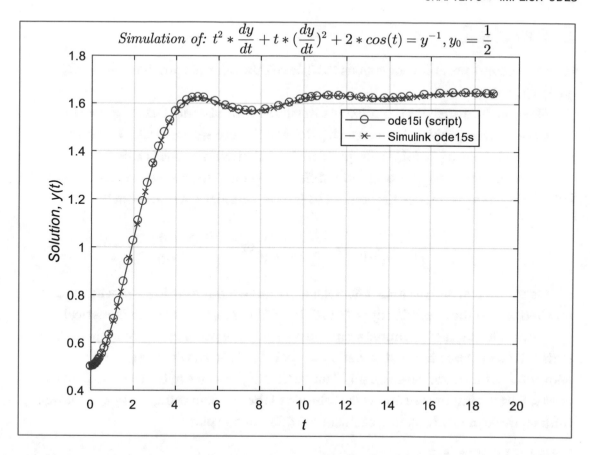

Figure 6-2. *Simulation results of the implicit ODE problem: $t^2 \dot{y} + t\dot{y}^2 + 2\cos(t) = \dfrac{1}{y}$ with ode15i and ode15s (Simulink)*

From the simulation results shown in Figure 6-2, we can conclude that both solver tools (MATLAB's ode15i and Simulink's ode15s) perform well-converged solutions. That means to solve implicitly defined ODEs, we can also employ Simulink modeling by adjusting the ODE solver options appropriately.

The second-order ODEs of the IVP (in Example 2 and Example 3) are meant to demonstrate the implementation procedures of ode15i by treating the given second-order homogeneous ODEs as implicit IVPs, even though they are explicitly defined.

Example 2

Given a second-order homogeneous ODE, here's the example problem: $\ddot{y} + y = 0$, $y(0) = y_0 = 0$, $\dot{y}(0) = 1$.

Note that the analytical solution of this simple second-order ODE is $y(t) = \sin(t)$. In fact, the given problem is an explicitly defined IVP of a second order. To better understand the use of ode15i, this problem is treated as an implicit ODE.

We rewrite the given second-order differential equation as a system of two first-order ODEs expressed as implicit first-order ODEs or differential algebraic equations.

$$\begin{cases} \dot{y}_1 = y_2 \\ \dot{y}_2 + y_1 = 0 \end{cases} \Rightarrow \begin{cases} \dot{y}_1 - y_2 = 0 \\ \dot{y}_2 + y_1 = 0 \end{cases} \text{ or } \begin{cases} dy_1 - y_2 = 0 \\ dy_2 + y_1 = 0 \end{cases}$$

Using the two derived first-order ODEs, an anonymous function is created and recalled within the created script, IMPLICIT_2ODE_EX2.m, to obtain the numerical solutions. The simulation model script consists of three parts. Part 1 computes solutions of the given problem by treating it as an implicit ODE solved with ode15i. Part 2 computes solutions of the given problem by treating it as an explicit ODE (as it is) with ode45. Part 3 computes an analytical solution of the problem using dsolve(). Outputs from all three parts are plotted against each other in one plot.

```
% IMPLICIT_2ODE_EX2.m
%{
EXAMPLE 2. IMPLICITLY expressed ODE problem: y"(t)+y(t)=0
We first define the given ODE as two 1st order ODEs:
F(t,y1,y2,dy1,dy2)=>dy1=y2;=>dy2+y1=0]=>{dy1-y2; dy2+y1};
NOTE: the difference(s) in defining functions (function file or anonymous
function (@)) between implicit ODEs and explicit ODEs.
%}
clearvars; close all
% Part 1. ODE15i

  Fex2=@(t, y, dy)([dy(1)-y(2); dy(2)+y(1)]);
  y0=[0, 1]; dy0=[1, 0];       % ICs
  F_aty0=[0, 0]; F_atydo=[];   % Function values at y0 and dy0
  Tspan=[0, 2*pi];             % Time span
```

```
% Find Consistent Initial Values:
[y0, dy0]=decic(Fex2, 0, y0, dy0, F_aty0, F_atyd0);

% Compute solutions of the problem with ode15i:
[t, y]= ode15i(Fex2, Tspan, y0, dy0);
 plot(t, y(:,1), 'bd-',t, y(:,2), 'rx-')
 hold on

% Part 2. ODE45
%%%% NOTE: differences between Fex2 and FEX2 anonymous.
% function handles:
FEX2=@(t, y)([y(2); -y(1)]);
y0=0; dy0=1;                              % ICs
Tspan=[0, 2*pi];                         % Time span
[time, Y]=ode45(FEX2,Tspan, [y0, dy0], []);
plot(time, Y(:,1), 'kh--',time, Y(:,2), 'go--')
title('\it ode15i vs. ode45: Solutions of y"(t)+y(t)=0')
xlabel('\it t'), ylabel('\it Solutions:  y(t), $$ \frac{dy}{dt}$$',
'interpreter', 'latex'), grid

% Part 3. DSOLVE: Analytic solution

yt=dsolve('D2y+y=0', 'y(0)=0','Dy(0)=1', 't');
Yt=vectorize(yt);
Yt=subs(Yt, t); % Alternative:  eval(Yt, t);
plot(t, Yt, 'c-', 'linewidth', 1.5)
legend('y(t) found via ode 15i',...
'dy found via ode15i','y(t) found via ode45',...
'dy found via ode45','y(t) Analytical solution', ...
 'location', 'best')
```

Finally, the simulation output of the script is displayed in Figure 6-3; the output contains the numerical results from the ode15i and ode45 solvers and analytical solution found from dsolve(). The results shown in Figure 6-3 demonstrate the correct use of the ODE solvers.

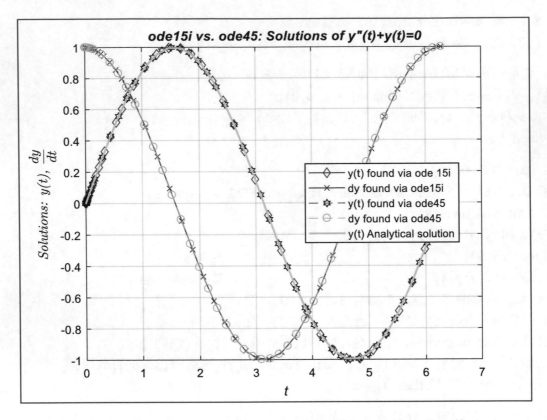

Figure 6-3. *Example 2:* $\ddot{y}+y=0$ *solved with* `ode15i`, `ode45`, `dsolve()`

Example 3

Here is the example problem: $m\ddot{x}+kx=0$, $x(0)=x_0=1$, $\dot{x}(0)=0$. Here are the parameters: $m=13$; $k=1213$. Note that the given second-order ODE is explicitly defined, but we intentionally treat it as an implicit ODE to demonstrate how to employ ODEx solvers.

First, we rewrite the given second-order ODE as two first-order ODEs.

$$\begin{cases} \dot{q}_1 = q_2 \\ m\dot{q}_2 + kq_1 = 0 \end{cases} => \begin{cases} \dot{q}_1 - q_2 = 0 \\ m\dot{q}_2 + kq_1 = 0 \end{cases} \text{ or } \begin{cases} dq_1 - q_2 = 0 \\ m\,dq_2 + kq_1 = 0 \end{cases}$$

Based on the derived two first-order differential algebraic equations, we write a function file called DAE_Fun.m. Another option is that, instead of a separate function m-file, we create an anonymous function (@).

```
function f=DAE_Fun(~,q, dq, k, m)
%{
A function file: DAE_Fun.m
NOTE: that following a sequence of arguments is IMPERATIVE. If the order
of k switched/swapped with m then totally different results are obtained.
Values for k and m (m=13, k=113) need to be assigned before ode solver in
the above simulation script.
%}
f = [dq(1)-q(2), m*dq(2) + k*q(1)]';
return
```

All initial conditions and the values of the input parameters such as $m = 13$ and $k = 1213$ are defined in the next script, named IMPLICIT_2ODE_EX3.m. This script is composed of two parts. Part 1 treats the given problem as an implicit ODE, and part 2 treats it as an explicit ODE.

```
%% EXAMPLE 3.  m*x"+k*x=0 with ICs: x(0)=1, x'(0)=0
%{
IMPLICITLY defined ode solved with ode15i
m=13; k=1213;
NOTE: that how the anonymous function (@) expression is defined in Fun to
solve implicit ODE problem. In this case, we have switched from a 2nd order
system into two 1st order ODE system of equations set equal to "0" as such:
dq(1)-q(2)=0 && m*dq(2)+k*q(1)=0. This is the very key point in solving
implicitly defined ODEs.
Fun = @(t, q, dq, k, m)([dq(1)-q(2); m*dq(2)+k*q(1)]);
NOTE: to call the anonymous function Fun in ode15i/ode45/etc, to use an
anonymous function name is needed without @ sign.
E.g. ode15i(Fun, t,[],[],[],...).
NOTE: in computing dq0, we plug in q0 into two equations of the function
Fun and by this way, we obtain the values of dq0 from the values of q0.
Similarly, the values of the Function (Fun) at q0 and dq0 are computed from
q0 and dq0 simultaneously.
%}
clearvars; close all
% PART 1.
```

```
m=13; k=113;
q0=[1,0]; dq0=[0,0];          % ICs
F_atq0=[0, k]; F_atdq0=[];   % Function values at q0 and dq0
% Consistent Initial values:
[q0, dq0]=decic(@Fun, 0, q0, dq0, F_atq0, F_atdq0,[], k, m);
% Compute solution values with ode15i:
[t, ft]=ode15i(@Fun, [0,pi], q0,[0, 0]', [], k, m);
plot(t, ft(:,1), 'bx--', t, ft(:,2), 'kd-.'); hold on
```

For comparison purposes, let's solve the previous problem with ode45 since it is an explicitly defined IVP.

```
% PART 2.
%{
EXPILICITLY defined ode solved with ode45.
GIVEN:  m*x"+k*x=0 with ICs: x(0)=1, x'(0)=0. The given 2nd order ODE can
be written also as: x"+w^2*x=0.
NOTE: that how two 1st order ODEs are defined in applying ode45 in
comparison with IMPLICITLY defined ODE solved with ODE15i in the above
case.
NOTE: Third input of the anonymous function FFunc is omegaSQ that is equal
to k/m
%}
FFunc=@(t, q, omegaSQ)([q(2); -omegaSQ*q(1)]);
M=13; K=113; omegaSQ=(K/M);
[t, FF]=ode45(FFunc, [0,pi], [1,0],[],  omegaSQ);
plot(t, FF(:,1), 'ro-',t, FF(:,2), 'm+-')
legend('ode15i x(t)','ode15i dx(t)','ode45 x(t)',...
'ode45 dx(t)', 'location', 'best')
title('ode15i vs. ode45: Solutions of  $$ m*\frac{d^2x}{dt^2}+k*x=0 $$ ',
'interpreter', 'latex'), shg
xlabel('\it t')
ylabel('Solutions: $$ x(t), \frac{dx}{dt} $$', 'interpreter', 'latex')
grid on; hold off
```

The numerical solution results from the two solvers, ode15i and ode45, are added to the plot, as shown in Figure 6-4.

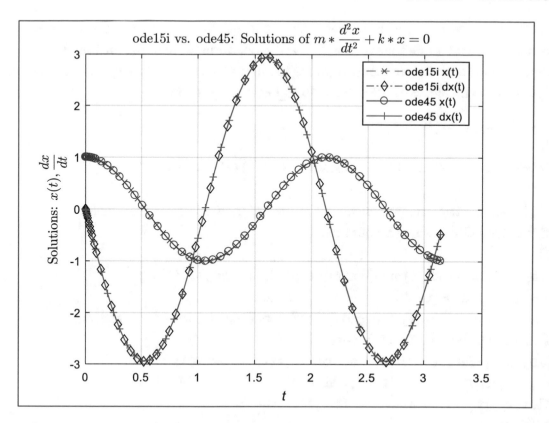

Figure 6-4. $m\ddot{x}+kx=0$ *solved with* `ode15i` *and* `ode45`

Well-converged results (shown in Figure 6-4) from both approaches (`ode15i` and `ode45`) demonstrate an appropriate use of ODE solvers and functions for implicit ODEs defined by differential algebraic equations.

Example 4

Given a second-order (truly) implicit ODE, here's the example problem: $e^{\ddot{x}} + \ddot{x} + x = 0$, $x(0) = 1$, $\dot{x}(0) = 0$.

This is a truly implicit IVP of an ODE that can be solved via the solver `ode15i`. In this example, like the previous examples, we rewrite it as two first-order differential algebraic equations.

$$\begin{cases} \dot{x}_1 = x_2 \\ e^{\dot{x}_2} + \dot{x}_2 + x_1 = 0 \end{cases} => \begin{cases} \dot{x}_1 - x_2 = 0 \\ e^{\dot{x}_2} + \dot{x}_2 + x_1 = 0 \end{cases} \text{ or } \begin{cases} dx_1 - x_2 = 0 \\ e^{dx_2} + dx_2 + x_1 = 0 \end{cases}$$

From the two previously derived differential equations, we create an anonymous function (@) and implement all the standard procedures for defining the initial conditions with decic and for computing the numerical solutions with the ode15i tools. The solution m-file, called IMPLICIT_2ODE_EX4.m, is as follows:

```
% IMPLICIT_2ODE_EX4.m
%{
EXAMPLE 4. Truly IMPLICIT problem: exp(x")+x"+x=0 with ICs: x(0)=1,
x'(0)=0.
NOTE: that Fimp anonymous function is defined as two first order
differential algebraic equations.
%}
Fimp = @(t, x, dx)([dx(1)-x(2); exp(dx(2))+dx(2)+x(1)]);
x0=[1, 0]; dx0=[0, 0];              % ICs
F_atx0=[0,exp(0)+1]; F_atdx0=[ ]; % Function values at x0 and dx0
% Find Consistent Initial values:
[x0, dx0]=decic(Fimp, 0, x0, dx0, F_atx0, F_atdx0);
[t, ft]=ode15i(Fimp, [0,13], x0,dx0);
plot(t, ft(:,1), 'bx-', t, ft(:,2), 'kd-')
hold on
Out=sim('IMPLICIT_2ODE_EX4_sim.mdl');
plot(Out.yout{1}.Values.Time, Out.yout{1}.Values.Data, 'r--',
'linewidth', 1.5)
plot(Out.yout{2}.Values.Time, Out.yout{2}.Values.Data, 'g--',
'linewidth', 1.5)
legend('x(t) with ode15i','dx(t) with ode15i', ...
  'x(t) with Simulink','dx(t) with Simulink','location', 'best')
title('\it ODE15i solutions of:  $$ e^{\frac{d^2x}{dt^2}}+\frac{d^2x}
{dt^2}+x=0 $$',  'interpreter', 'latex')
 xlabel('\it t')
ylabel('\it $$ x(t), \frac{dx}{dt} $$', 'interpreter', 'latex')
grid on;
axis tight; shg
```

Figure 6-5 shows the Simulink model solution (IMPLICIT_2ODE_EX4_sim.mdl) for the problem. Both of the initial conditions are set internally in the [Integrator] blocks, and the ODE solver selection is set to auto.

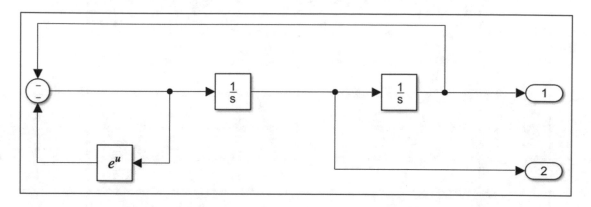

Figure 6-5. *Simulink model, IMPLICIT_2ODE_EX4_sim.mdl, of* $e^{\ddot{x}} + \ddot{x} + x = 0$

Note   The latest versions of Simulink have a DAE solver called DAESSC that is meant to be used with the Simscape Toolbox of Simulink.

The simulation results (shown in Figure 6-6) of ode15i and the Simulink model show that the computed solutions are converged perfectly well.

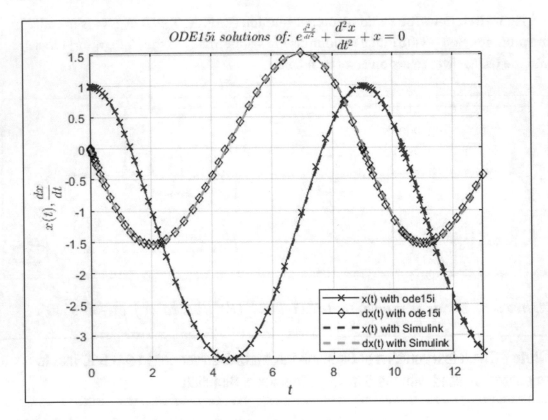

Figure 6-6. *The $e^{\ddot{x}} + \ddot{x} + x = 0$ problem is solved with* `ode15i`.

Example 5

Given an implicitly defined second-order ODE, here is the example problem:
$y(\ddot{y})^2 = e^{2x}$, $y(0) = 0$, $y'(0) = 0$.

This is a differential equation that describes the space charge current in a cylindrical capacitor. The problem is taken from [2]. This is a special exercise; y' near the center of $x = 0$ increases substantially. Thus, we need to approximate $x(0) \rightarrow 0.001$ and integrate the first-order system over $[x(0), 1]$ with default tolerances of the solvers –`ode15i` and `ode45`. The solution algorithm of this problem is similar to Example 2 and Example 3, and the rewriting procedures are similar; thus, we have not put any explanatory notes here. Note that in the given solution script, `IMPLICIT_2ODE_EX5.m`, now the anonymous functions (@) are created in the implicit case (`Fimp`) and the explicit case (`Fexp`).

```
%{
EXAMPLE 5. Implicitly defined problem:  y*(y")^2=exp(2*x) with ICs:y(0)=0,
y'(0)=0. In this case, we can define the given problem in the explicit
form, but for demonstration purposes we define this problem as an implicit
problem and solve it by using ode15i. Then we define it explicitly and
solve it by using ode45.
NOTE: Fimp anonymous function is defined as two 1st order ODEs. Compare
Fimp with Fexp that defines DEs for an explicit ODE solver ode45.
%}
Fimp = @(t, y, dy)([dy(1)-y(2); y(1)*dy(2).^2-exp(2*t)]);
y0=[0.001, 0]; dy0=[0,0];          % ICs
F_aty0=[0,sqrt(1e3)]; F_atdy0=[]; % Function values at y0 and dy0
% Consistent Initial values
[y0, dy0]=decic(Fimp, 0, y0, dy0, F_aty0, F_atdy0);
[t, ft]=ode15i(Fimp, [0.001,1], y0,dy0);
plot(t, ft(:,1), 'bx-', t, ft(:,2), 'kd-')
title('Solutions of: $$ y*(\frac{d^2y}{dx^2})^2=e^{2*x} $$', 'interpreter',
'latex')
xlabel('\it x'), ylabel('\it y(x), dy(x)')
grid on, hold on
Fexp=@(t, y)([y(2); sqrt(exp(2*t)./y(1))]);
ICs=[0.001, 0];
timespan=[0, 1];
[t, yt]=ode45(Fexp, timespan, ICs, []);
plot(t, yt(:,1),'ro-', t, yt(:,2), 'g+-')
Out=sim('IMPLICIT_2ODE_EX5_sim.slx', [1e-3, 1]);
plot(Out.yout{1}.Values.Time, Out.yout{1}.Values.Data, 'r--',
'linewidth', 1.5)
plot(Out.yout{2}.Values.Time, Out.yout{2}.Values.Data, 'k--',
'linewidth', 1.5)
legend('y(x) with ode15i','dy(x) with ode15i','y(x) with ode45', ...
    'dy(x) with ode45', 'y(x) with Simulink', ...
    'dy(x) with Simulink', 'location', 'best');
hold off; shg
```

In Simulink, a model called `IMPLICIT_2ODE_EX5_sim.slx` (shown in Figure 6-7) takes a variable step solver by default, and the solver selection is set to `auto`. The initial condition on [Integrator1] (before $y(t)$) is set to 0.001 to make a good approximation of the initial condition and to avoid singularity problems.

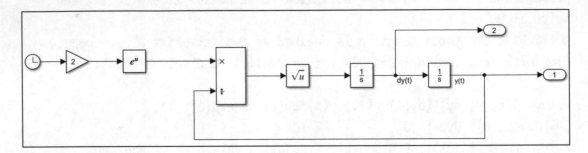

Figure 6-7. *Simulink model of the implicitly defined second-order ODE:* $y(\ddot{y})^2 = e^{2x}$, $y(0) = 0$, $y'(0) = 0$

The simulation results (shown in Figure 6-8) show that all chosen solvers, such as `ode15i`, `ode45`, and `ode2`, have resulted in well-converged numerical solutions that prove the correctness of the chosen approaches. Note that in all the approaches, ODEx solvers are selected with variable step sizes.

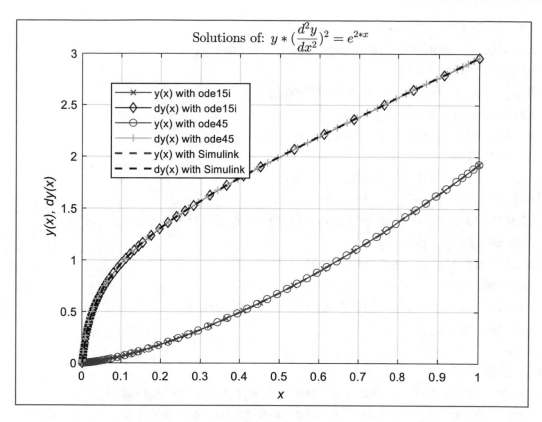

Figure 6-8. *Simulation of the problem:* $y(\ddot{y})^2 = e^{2x}$, $y(0) = 0$, $y'(0) = 0$

Example 6

Here is the example problem: $e^{0.01t} + \dot{y}e^{y} = 2$, $y(0) = 2$.

We start by rewriting the equation in the form of an implicit or differential algebraic equation.

$$2 - \left(e^{0.01t} + \dot{y}e^{y} \right) = 0$$

This can be also formulated as follows:

$$2 - \left(e^{0.01t} + \frac{dy}{dt}e^{y} \right) = 0$$

This equation contains three arguments, which are t, y, and $\dfrac{dy}{dt}$, which can be expressed via an anonymous function (@). The complete simulation model is in a file called `IMPLICIT_1ODE_EX6.m` that calls the Simulink model - `IMPLICIT_1ODE_Ex6_sim.mdl` shown in Figure 6-9. The simulation results of the exercise are shown in Figure 6-10.

```
% IMPLICIT_1ODE_EX6.m
% EXAMPLE 6. exp(.01*t)+y'*exp(y)=2 with ICs: y(0)=2
clearvars; close all
Fun=@(t, y, dy)(2-(exp(.01*t)+dy*exp(y)));
y0=2; dy0=1/2; t0=0; F_aty0=1;
[y0, dy0]=decic(Fun,t0, y0, dy0, F_aty0, [] );
tspan=[0, 130];
[time, y]=ode15i(Fun, tspan, y0, dy0, []);
plot(time, y, 'bd-'), grid minor
title('Solutions of: $$ e^{0.01*t}+\frac{dy}{dt}*e^y=2 $$', ...
'interpreter', 'latex')
xlabel '\it t', ylabel '\it y(t)'
% dsolve
yt=dsolve('2-(exp(.01*t)+Dy*exp(y))=0', 'y(0)=2');
Yt=vectorize(yt);
t=time;
Yt=eval(Yt);
hold on; plot( t, Yt, 'mx--', 'linewidth', 1.5), ylim([-.4 4])
% Simulink model executed for time span of 0...130 sec.
[time,y]=sim('IMPLICIT_1ODE_EX6_sim', [0, 130]);
plot(time, y, 'k-', 'linewidth', 1.5)
legend('ode15i','dsolve: Analytical solution','SIMULINK', ... 'location',
'best'), shg
```

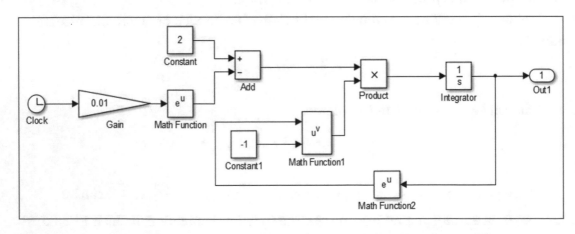

Figure 6-9. *Simulink model* IMPLICIT_1ODE_Ex6_sim.mdl *of the problem:*
$e^{0.01t} + \dot{y}e^y = 2$

The Simulink model (IMPLICIT_1ODE_Ex6_sim.mdl) is built based on the problem

formulation of $\frac{dy}{dt} = \left(2 - e^{0.01t}\right)e^{y}$. Note that the initial condition $y(0) = 2$ is implemented

via the Integrator block as an internal source, and all error tolerances are left unchanged.

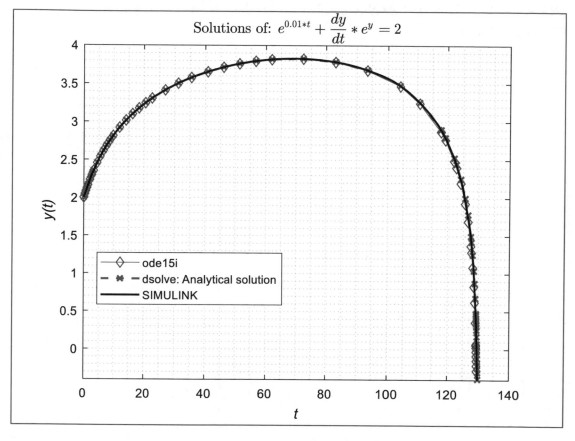

Figure 6-10. *Simulation of the problem: $e^{0.01t} + \dot{y}e^{y} = 2$, $y(0) = 2$*

The previous procedures and scripts that use ode15i and decic, as well as the
Simulink modeling approach with tuned initial conditions to compute numerical
solutions of implicitly defined (truly implicit or "intentionally" treated as implicit ODEs
for demonstration purposes) first- and second-order ODEs, are valid for higher-order
implicit ODEs (or DAEs) and systems of ODEs. From the previous examples, we can
conclude that Simulink modeling can be a good alternative to solving implicitly defined
ODEs or DAEs, with the necessary adjustments in the initial condition settings and in the
solver settings if required.

References

[1] Ascher, U. M., Petzold, L. R., *Computer Methods for Ordinary Differential Equations and Differential-Algebraic Equations*. Society for Industrial and Applied Mathematics (1998).

[2] Kamke, E., *Differential gleichungen Lösungsmethoden und Lösungen*, vol. I. New-York: Chelsea (1971), p. 589.

CHAPTER 7

Comparative Analysis of ODE Solution Methods

In the previous chapters, we discussed a few different methods and MATLAB/Simulink built-in functions to solve IVPs of first-, second-, higher-order, and coupled ODEs defined explicitly and implicitly, and we showed examples of them. Summarizing all discussed methods, we can generalize that there are six different approaches to solving IVPs with MATLAB/Simulink tools, functions, and toolboxes and with script writing.

- Analytic solution search with `dsolve()` from the Symbolic Math Toolbox

- Analytic solution search with the Laplace transforms called `laplace` and `ilaplace` from the Symbolic Math Toolbox

- Analytic solution search with MuPAD notes (Symbolic MATH Toolbox) with `ODE` and `ODE::SOLVE`

- Numerical search solution with the built-in ODEx solvers: `ode15s`, `ode23`, `ode45`, `ode113`, and so on

- Numerical search solution with Simulink models with a few different fixed and variable step solvers: `ode2`, `ode4`, `ode8`, `ode15s`, `ode45`, and so on

- Numerical search solution by writing scripts based on Euler, Heun, Milne, Adams-Bashforth, Adams-Moulton, Runge-Kutta, Taylor series expansion, and other methods

Let's demonstrate all six approaches by solving the next second-order explicitly defined nonhomogeneous ODE and then compare the accuracy and efficiency of the results.

© Sulaymon L. Eshkabilov 2020
S. L. Eshkabilov, *Practical MATLAB Modeling with Simulink*, https://doi.org/10.1007/978-1-4842-5799-9_7

Example 1

Here's the example problem: $\ddot{y} + \dot{y} + 3y = \sin(5t)$, $\dot{y}(0) = -2; y(0) = 3$.

We start with two approaches: analytic solutions computed with dsolve() and the laplace and ilaplace tools of the Symbolic Math Toolbox. The solutions of the exercise are given in COMPARE_ODE_Solvers.m, COMPARE_ODE_Solvers_MUPAD.mn, and COMPARE_ODE_Solvers_sim.slx. The script COMPARE_ODE_Solvers.m computes using these approaches: dsolve(), laplace/ilaplace, ode45/ode113, and the Euler/Adams-Moulton methods. The script calls the numerical solutions from the Simulink model (COMPARE_ODE_Solvers_sim.slx shown in Figure 7-5) with ode8. The numarical solutions computed from the evaluated analytical formulations are displayed in Figure 7-1 and 7-2.

```
% COMPARE_ODE_Solvers.m
close all; clearvars
% Part 1
%% 1st Solution Approach: DSOLVE
syms t s Y y(t)
Dy=diff(y, t);
D2y=diff(y, t, 2);
Equation = D2y==sin(5*t)-Dy-3*y;
ICs = [y(0)==3, Dy(0)==-2];
Solution_dsolve=dsolve('D2y+Dy+3*y-sin(5*t)=0', 'y(0)=3','Dy(0)+2=0');
disp('Symbolic MATH dsolve function output is: ')
pretty(Solution_dsolve)
% display plot of the solution of dsolve and compare it with LT
figure;
subplot(211)
fplot(Solution_dsolve, [0, 2*pi], 'b-', 'linewidth', 1.5)
title('DSOLVE solution of: $$ \frac{d^2y}{dt^2}+\frac{dy}{dt}+3*y=sin(5*t)
$$', 'interpreter', 'latex')
grid on
ylabel('\it y(t)'), xlabel('\it t')
%% 2nd Solution Approach: Laplace Transforms laplace/ilaplace
% Define symbolic variables' names
syms t s Y y(t)
```

```
assume([Y, t]>0)
Dy=diff(y, t);
D2y=diff(y, t, 2);
Equation = D2y==sin(5*t)-Dy-3*y;
% Laplace Transforms
LT_Y=laplace(Equation, t, s);
% Substitute the arbitrary unknown Yand Initial Conditions
LT_Y=subs(LT_Y,laplace(y(t),t, s),Y);
LT_Y=subs(LT_Y,y(0),3);
LT_Y=subs(LT_Y,subs(diff(y(t), t), t),-2);
% Solve for Y unknown
disp('Laplace Transforms of the given 2nd Order ODE with ICs')
Y=solve(LT_Y, Y)   %#ok
Solution_Laplace=ilaplace(Y);
disp('Solution found using Laplace Transforms is:')
Solution_Laplace=simplify(Solution_Laplace);
pretty(Solution_Laplace)
subplot(212)
fplot(Solution_Laplace, [0, 2*pi], 'r-', 'linewidth', 1.5); grid on
title('LAPLACE/iLAPLACE solution of: $$ \frac{d^2y}{dt^2}+\frac{dy}
{dt}+3*y=sin(5*t) $$', 'interpreter', 'latex')
ylabel('\it y(t)'), xlabel('\it t')
```

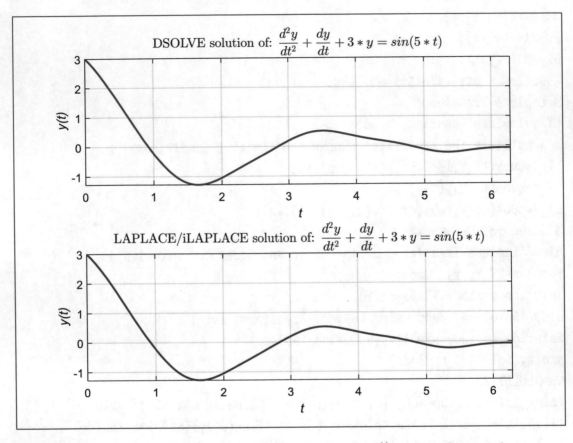

Figure 7-1. *Analytic solutions computed with* dsolve() *and laplace and ilaplace*

Symbolic MATH dsolve function output is:

```
          /    1/2    \
          | 11     t |
1532 cos| ------- |
          \   2    /       /   22 sin(5 t)    5 cos(5 t)
------------------- - ----------- - ---------- -
          / t \              509            509
  509 exp| - |
          \ 2 /

                     /    1/2    \
          1/2    | 11     t |
    284 11     sin| ------- |
```

```
                \     2    /
       ----------------------
                / t \
       5599 exp| - |
                \ 2 /
```

Laplace Transforms of the given 2nd Order ODE with ICs
(3*s + 5/(s^2 + 25) + 1)/(s^2 + s + 3)

Solution found using Laplace Transforms is:
(1532*exp(-t/2)*(cos((11^(1/2)*t)/2) - (71*11^(1/2)*sin((11^(1/2)
*t)/2))/4213))/509 - (22*sin(5*t))/509 - (5*cos(5*t))/509

```
     /                                    /   1/2    \ \
     |                          1/2       | 11     t | |
     |      /   1/2   \    71 11     sin| ------- | |
     |      | 11     t |                  \    2    / |
1532 | cos| ------- | - ----------------------- |
     \     \    2    /              4213                 /   22 sin(5 t)
------------------------------------------------------ - ----------- -
                    / t \                                     509
            509 exp| - |
                    \ 2 /

  5 cos(5 t)
  ----------
     509
```

Thus, the analytic solution of the problem is as follows:

$$y(t) = \frac{1532\cos\left(\dfrac{\sqrt{11}t}{2}\right)}{509e^{\frac{t}{2}}} - \frac{22\sin(5t)}{509} - \frac{5\cos(5t)}{509} - \frac{284\sqrt{11}\sin\left(\dfrac{\sqrt{11}t}{2}\right)}{5599e^{\frac{t}{2}}}$$

An analytical solution of the given ODE can also be solved in MuPAD note (COMPARE_
ODE_Solvers_MUPAD.mn) with the following MuPAD commands:

Third Approach with MuPAD:
[ODE_eqn:=ode({y"(t)+y'(t)+3*y(t)=sin(5*t),y(0)=3,y'(0)=-2}, y(t))

$$\text{ode}\left(\left\{y'(0) = -2,\ y(0) = 3,\ -\sin(5\,t) + 3\,y(t) + \frac{\partial}{\partial t}\,y(t) + \frac{\partial^2}{\partial t^2}\,y(t)\right\},\ y(t)\right)$$

```
[Solution:=ode::solve(ODE_eqn)
```

$$\left\{\frac{1532\cos(\sigma_2)}{509\,e^{\frac{t}{2}}} - \frac{284\,\sqrt{11}\,\sin(\sigma_2)}{5599\,e^{\frac{t}{2}}} + \frac{\cos(\sigma_2)\left(\frac{\sqrt{11}\,e^{\frac{t}{2}}\sin(\sigma_2)\,\sigma_3}{11} - \frac{\sqrt{11}\,e^{\frac{t}{2}}\cos(\sigma_2)\,\sigma_1}{11}\right)}{e^{\frac{t}{2}}} - \frac{\sin(\sigma_2)\left(\frac{\sqrt{11}\,e^{\frac{t}{2}}\cos(\sigma_2)\,\sigma_3}{11} + \frac{\sqrt{11}\,e^{\frac{t}{2}}\sin(\sigma_2)\,\sigma_1}{11}\right)}{e^{\frac{t}{2}}}\right\}$$

where

$$\sigma_1 = \frac{22\,\sqrt{11}\,\sin(5\,t)}{509} + \frac{5\,\sqrt{11}\,\cos(5\,t)}{509}$$

$$\sigma_2 = \frac{\sqrt{11}\,t}{2}$$

$$\sigma_3 = \frac{225\,\cos(5\,t)}{509} - \frac{28\,\sin(5\,t)}{509}$$

```
[simplify(Solution)
```

$$\left\{\frac{1532\cos\left(\frac{\sqrt{11}\,t}{2}\right)}{509\,e^{\frac{t}{2}}} - \frac{22\,\sin(5\,t)}{509} - \frac{5\,\cos(5\,t)}{509} - \frac{284\,\sqrt{11}\,\sin\left(\frac{\sqrt{11}\,t}{2}\right)}{5599\,e^{\frac{t}{2}}}\right\}$$

```
[plot(Solution, #G, #L, t=0..2*PI)
```

238

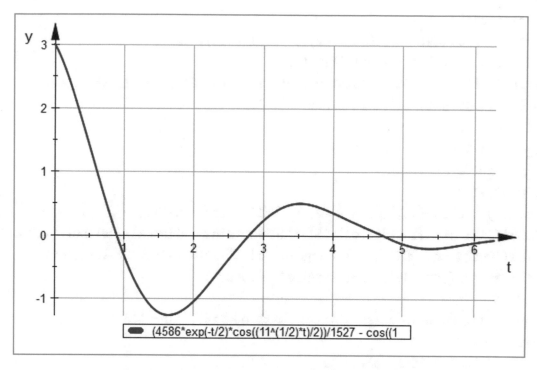

Figure 7-2. *The solution of the exercise computed in MuPAD notes*

Note that the analytical solution expressions of the given second-order IVP found via the three approaches (dsolve(), laplace/ilaplace, and MuPAD notes of the Symbolic Math Toolbox) differ in formulation; however, with some algebraic changes and simplifications, it is feasible to obtain the same format from the solutions of all three approaches.

The fourth approach is to compute numerical solutions with the built-in ODEx solvers (ode15s, ode23, ode23tb, ode45, and ode113). Let's look at the ode45 and ode113 solvers to compute numerical solutions of the given problem, as shown here:

```
%% Part 3. 4th Solution Approach: ODE45, ODE113 solvers
Ys=@(t, y)([y(2);-(y(2)+3*y(1))+sin(5*t)]);
ICs=[3, -2];
timeSPAN=0:pi/100:2*pi;
options_ODE=odeset('RelTol', 1e-6, 'AbsTol', 1e-8);
[t1, y_ODE45]=ode45(Ys, timeSPAN, ICs, options_ODE);
[t2, y_ODE113]=ode113(Ys, timeSPAN, ICs, options_ODE);
figure('name', 'ODEx solvers 1')
```

```
subplot(211)
plot(t1, y_ODE45(:,1), 'bo', t2, y_ODE113(:,1), 'k-')
ylabel('\it y(t)'), xlabel('\it t')
title('ODE45 and ODE113 solutions of: $$ \frac{d^2y}{dt^2}+\frac{dy}
{dt}+3*y=sin(5*t) $$', 'interpreter', 'latex')
grid on
legend('\it ode45', 'ode113')
axis tight
subplot(212)
plot(t1, y_ODE45(:,2), 'bo', t2, y_ODE113(:,2), 'k-')
ylabel('\it $$ \frac{dy}{dt} $$', 'interpreter', 'latex'), xlabel('\it t')
title('ODE45 and ODE113 solutions of: $$ \frac{d^2y}{dt^2}+\frac{dy}
{dt}+3*y=sin(5*t) $$', 'interpreter', 'latex')
grid on
legend('\it dy with ode45','\it dy with ode113', 'location', 'best')
axis tight
figure('name', 'ODEx solvers 2')
plot(y_ODE45(:,1), y_ODE45(:,2),'bo', ...
y_ODE113(:,1), y_ODE113(:,2), 'r-')
title('Phase plot of:  $$ \frac{d^2y}{dt^2}+\frac{dy}{dt}+3*y=sin(5*t) $$',
'interpreter', 'latex')
legend('\it ode45', 'ode113'), grid on
axis tight; shg
xlabel('\it y(t)')
ylabel('\it $$ \frac{dy}{dt} $$', 'interpreter', 'latex')
```

Figure 7-3 and Figure 7-4 show the numerical solutions in plots obtained with the ode45 and ode113 solvers.

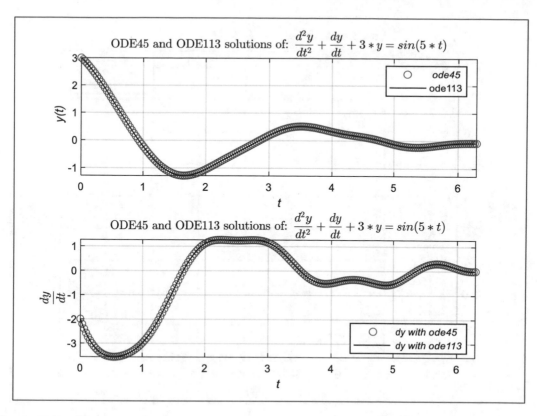

Figure 7-3. *Numerical solutions computed with* ode45 *and* ode113

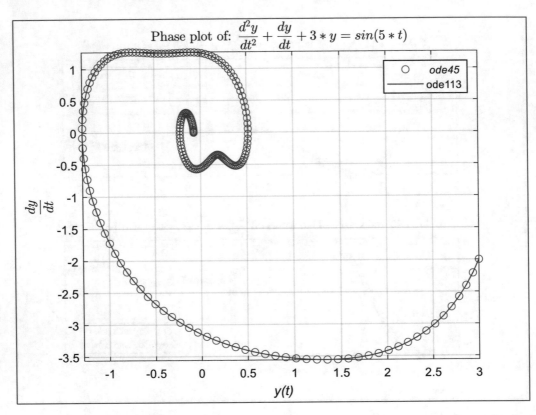

Figure 7-4. *Numerical solutions (phase plot) computed with ode45 and ode113*

The next approach is Simulink modeling. The model of the given exercise is shown in Figure 7-5. Note that two initial conditions are input via the [Integrator] and [Integrator1] blocks. Moreover, a fixed step solver (ode8, Dormand-Prince algorithm) with a fixed step size of [pi/100] is set in the model settings.

```
%% Part 4. 5th Solution Approach: Simulink modeling
%{
NB: This model uses ODE solver ODE8 in simulation.
A solver type can be altered if needed for a better/faster solution search.
This cell first recalls and then executes the Simulink model
COMPARE_ODE_Solvers_sim.slx that is to be put in a current directory
of MATLAB from the archived folder.
A fixed step is set to pi/100.
%}
%set_param('COMPARE_ODE_Solvers_sim','Solver','ode8','StopTime','2*pi');
[tout, SIMout]=sim('COMPARE_ODE_Solvers_sim.slx');
```

242

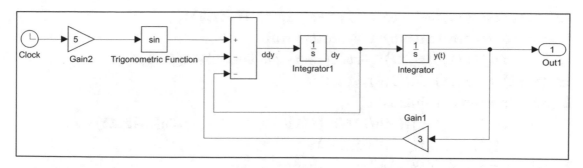

Figure 7-5. *Simulink model* COMPARE_ODE_Solvers_sim.slx *of the given problem*

The last approach is script writing based on the Euler and Adams-Moulton methods.

```
%% Part 5. 6th Solution Approach: scripts using Euler's Method
t0=0;                   % start time
tend=2*pi;              % end time
h=pi/100;               % time step
t=t0:h:tend;            % time space
steps=length(t);        % # of steps
y=[3, -2];              % ICs: u0 at t0
for k=2:steps
    f=[y(k-1,2), sin(5*t(k-1))-3*y(k-1,1)-y(k-1,2)];
    y(k,:)=y(k-1,:)+h*f;
end

%% Part 6. 6th solution approach: scripts - 5-step Adams-Moulton
% method
t0=0;                   % start time
tend=2*pi;              % end time
h=pi/100;               % time step
t=t0:h:tend;            % time space
steps=length(t);        % # of steps
Fn(1,:)=[3, -2];        % ICs: u0 at t0
% function is defined via anonymous function handle:
Fcn =@(t, y1, y2)([y2, sin(5*t)-y2-3*y1]);
for ii=1:steps-4
% 1st: Predicted value by EULER's forward method
```

243

```
Fn(ii+1,:)=Fn(ii,:)+h*Fcn(t(ii), Fn(ii, 1),Fn(ii,2));
% 2nd: Corrected value by trapezoidal rule
Fn(ii+1,:)=Fn(ii,:)+(h/2)*(Fcn(t(ii+1),Fn(ii+1,1),Fn(ii+1,2))+...
Fcn(t(ii),Fn(ii,1),Fn(ii,2)));
% 3rd: Predicted solution:
Fn(ii+2,:)=Fn(ii+1,:)+(3*h/2)*Fcn(t(ii+1),Fn(ii+1,1),Fn(ii+1,2))-...
(h/2)*Fcn(t(ii),Fn(ii,1),Fn(ii,2));
% 3rd: Corrected ADAMS-Moulton method:
Fn(ii+2,:)=Fn(ii+1,:)+h*((5/12)*Fcn(t(ii+2),Fn(ii+2,1),Fn(ii+2,2))+...
(2/3)*Fcn(t(ii+1),Fn(ii+1,1),Fn(ii+1,2))-...
(1/12)*Fcn(t(ii),Fn(ii,1),Fn(ii,2)));
% Predicted solution:
Fn(ii+3,:)=Fn(ii+2,:)+h*((23/12)*Fcn(t(ii+2),Fn(ii+2,1),Fn(ii+2,2))...
-(4/3)*Fcn(t(ii+1),Fn(ii+1,1),Fn(ii+1,2))+...
(5/12)*Fcn(t(ii),Fn(ii,1),Fn(ii,2)));
% Corrected solution by ADAMS-Moulton method 4-step from here:
Fn(ii+3,:)=Fn(ii+2,:)+h*((3/8)*Fcn(t(ii+3),Fn(ii+3,1),Fn(ii+3,2))+...
(19/24)*Fcn(t(ii+2),Fn(ii+2,1),Fn(ii+2,2))-...
(5/24)*Fcn(t(ii+1),Fn(ii+1,1),Fn(ii+1,2))+...
(1/24)*Fcn(t(ii),Fn(ii,1),Fn(ii,2)));
% Predicted solution:
Fn(ii+4,:)=Fn(ii+3,:)+h*((55/24)*Fcn(t(ii+3),Fn(ii+3,1),Fn(ii+3,2)) ...
-(59/24)*Fcn(t(ii+2),Fn(ii+2,1),Fn(ii+2,2))+...
(37/24)*Fcn(t(ii+1),Fn(ii+1,1),Fn(ii+1,2))-...
(3/8)*Fcn(t(ii),Fn(ii,1),Fn(ii,2)));
% Corrected solution by ADAMS-Moulton 5-step method from here:
Fn(ii+4,:)=Fn(ii+3,:)+ ...
    h*((251/720)*Fcn(t(ii+4),Fn(ii+4,1),Fn(ii+4,2))+...
(646/720)*Fcn(t(ii+3),Fn(ii+3,1),Fn(ii+3,2))-...
(264/720)*Fcn(t(ii+2),Fn(ii+2,1),Fn(ii+2,2))+...
(106/720)*Fcn(t(ii+1),Fn(ii+1,1),Fn(ii+1,2))-...
(19/720)*Fcn(t(ii),Fn(ii,1),Fn(ii,2)));
end

%% Part 6. Numerical values from DSOLVE and LAPLACE/iLAPLACE
SDsolve  = eval(Solution_dsolve);
```

```matlab
SLaplace = eval(Solution_Laplace);
figure('name', 'Compare ALL 1')
plot(t, SDsolve, 'bx-'); grid on, hold on,
xlabel('\it t'), ylabel('\it y(t)')
plot(t, SLaplace, 'ro-')
plot(t, y_ODE45(:,1), 'm^--')
plot(t, y_ODE113(:,1), 'k<-')
plot(tout, SIMout(:,1), 'gd-')
plot(t, y(:,1), 'linewidth', 2)
plot(t, Fn(:,1), 'b--', 'linewidth', 1.5), grid minor
title('Solutions of: $$ \frac{d^2y}{dt^2}+\frac{dy}{dt}+3*y=sin(5*t) $$',
'interpreter', 'latex')
legend('dsolve','Laplace Transforms','ODE45','ODE113',...
'Simulink', 'Euler','Adams-Moulton'),
xlim([0, 2*pi]), hold off

figure('name', 'Compare ALL 2')
plot(y_ODE45(:,1), y_ODE45(:,2), 'mo'); grid on, hold on,
xlabel('\it y(t)'), ylabel('\it $$ \frac{dy}{dt} $$', ... 'interpreter', 'latex')
plot(y_ODE113(:,1), y_ODE113(:,2), 'k<')
plot(SIMout(:,1), SIMout(:,2), 'gd')
plot(y(:,1), y(:,2), 'linewidth', 2)
plot(Fn(:,1), Fn(:,2),'b--', 'linewidth', 1.5), grid minor
title('Phase plot of: $$ \frac{d^2y}{dt^2}+\frac{dy}{dt}+3*y=sin(5*t) $$',
...
'interpreter', 'latex')
legend('ODE45','ODE113', 'Simulink', 'Euler','Adams-Moulton'),
axis tight, hold off
%% Part 7. Residuals are computed
Res_ode45=abs(SDsolve)-abs(y_ODE45(:,1)');
Res_ode113=abs(SDsolve)-abs(y_ODE113(:,1)');
Res_sim=abs(SDsolve)-abs(SIMout(:,1)');
Res_Euler=abs(SDsolve)-abs(y(:,1)');
Res_AM=abs(SDsolve)-abs(Fn(:,1)');
Res_LTSD=abs(SLaplace)-abs(SDsolve);
figure('name', 'Residuals')
```

```
subplot(311)
plot(t, Res_ode45, 'r-',t,Res_ode113, 'k-.', t, Res_sim, 'b-')
hold on
plot(t, Res_LTSD, 'g--', 'linewidth', 1.5)
title('\it Residuals'), ylabel('\it Residual values')
legend('ode45','ode113','Simulink', 'Laplace', 'location', 'best')
subplot(312)
plot(t, Res_Euler, 'k-'), hold on
plot(t, Res_LTSD, 'g--', 'linewidth', 1.5)
ylabel('\it Residual values')
legend('Euler', 'Laplace', 'location', 'best')
subplot(313)
plot(t, Res_AM, 'm-', 'linewidth', 1.5), hold on
plot(t, Res_LTSD, 'g--', 'linewidth', 1.5)
ylabel('\it Residual values')
xlabel('\it t')
legend('Adams-Moulton', 'Laplace', 'location', 'best')
xlim([0, 2*pi]), shg
```

By executing the script COMPARE_ODE_Solvers.m and the Simulink model, (COMPARE_ODE_Solvers_sim.mdl, shown in Figure 7-5), the plots shown in Figure 7-6, Figure 7-7, and Figure 7-8 are obtained. They display numerical solutions and residuals of the numerical approaches from analytical solution values.

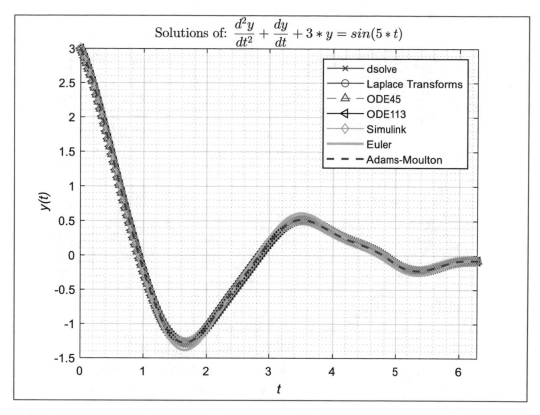

Figure 7-6. *All numerical solutions found with the* `dsolve()`*, laplace/ilaplace,* `ode45`*/*`ode113`*, Simulink (*`ode8`*), Euler, and Adams-Moulton methods*

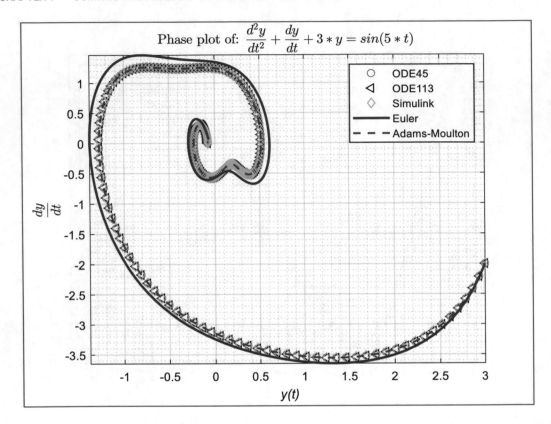

Figure 7-7. *Phase plot (y(t) vs. $\frac{dy}{dt}$) of solutions computed by the ode45, ode113, Simulink, Euler, and Adams-Moulton methods*

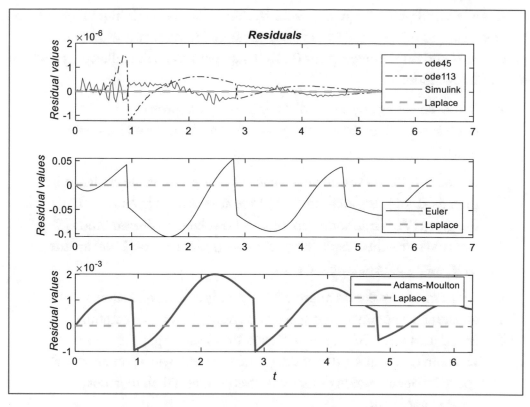

Figure 7-8. *Residuals of numerical solutions from the analytical solution*

In this exercise of a comparative analysis of IVP-solving methods, are obtained simulation results from six approaches converge relatively well (Figures 7-6 and 7-7). At the same time, there are certain error margins, which are displayed in the plot of residuals shown in Figure 7-8. From the residuals, it is clear that the Euler method produces the largest error margins, but they tend to decrease over a longer simulation time. The most accurate numerical solution is obtained from the Simulink model (`COMPARE_ODE_Solvers_sim.mdl`). Results of other numerical methods are so well converged with the analytical solution that error margins smaller than $2 * 10^{-3}$ can be reduced even further by taking smaller time steps; on the other hand, decreasing the step size increases the computation time substantially.

Generalizing all points from the demonstrated methods of solving IVPs analytically and numerically in the examples of specific linear and nonlinear, stiff and nonstiff, homogeneous and nonhomogeneous ODEs (Chapters 1 to 7), the following concluding remarks can be made:

- Before selecting any method to apply to a given problem, it is necessary to identify what the given ODE problem type is, i.e., stiff/nonstiff.

- Not all IVP can be solved analytically. If they are solvable analytically, then `dsolve()`, `laplace`, and `ilaplace` are the tools to employ. If the given IVP is nonhomogeneous and is subject to discontinuous external force, then Laplace transforms (`laplace/ ilaplace`) are the most adequate approach to take.

- If a numerical solution approach is the only option, one should consider a solution search space and choose a step size carefully based on a given ODE type. If the solution search space of a given problem is very large, it is advised to start working in a smaller search space by benchmarking numerical methods and then to choose a most suitable one.

- Consider what the priority is for a given task: accuracy versus speed.

- While working with differential algebraic equations, rewriting a given implicit ODE and defining its consistent initial conditions are first imperative steps. In addition, modeling in Simulink and then comparing the solutions computed by both approaches (`ode15i` versus Simulink) is recommended.

- It is advised to compute numerical solutions with several different methods. We have even seen that some recommended solvers of MATLAB may produce misleading numerical solutions in stiff problems, for instance. In addition, it is recommended to tune the settings of built-in ODEx (`ode15s`, `ode23`, `ode45`, `ode113`, etc.) solvers accordingly and provide Jacobian matrices to improve the efficiency of a chosen solver for stiff ODEs.

Drill Exercises

The following are some exercises to try.

Exercise 1

The following are IVPs of second-order nonhomogeneous ODEs:

- $\ddot{y} + 9y^2 = \sin(2t)$, $y(0) = 0$ and $\dot{y}(0) = 6$ for $t \in [0,\ 3\pi]$

- $\ddot{y} + 4\dot{y} + 104y = 2\cos(10t)$, $y(0) = 0$ and $\dot{y}(0) = 0$ for $t \in [0,\ 5\pi]$

- $\ddot{y} + y = e^x + 2$, $y(0) = 0$ and $\dot{y}(0) = 0$ for $x \in [0,\ 13]$

- $\ddot{y} + 2\dot{y} + y = 2x$, $\dot{y}(0) = 0$ and $\dot{y}(0) = 6$ for $x \in [0,\ 15]$

- $\ddot{y} + 2\dot{y} + 101y = 5\sin(10t)$, $y(0) = 0$ and $\dot{y}(0) = 20$ for $t \in [0,\ 5\pi]$

First, solve each of the previous second-order ODEs with the following methods:

a) Writing a script based on the forward Euler method

b) Writing a script based on the fourth-order Adams-Moulton method

c) Writing a script based on the 4th-order Runge-Kutta method

d) Using the MATLAB built-in ODE solvers ode23, ode45, and ode113, and adjusting their settings, namely, relative and absolute error tolerances, accordingly

e) Building a Simulink model and using the solver ode3

Next, compare all the solutions found from (a) to (e) and figure out which approach is the most efficient and accurate (correct and has smallest error margins).

Finally, is it possible to compute an analytical solution of the given problems by using dsolve() and the Laplace transforms (laplace and ilaplace)? If yes, plot the analytical solutions against the numerical solutions found from (a) to (e).

251

Exercise 2

First, solve the second-order nonhomogeneous ODE problem given here:
$x^2\ddot{y} - 5x\dot{y} + 8y = e^{3x}$, $y(1) = 0$, and $\dot{y}(1) = 24$ for $x \in [1, 15]$. Note that the initial point is at $x = 1$. Use the following methods:

a) Writing a script based on the backward Euler method

b) Writing a script based on the five-step Adams-Bashforth method

c) Writing a script based on Milne's method

d) Writing a script based on Euler's improved method

e) Using MATLAB's built-in ODE solvers: ode23, ode45, ode113

f) Building a Simulink model and employing the solver ode2

g) Comparing all solutions found from (a) to (e) and pinpointing which approach is the most adequate and efficient

Next, is it possible to compute an analytical solution of the given problem by using dsolve() and the Laplace transforms (laplace and ilaplace)? If yes, plot the analytical solutions against the numerical solutions found from (a) to (e).

Finally, is it possible to compute an analytical solution of the given problem by using MuPAD? If yes, compare it with the analytical solutions found with dsolve() and the Laplace transforms (laplace and ilaplace).

Exercise 3

First, solve the second-order nonhomogeneous and nonlinear ODE given here:
$\ddot{y} + 16|y|\dot{y} + 12y = 3t^3 + 12\cos(3t)$, $y(0) = -1$ and $\dot{y}(0) = 0$ for $x \in [0,\ 13]$. Use the following methods:

a) Writing a script based on the Ralston and Milne methods

b) Writing a script based on the fourth-order Adams-Moulton method

c) Writing a script based on the Runge-Kutta-Gill method

d) Using MATLAB's built-in ODE solvers: ode23, ode45, ode113

e) Building a Simulink model and employing the solver ode4

Next, compare all the solutions found from (a) to (e) by plotting t versus $y(t)$ and t versus $\dot{y}(t)$, and find out which approach is the most adequate (correct and has smallest error margins) and efficient.

Exercise 4

First, solve the given IVP of the second-order nonhomogeneous and nonlinear ODE given here: $\ddot{y} + 2y^2\dot{y}^2 + 101y = e^{5t} + 2t^2 + 5\sin(10t)$, $y(0) = 0$ and $\dot{y}(0) = 20$ for $t \in [0, \ 6\pi]$. Use the following methods:

   a) Writing a script based on Heun's method

   b) Writing a script based on the fifth-order Adams-Moulton method

   c) Writing a script based on the Runge-Kutta-Gill method

   d) Using MATLAB's built-in ODE solvers: `ode23`, `ode45`, `ode113`

   e) Building a Simulink model and employing the solver `ode14x`

Next, compare all the solutions found from (a) to (e) and find out which approach is the most efficient. Take smaller time steps if necessary.

Finally, is it possible to compute an analytical solution of the given problem by `dsolve()` and Laplace transforms (`laplace` and `ilaplace`)?

Exercise 5

First, solve the second-order nonhomogeneous and nonlinear ODE given here: $\ddot{y}^2 - 5y^2|\dot{y}| = e^{2t} + 2t$, $y(0) = 1$ and $\dot{y}(0) = 0$ for $t \in [0, \ 13]$. Use the following methods

   a) Writing a script based on the Milne method

   b) Writing a script based on the five-step Adams-Bashforth method

   c) Writing a script based on the Runge-Kutta-Fehlberg method

   d) Using MATLAB's built-in ODE solvers: `ode23`, `ode45`, `ode113`

   e) Building a Simulink model and employing the solver `ode1`

Next, compare all the solutions found from (a) to (e) and find out which approach is the most efficient.

Exercise 6

Here is an equation of charge in a resistor-inductance-capacitor (RLC) circuit in a series by Kirchhoff's law: $L\ddot{q} + R\dot{q} + \dfrac{q}{C} = \mathcal{E}_{max}\cos\omega t$.

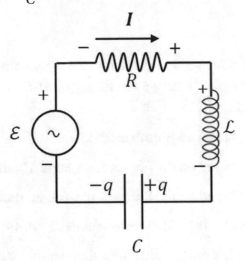

where

$q(0) = \dot{q}(0) = 0$ for $t \in [0, 4\pi]$

EMF: $\varepsilon_{max} = 110$ $[V]$

Resistance: $R = 7.17$ $[\Omega]$

Capacitor: $C = 50 * 10^{-3}[F]$

Armature inductance: $\mathcal{L} = 9.53 * 10^{-4}[H]$

Frequency: $\omega = 60$ $[Hz]$

First, do the following:

a) Find the numerical solutions of $q(t)$ by writing a script based on Euler's improved method and the Ralston method.

b) Find the numerical solutions of $q(t)$ by writing a script based on the fifth-order Adams-Moulton method.

c) Find the numerical solutions of $q(t)$ by writing a script based on the 4th-order Runge-Kutta-Gill method.

d) Find the numerical solutions of $q(t)$ using MATLAB's built-in ODE solvers: ode23, ode45, ode113.

e) Find the numerical solutions of $q(t)$ by building a Simulink model and employing the solver ode8.

f) Compare all solutions found from (a) to (e) and find out which approach is the most efficient and correct/appropriate. If necessary, take reasonably smaller time steps and specify (appropriate) initial step size, and relative and absolute tolerances, as well.

Then, is it possible to compute the analytical solution of the problem using dsolve() and the Laplace transforms (laplace and ilaplace)? If yes, plot the analytical solutions against the numerical solutions found from (a) to (e).

Finally, is it possible to compute the analytical solution of the problem by using MuPAD notes? If so, compare it with the analytical solutions found with dsolve() and the Laplace transforms (laplace and ilaplace).

Exercise 7

First, solve the given IVP of the fourth-order nonhomogeneous ODE given here: $y^{iv} + 3\ddot{y}^3 - \cos(100t)\dot{y} + 8y = t^2 + 10\sin(100t)$ with $y(0) = 0, \dot{y}(0) = 1,\ \ddot{y}(0) = 2,$ and $\dddot{y}(0) = -3$. For $t \in [0,\ 3\pi]$. Use the following methods:

a) Solve the problem by writing a script based on Heun's method.

b) Solve the problem by writing a script based on the fifth-order Adams-Moulton method.

c) Solve the problem by writing a script based on the Runge-Kutta-Gill method.

d) Solve the problem by using MATLAB's built-in ODE solvers, specifically, ode23, ode45, and ode113. Also set up the relative and absolute error tolerances.

e) Solve the problem by using MATLAB's built-in ODE solvers, specifically, ode23s, ode15s, and ode23tb. Also obtain the numerical solution of the problem in the plot only. (Hint: set up OutputFcn for @odeplot with odeset.)

255

f)   Solve the problem by building a Simulink model with the solver ode2.

Finally, compare all the solutions found from (a) to (f) and find out which approach is the most efficient and adequate.

Exercise 8

Solve the given IVP of the fourth-order nonhomogeneous ODE given here:

$y^{iv} + 2\ddot{y} + \ddot{y} + 8\dot{y} - 12y = 12\sin(25t) - e^{-5t}$, $y(0) = 3, \dot{y}(0) = 0, \ddot{y}(0) = -1$ and $\dddot{y}(0) = 2$, for $t \in [0, \ 5\pi]$. Use the following methods:

a)   Solve the problem by writing a script based on Euler's forward method.

b)   Solve the problem by writing a script based on the fifth-order Adams-Bashforth method.

c)   Solve the problem by writing a script based on the Runge-Kutta-Fehlberg method.

d)   Solve the problem by using MATLAB's built-in ODE solvers, specifically, ode23s, ode15s, and ode113. Also set up relative and absolute tolerances.

e)   Solve the problem by using MATLAB's built-in ODE solvers, specifically, ode23, ode45, and ode23tb. Also obtain the numerical solutions of the problem in the plot only. (Hint: set up OutputFcn for @odeplot with odeset.)

f)   Solve the problem by building a Simulink model with the solver ode8.

g)   Compare all the solutions found from (a) to (f) and pinpoint which approach is the most efficient and appropriate.

Exercise 9

Find the numerical solutions of the following systems of coupled ODEs defined here:

1. $\begin{cases} \dfrac{dx_1}{dt} = -x_2 + \cos(t) \\ \dfrac{dx_2}{dt} = x_1 + \sin(t) \end{cases}$ with ICs: $x_1(1) = 2.5$, $x_2(1) = 3.5$, $t \in [1, 13]$.

2. $\begin{cases} \dfrac{dx}{dt} = -3x + 5y + 2ye^x \\ \dfrac{dy}{dt} = -13x - x^2 - y^2 \end{cases}$ with ICs: $x(0.5) = 2$, $y(0.5) = -2$, $t \le 5.55$.

3. $\begin{cases} \dfrac{dx}{dt} = (1+x)\sin(y) \\ \dfrac{dy}{dt} = 1 - x - \cos(y) \end{cases}$ with ICs: $x\left(\dfrac{\pi}{4}\right) = 1.25$, $y\left(\dfrac{\pi}{4}\right) = 0.75$, $t \in \left[\dfrac{\pi}{4}, \dfrac{7\pi}{2}\right]$.

For each of the systems, perform the following tasks:

- Write an anonymous function of the coupled system.

- Write an inline function of the coupled system.

- Create a function file called, e.g., CoupleODE.m.

- Solve the problem by building a Simulink model called, e.g., CoupledODEsim.mdl, with the fixed step solver ode3.

- Find the numerical solutions of the problem by employing ode23, ode45, and ode113. Compare the solutions from ODEx solvers to the Simulink model and check the efficiency of each approach. Take smaller time steps, adjust the relative and absolute tolerances, and simulate your created Simulink model (CoupledODEsim.mdl) from an m-file. (Hint: use sim() and simset().)

- Find the numerical solutions of the given coupled system by writing a script based on the midpoint rule method and compare its numerical solutions with the ones found from ode23, ode45, and ode113 and the Simulink model.

Exercise 10

By using the Laplace transforms (`laplace` and `ilaplace`), solve the following second-order nonhomogeneous ODEs subject to the discontinuous forcing function:

1. $\ddot{y} + 5y = h(t)$, $y(0) = 0, \dot{y}(0) = 0$, $h(t) = \begin{cases} 0, & 0 \le t < 3 \\ (t-3)/3 & 3 \le t < 11 \\ 13 & t \ge 11 \end{cases}$

2. $\ddot{y} + 5\dot{y} + 5y = g(t)$, $y(0) = 0, \dot{y}(0) = 2$,

   $g(t) = \begin{cases} 5 & \pi \le t < 3\pi \\ 0 & 0 \le t < \pi \text{ and } t \ge 3\pi \end{cases}$

3. $\ddot{x} + 5\dot{x} + \dfrac{5}{6}x = u(t)$, $x(0) = 0, \dot{x}(0) = 0$, $u(t) = \begin{cases} \sin(t) & 0 \le t < 2\pi \\ 0 & t \ge 2\pi \end{cases}$

Plot the numerical values of the analytical solutions for a sufficient time period.

Exercise 11

Simulate Robertson's problem with these approaches:

1) Simulate the Robertson's problem model by using `ode15s` and `ode113` and by writing a script based on the Adams-Moulton method, and compare the numerical solutions. Which method is most appropriate for this problem?

2) Simulate the Akzo-Nobel problem by writing a script based on Euler's improved method and the Runge-Kutta-Fehlberg method. Compare the simulation results from the two methods.

3) Simulate the HIRES problem by building a Simulink model with the variable step-size solver `ode23` and by using the solver `ode15s`. If necessary, tune the relative and absolute error tolerances in both models and compare their numerical solutions. Which approach is more appropriate and why?

Exercise 12

Solve the given implicitly defined IVPs of the first-order ODEs:

1) $(y')^2 + xy' - y = 2x$, $y(2) = -1$ for $x \in [2, 15]$. Note an initial point is not at zero but at 2.

2) $y(y')^3 + 5ty' - 3y = 2t$, $y(1) = -2$ for $t \in [1, 15]$. Note an initial point is not at zero but at 1.

   - Use the MATLAB solver ode15i.

   - Build a Simulink model.

   - Compare the results from ode15i and a Simulink model.

Exercise 13

Solve the given implicitly defined IVPs of the second-order ODE defined by these:

- $3\ddot{y}^2 - 2\ddot{y} + e^{2t}\dot{y} - 12y = 0$, $y(0) = 2$, $\dot{y}(0) = 2e^2$, $t \in [0,25]$
- $3\ddot{y}^3 - 2y\ddot{y} + \dot{y} - 13y = e^{3t}$, $y(0) = 5$, $\dot{y}(0) = 3e^4$, $t \in [0,15]$

a) Create a Simulink model of the problem.

b) Use MATLAB's built-in ODE solvers ode15i and decic.

c) Compare numerical solutions found from (a) and (b) and find out which approach is most efficient. Take smaller time steps if necessary.

d) Is it feasible to evaluate an analytical solution of the problem using dsolve() or Laplace transforms (laplace and ilaplace)?

PART II

Boundary Value Problems in Ordinary Differential Equations

CHAPTER 8

Boundary Value Problems

In this chapter, we'll discuss the essential steps of solving boundary value problems (BVPs) of ordinary differential equations (ODEs) using MATLAB's built-in solvers. The only difference between BVPs and IVPs is that the given differential equation in a BVP is valid within two boundary conditions, which are the initial and end conditions. A BVP can be formulated with the following expressions:

$$\frac{\partial u}{\partial t}(y,t) = \beta \frac{\partial^2 y}{\partial y^2}(y,t), \quad a < y < b, \ 0 < t \tag{8-1}$$

There are three main types (plus one other kind) of boundary conditions (BCs) in a BVP, as listed here:

- Dirichlet BCs:

$$u(a,t) = 0; \quad u(b,t) = 0 \tag{8-2}$$

- Neumann BCs:

$$\frac{\partial u}{\partial y}(a,t) = 0; \quad \frac{\partial u}{\partial y}(b,t) = 0 \tag{8-3}$$

- Robin or mixed BCs:

$$u(a,t) = 0; \quad \frac{\partial u}{\partial y}(b,t) = 0 \tag{8-4}$$

© Sulaymon L. Eshkabilov 2020
S. L. Eshkabilov, *Practical MATLAB Modeling with Simulink*, https://doi.org/10.1007/978-1-4842-5799-9_8

or

$$\frac{\partial u}{\partial y}(a,t)=0; \quad u(b,t)=0 \tag{8-5}$$

- Periodic BCs:

$$u(a,t)=u(b,t); \quad \frac{\partial u}{\partial y}(a,t)=\frac{\partial u}{\partial y}(b,t); \tag{8-6}$$

Most BVPs are second-order problems; thus, we will focus on solving or simulating second-order BVP equations defined as follows:

$$\ddot{y}+p(t)\dot{y}+q(t)y=r(t), \; a<t<b \tag{8-7}$$

where we let $I=(a,b)\subseteq \mathbb{R}$ be an interval and let $p, q, r:(a,b)\to \mathbb{R}$ be continuous functions.

The ODEs in a BVP might have one, several, or an infinite number of solutions depending on the boundary conditions (the boundary values and boundary kinds). For example, let's consider the second-order differential equation shown here:

$$\ddot{y}+y=0 \tag{8-8}$$

Here are three cases of this equation:

- **Case #1:** Equation (8-8) with the boundary conditions $y(0)=1$ and $y\left(\frac{\pi}{2}\right)=1$ has a unique solution given by $y(t)=\ sin\ (t)+\ cos\ (t)$.

- **Case #2:** Equation (8-8) with the boundary conditions $y(0)=1$ and $y(\pi)=1$ has no solutions.

- **Case #3:** Equation (8-8) with the boundary conditions $y(0)=1$ and $y(2\pi)=1$ has an infinite number of solutions.

Similarly, we can generate a few more other cases with other boundary conditions if necessary.

For Case #1. With $y(0)=1$ and $y\left(\dfrac{\pi}{2}\right)=1$, we assume that a general solution of Equation (8-8) will be $y(t) = C_1\,sin\,(t) + C_2\,cos\,(t)$. C_1 and C_2 can be found by using the following expressions and applying the boundary conditions to the problem's general solution:

$$1 = y(0) = C_1\sin(0) + C_2cos(0) = C_1 * 0 + C_2 * 1 = C_2 \Rightarrow C_2 = 1$$

$$1 = y\left(\frac{\pi}{2}\right) = C_1\sin\left(\frac{\pi}{2}\right) + C_2cos\left(\frac{\pi}{2}\right) = C_1 * 1 + C_2 * 0 = C_1 \Rightarrow C_1 = 1$$

Now, the solution is as follows:

$$y(t) = \sin(t) + \cos(t)$$

For Case #2. With $y(0) = 1$ and $y(\pi) = 1$, we assume that a general solution of Equation (8-8) is $y(t) = C_1\,sin\,(t) + C_2\,cos\,(t)$. C_1 and C_2 can be found by using the following expressions and applying the boundary conditions to the problem's general solution:

$$1 = y(0) = C_1\sin(0) + C_2cos(0) = C_1 * 0 + C_2 * 1 = C_2 \Rightarrow C_2 = 1$$

$$1 = y(\pi) = C_1\sin(\pi) + C_2cos(\pi) = C_1 * 0 + C_2 * (-1) = C_2 \Rightarrow C_2 = -1$$

$C_2 = 1$ for $a = 0$, and $C_2 = -1$ for $b = \pi$. This is contradictory, which means that there is no solution for these boundary conditions.

For Case #3. With $y(0) = 1$ and $y(2\pi) = 1$, we assume that a general solution for Equation (8-8) will be $y(t) = C_1\,sin\,(t) + C_2\,cos\,(t)$. C_1 and C_2 can be found by using the following expressions and applying the boundary conditions to the problem's general solution:

$$1 = y(0) = C_1\sin(0) + C_2cos(0) = C_1 * 0 + C_2 * 1 = C_2 \Rightarrow C_2 = 1$$

$$1 = y(2\pi) = C_1\sin(2\pi) + C_2cos(2\pi) = C_1 * 0 + C_2 * (1) = C_2 \Rightarrow C_2 = 1$$

C_1 is an arbitrary number that can have any value, which means the given problem has infinitely many solutions for the given boundary conditions in the form of $y(t) = C_1 \sin(t) + \cos(t)$ for $C_1 \subseteq \forall$.

Let's explore the MATLAB tools and functions along with the Symbolic Math Toolbox's MuPAD note to solve a few different BVPs. The following built-in functions of MATLAB are used to solve BVPs: `bvpinit`, `bvp4c`, `bvp5c`, `bc`, `odeset`, and `deval`. In the Symbolic Math Toolbox, the command `ode::solve` is employed to compute an analytical solution of a given problem.

Now let's consider a few different second-order ODEs of BVPs to find numerical solutions and simulate them using MATLAB and the Symbolic Math Toolbox.

There are a few general "must-follow" steps when computing the numerical solutions of the BVPs in MATLAB.

- Step #0. Rewrite a given higher-order differential equation via a system of first-order differential equations. This step is like in the IVP.

- Step #1. Define a given problem differential equation via an inline function or function handle (@) or function file (m-file).

- Step #2. Define residues of boundary conditions via an anonymous function (function handle, @) or a separate function file (m-file) or a nested function.

- Step #3. Define a simulation range/space (e.g., time space).

- Step #4. Guess at the solution structure by using `bvpinit()`. Note that initial guesses can be constant values or functions to be computed on specific (constant) values.

- Step #5. Obtain solutions by using `bvp4c()` or `bvp5c()`.

- Step #6. Compute numerical solutions by using `deval()`.

Let's consider a few different problems to show how to employ MATLAB tools according to the previous procedure steps used to solve the BVP.

Dirichlet Boundary Condition Problem
Example 1

Here's the example:

$$\ddot{y} + 2y = e^t \text{ with } y = [y_0 \; y_{end}]$$

where $y_0 = y(0) = 0$ and $y_{end} = y(\pi) = 1$ are Dirichlet boundary conditions.

The solution of the given problem $f(t, y)$ is valid only within $t \in [0, \pi]$ and beyond the boundaries. The validity of $f(t, y)$ is not of any interest.

We first need to rewrite this as two first-order ODEs: $\begin{cases} y_2 = \dot{y}_1 \\ \dot{y}_2 = e^t - 2y_1 \end{cases}$.

This is similar to the IVP-solving approach for second-order ODEs. Now we write an m-file to compute its numerical solutions. Here is the solution script (BVP_EX1.m) to compute numerical solutions of the given differential equation. The computed solutions of the exercise are shown in Figure 8-1.

```
clc; clearvars
% BVP_EX1.m
%{
EXAMPLE 1. Given: y"+2y=exp(t) with BCs: y(0)=0, y(pi)=1.
HELP: 1st we need to rewrite the given problem equation
as a set of 1-st order ODEs likewise to 2nd order IVP problems.
Thus, we re-write it as:  dy/dt=[y2, exp(t)-2*y1];
dy/dt can be defined via a function file or anonymous function.
%}
% Defined by anonymous function:
dy=@(t, y)([y(2); exp(t)-2*y(1)]);
% Define residues of BCs (Dirichlet BCs) by anonymous function:
Res=@(yl, yr)([yl(1), yr(1)-1]);
% Create time space for simulation:
t = linspace(0, pi, 40);
% Create a guess structure consisting of an initial mesh
% of 40 (opts) equally spaced points within [0,pi] and
% a guess of constant values y1=0  and y2=1 with this command:
SOLin1 = bvpinit(t,[0 1]);
```

```
% Another initial guess (for fun): y1=2 and y2=1
SOLin2 = bvpinit(t,[2 1]);
% Obtain numerical solutions of the problem:
SOL1 = bvp4c(dy, Res, SOLin1);
SOL2 = bvp4c(dy, Res, SOLin2);
y1 = deval(SOL1,t);
y2 = deval(SOL2,t);
figure
plot(t, y1(1,1:end), 'ro-', t, y2(1, :),'bx-', 'linewidth', 1.5),
grid minor;
legend('\it Guess #1: y_1=1  & y_2=0 ',...
'\it Guess #2: y_1=2  & y_2=1', 'location', 'SouthWest')
xlabel( '\it t, time')
ylabel( '\it y(t) solution values')
axis tight
syms y t d dt e
EQN = d^2*y/dt^2+2*y==e^t;
title('\it Solution of BVP: $\frac{d^2y}{dt^2}+2y=e^t$' ,
'Interpreter',  'latex')
```

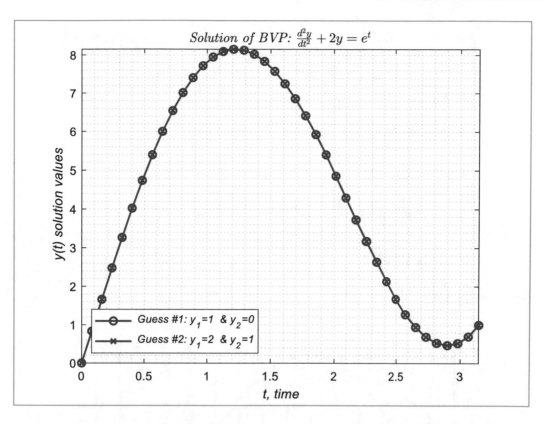

Figure 8-1. *Simulation results of the second-order BVP:* $\ddot{y}+2y=e^{t}$

An alternative way to solve this exercise is to use the Symbolic Math Toolbox that evaluates the analytical solution of the problem with two lines of commands (in a MuPAD notes) that have the following syntaxes:

```
[ Sol1:=ode::solve({u"(t)=exp(t)-2*u(t),
 u(0) = 0, u(PI) = 1}, u(t));
```

$$\left\{ \frac{e^t \sigma_2 \left(2\sigma_2 - \sqrt{2}\,\sigma_1\right)}{6} - \frac{\sigma_2}{3} + \frac{\sigma_1 \left(-e^\pi \sigma_3{}^2 + \sigma_3 - e^\pi \sigma_4{}^2 + 3\right)}{3\,\sigma_4} + \frac{e^t \sigma_1 \left(2\sigma_1 + \sqrt{2}\,\sigma_2\right)}{6} \right\}$$

where

$$\sigma_1 = \sin(\sqrt{2}\ t)$$

$$\sigma_2 = \cos(\sqrt{2}\ t)$$

$$\sigma_3 = \cos(\pi\ \sqrt{2})$$

$$\sigma_4 = \sin(\pi\ \sqrt{2})$$

```
[ plot(Sol1, t=0..PI, #G, #L)
```

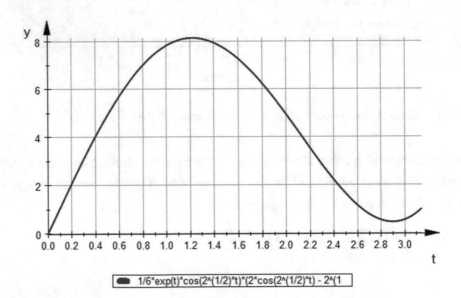

1/6*exp(t)*cos(2^(1/2)*t)*(2*cos(2^(1/2)*t) - 2^(1

The numerical simulations of this BVP with two different numerical initial guesses with relatively close values to each other have shown that initial numerical guesses for such BVPs do not have any effect on the final numerical solutions of a given problem. This exercise can also be tested for a second initial guess of $[10^7, -10^7]$ or $[-10^7, 10^7]$, and the previously displayed results in the plot figure will be repeated in the same way. However, these initial numerical guesses cannot have the value of infinity (∞).

If the initial guess values are set to be `inf`, then the solver `bvp4c` can't solve the problem because of a singularity issue in the Jacobian matrices of the problem.

The analytical solution evaluation with the Symbolic Math Toolbox (MuPAD note) has a simple syntax. However, not all BVPs can be solved with the Symbolic Math Toolbox commands and functions. This issue is similar to IVPs.

Example 2

Here is another Dirichlet BC problem: $t^2\ddot{y} - 5t\dot{y} + 8y = 0, y(1) = 0, y(2) = 24.$ We rewrite

the given problem as two first-order ODEs: $\begin{cases} y_2 = \dot{y}_1 \\ \dot{y}_2 = \dfrac{5ty_2 - 8y_1}{t^2} \end{cases}.$

Now, we implement the previous system of equations via a nested function called dy within our script function file, named BVP_EX2.m. Note that this problem has been implemented by employing a function file, and several nested functions defining BCs and initial guess structures are embedded within the function file. The simulation results are shown in Figure 8-2.

```
function BVP_EX2
% BVP_EX2.m solves the given BVP: t^2*y"-5t*y'+8y=0, y(1)=0,      %y(2)=24
t = linspace(1, 2, 30);
% A guess structure is created that consists of time mesh
% in the range of BCs and a guess function (Gsol1):
SOLin1 = bvpinit(t,@Gsol1);
% Another initial guess (with a function) Gsol2
SOLin2 = bvpinit(t,@Gsol2);
% Another initial guess (with numerical values) y1=0 and y2=24
SOLin3 = bvpinit(t,[0,24]);
% Numerical solutions of the problem are obtained with:
SOL1 = bvp4c(@dy,@Res,SOLin1);
SOL2 = bvp4c(@dy,@Res,SOLin2);
SOL3 = bvp4c(@dy,@Res,SOLin3);
y1 = deval(SOL1,t);
y2 = deval(SOL2,t);
y3 = deval(SOL3,t);
figure
```

```matlab
plot(t, y1(1,1:end), 'ro-', t, y2(1, :), 'bx-',...
    t, y3(1,:),'kd:','linewidth', 1.5), grid on
legend('\it Guess #1: Polynom', '\it Guess #2: cos&sin fcns ',...
'\it Guess #3: numerical values', 'location', 'SouthEast')
title('\it Solution of BVP:  $\frac{d^2y}{dt^2}*t^2-5t*\frac{dy}
{dt}+8y=0$', 'Interpreter', 'latex')
xlabel '\it t',
ylabel '\it y(t) solution values'
function Ft1=Gsol1(t,~)
% 1st guess solution function:
Ft1=([8*t.^2-8*t, 16*t-8]);
end
function Ft2=Gsol2(t,~)
% 2nd guess solution function:
        Ft2=([cos(t), -sin(t)]);
end
function f=dy(t,y)
% Given problem formulation:
        f=[y(2), (5*t.*y(2)-8*y(1))./t.^2];
end
function F=Res(yl,yr)
% Now, we define residues of BCs as a nested function.
        F=[yl(1), yr(1)-24];
end
end
```

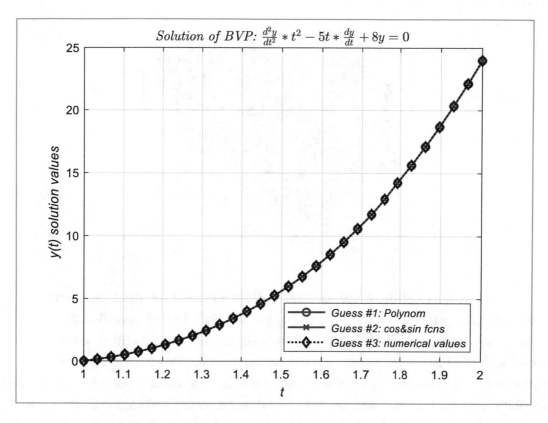

Figure 8-2. *Simulation of the problem:* $t^2\ddot{y} - 5t\dot{y} + 8y = 0$

The solution commands of the Symbolic Math Toolbox in the MuPAD note are as follows:

```
[ Sol1:=ode::solve({u"(t) =(5*t*u'(t)-8*u(t))/t^2,
u(1) = 0, u(2) = 24}, u(t));
```

$$\{2\,t^4 - 2\,t^2\}$$

```
[ plot(Sol1, t=1..2, #G, #L, #3D)
```

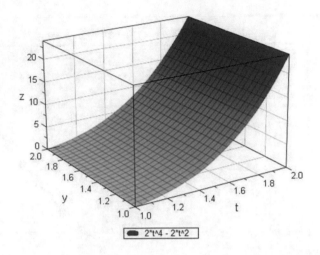

For numerical solutions of this second-order nonlinear BVP (Example 2) with Dirichlet BCs, we have made three different guesses (BVP_EX2.m), specifically, polynomials (quadratic and linear equations), sine and cosine functions, and numerical guesses, all of which have resulted in well-converged solutions. Moreover, we have explored in this example ways of using nested functions and a function file to speed up a simulation process. In addition, we have computed an analytical solution of this exercise in the Symbolic Math Toolbox in a much simpler way with two lines of commands and plotted numerical values.

Robin Boundary Condition Problem

Robin boundary conditions are mixed boundary conditions.

Example 3

In this example, we look at two BVPs with the same differential equation but use two different boundary conditions, Robin boundary conditions and Dirichlet boundary conditions, for comparison purposes. The two different boundary conditions of the BVPs are compared.

- **Problem 1:** $\ddot{y} + 2y = 0$ with BCs: $y(0) = 1, \dot{y}(\pi) = 0$ (Robin or mixed boundary conditions)

- **Problem 2:** $\ddot{y} + 2y = 0$ with BCs: $y(0) = 1, y(\pi) = 0$ (Dirichlet boundary conditions)

Note that the only difference in these two problems is in their BCs.

First, we rewrite the given problem equation as a system of first-order ODEs: $\begin{cases} y_2 = \dot{y}_1 \\ \dot{y}_2 = -2y_1 \end{cases}$.

Now, we simulate the given problem with the following script (BVP_EX3.m), which consists of two parts; part 1 solves problem 1 (with Dirichlet boundary conditions), and part 2 solves problem 2 (Robin or mixed boundary conditions). Solving the two problems is similar except for how residues of their BCs are defined.

```
% BVP_EX3.m
% Problem 1. Given: y"+2y=0
% Part 1. BCs: y(0)=1, dy(pi)=0. Robin BC.
clearvars; clc; close all
dy=@(t, y)([y(2), -2*y(1)]);
% Define residues of BCs:
Res=@(yl, yr)([yl(1)-1, yr(2)]);
% Create guess structures:
SOLin1 = bvpinit(linspace(0,pi,40),[1 0]);
% Another initial guess:
SOLin2 = bvpinit(linspace(0,pi,40),[2 1]);
% Obtain the solutions of the problem:
SOL1 = bvp4c(dy,Res,SOLin1);
SOL2 = bvp4c(dy,Res,SOLin2);
t = linspace(0,pi, 40);
y1 = deval(SOL1,t);
y2 = deval(SOL2,t);
plot(t, y1(1,1:end), 'mp', t, y2(1, :),'k-', 'linewidth', 2.0),
grid on; axis tight; hold on

% Part 2. Problem 2. y"+2y=0, BCs: y(0)=1, y(pi)=0. Dirichlet BC
%{
Rewrite the given problem equation as a set of 1st order ODEs:
```

```
dydt=[y2, 2*y1];
Define dydt with an anonymous function:
%}
dy=@(t,y)([y(2), -2*y(1)]);
% Define residues of BCs:
Res=@(yl, yr)([yl(1)-1, yr(1)]);
%{
Create a guess structure consisting of an initial mesh
of 40 (opts) equally spaced points in [0,pi] (in the range of BCs)
and a guess of constant values y1=1  and y2=0 as follows:
%}
SOLin1 = bvpinit(linspace(0,pi,40),[1 0]);
% Another initial guess is taken: y1=2 and y2=1
SOLin2 = bvpinit(linspace(0,pi,40),[2 1]);
% Obtain solutions of the problem:
SOL1 = bvp4c(dy,Res,SOLin1);
SOL2 = bvp4c(dy,Res,SOLin2);
t   = linspace(0,pi, 40);
y1 = deval(SOL1,t);
y2 = deval(SOL2,t);
plot(t, y1(1,1:end), 'bo', t, y2(1, :), 'r', 'linewidth',2)
legend('\it Prob1: Guess #1 [y1=1 & y2=0] ',...
'\it Prob1: Guess #2 [y1=2 & y2=1]', ...
'\it Prob2: Guess #1 [y1=1 & y2=0] ', ...
'\it Prob2: Guess #2 [y1=2  & y2=1]', ...
'location', 'southwest')
xlabel '\it t', ylabel '\it y(t) solution values'
title('Solution of BVP: $\frac{d^2y}{dt^2}+2y=0$', 'interpreter', 'latex')
gtext('Problem 1: y(0)=1, y(\pi)=0')
gtext('Problem 2: ; y(0)=1,dy(\pi)=0')
hold off
```

Figure 8-3 shows the simulation results of the script. This exercise demonstrates clearly how much the BCs affect the final solutions of BVP problems.

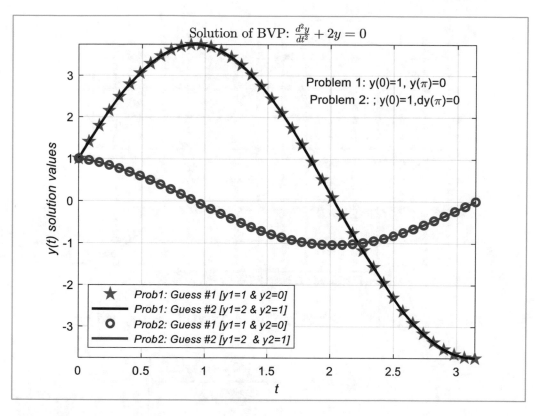

Figure 8-3. *Simulation of $\ddot{y}+2y=0$ with two different boundary conditions, Robin BC (problem 1) and Dirichlet BC (problem 2)*

An alternative solution can be computed via the Symbolic Math Toolbox in MuPAD notes.

```
[ SOL1:=ode::solve({y"(t) =-2*y(t),y(0) = 1, y'(PI) = 0}, y(t));
```

$$\left\{ \cos(\sqrt{2}\, t) + \frac{\sin(\pi\,\sqrt{2})\,\sin(\sqrt{2}\, t)}{\cos(\pi\,\sqrt{2})} \right\}$$

```
[ plot(SOL1, t=0..PI, #G, #L)
```

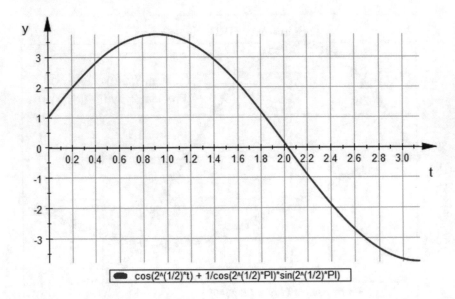

cos(2^(1/2)*t) + 1/cos(2^(1/2)*PI)*sin(2^(1/2)*PI)

[SOL2:=ode::solve({y"(t) =-2*y(t), y(0) = 1, y(PI) = 0}, y(t));

$$\left\{ \cos(\sqrt{2}\, t) - \frac{\cos(\pi\, \sqrt{2})\, \sin(\sqrt{2}\, t)}{\sin(\pi\, \sqrt{2})} \right\}$$

[plot(SOL2, t=0..PI, #G, #L)

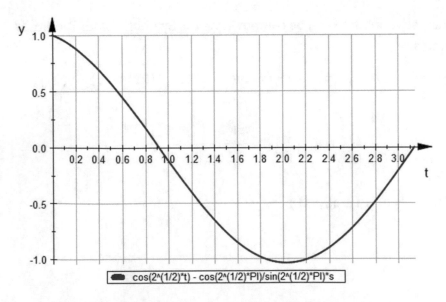

cos(2^(1/2)*t) - cos(2^(1/2)*PI)/sin(2^(1/2)*PI)*s

[plot(SOL1,SOL2, t=0..PI, #G, #L)

278

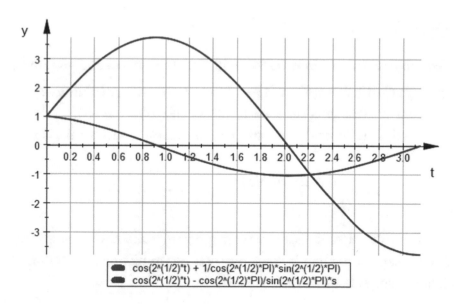

The simulation results of the exercise demonstrate how two different guesses of boundary values yield well-converged solutions. In this example, like with the previous example (Example 1), numerical guesses of boundary values can be tested for a wide range of values, for instance, $[-10^8, 10^{10}]$ or vice versa.

Sturm-Liouville Boundary Value Problem

There are a number of BVPs with many applications in scientific computing and engineering studies. One type of BVP is the Sturm-Liouville boundary value problem named after two French mathematicians, Jacques Sturm and Joseph Liouville. Sturm-Liouville BVPs [1] have a broad range of application areas, for instance:

— Heat conduction in nonuniform bars

— Quantum mechanics, the one-dimensional Schrödinger equation

— Vibrations of nonuniform bars and strings

— Scattering theory, Hill's equation

— Oscillations of a sphere, Legendre's equation

— Thermodynamics of cylindrical and spherical bodies

Example 4

Here's an example of Sturm-Liouville BVP with the constant coefficient values:

$-\ddot{u} + \omega^2 u = f(t)$ with BCs $u(0) = u(1) = 0$ (periodic boundary conditions)

Let's consider the following case: $\omega = 0, 3, 7, 13$ and $f(t) = sinh(t)$.

Now, we rewrite the Sturm-Liouville equation with our considered case.

$$-\ddot{u} + \omega^2 u = sinh(t) \Rightarrow \begin{cases} u_2 = \dot{u}_1 \\ \dot{u}_2 = \omega^2 u_1 - sinh(t) \end{cases}$$

In MATLAB's BVP solver tools, there is an option to include parameters within bvpinit that we can use to include the values of ω for simulations and write BVP_EX4.m. The simulation results are shown in Figure 8-4.

```
clc; clearvars; close all
% BVP_EX4.m
% Sturm_Liouville_bvp. Given: -u"+w^2*u=sinh(t)
% periodic boundary conditions: u(0)=u(1)=0
% -----------------------------------------------------------
% Define residues of BCs via a function handle
Res=@(yl,yr)([yl(1) yr(1)]);
% 1st guess solution function defined with function handle:
Gsol=@(t)([sin(-0.540302305868140*t),cos(t)-0.540302305868140]);
close all; clc
omega=[0, 3, 5, 7, 13];
t = linspace(0, 1, 500);
% A guess structure consisting of time mesh within [0, 1]
% in the range of BCs and a guess function (Gsol):
SOLin1 = bvpinit(linspace(0,1, 10),Gsol);
% Another numeric value based initial guess y1=0 and y2=0
SOLin2 = bvpinit(linspace(0,1, 10),[0,0]);

% Plotting options:
Labelit = { };
Colorit1 = 'krgbm';
Colorit2 = 'krgbm';
Lineit1  = '---:--:---:';
```

```
Lineit2  = '---:--:---:';
for ii=1:numel(omega)
    Styleit1=[Colorit1(ii), Lineit1(ii)];
    Styleit2=[Colorit2(ii), Lineit2(ii)];
    % Given problem formulation:
    dy=@(t,y)([y(2), omega(ii)^2*y(1)-sinh(t)]);
    %Obtain the solutions of the problem for two initial guesses:
    SOL1 = bvp4c(dy,Res,SOLin1);
    SOL2 = bvp4c(dy,Res,SOLin2);
    % Compute numeric solutions of the problem within time-space:
    y1 = deval(SOL1,t);
    y2 = deval(SOL2,t);
    subplot(211)
    plot(t, y1(1,1:end), Styleit1, 'linewidth', 1.5), grid on
    hold on
    Labelit{ii}=['\omega = ' num2str(omega(ii))];
    legend(Labelit{:})
    title('Sturm Liouville BVP: $\frac{-d^2u}{dt^2}+\omega^2*u=sinh(t)$',
    'interpreter', 'latex')
    ylabel('\it y(t) solution values')
    subplot(212)
    plot(t,y2(1, :), Styleit2, 'linewidth', 1.5), grid on
    hold on
    Labelit{ii}=['\omega = ' num2str(omega(ii))];
    legend(Labelit{:})
    xlabel '\it t'
    ylabel('\it y(t) solution values')
end
gtext('Guess 1: sin&cos fcns', 'background', 'y')
gtext('Guess 2: Numerical guess values', 'background', 'y')
hold off
```

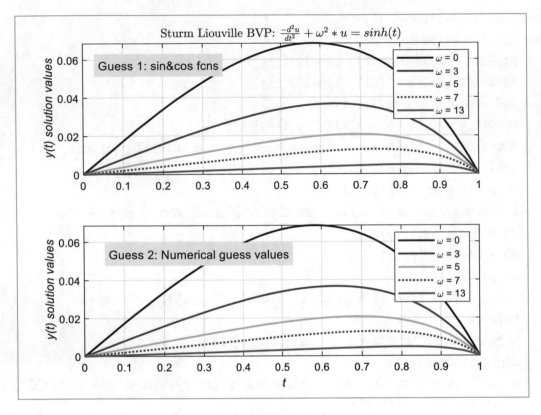

Figure 8-4. *Simulation of Sturm-Liouville BVP for ω = 0, 3, 5, 7, 13.*

An alternative solution of the Sturm-Liouville BVP is obtained in the Symbolic Math Toolbox script in a MuPAD note (BVP_EX4_MuPAD.mn).

```
[ Omega:=([0, 3, 5, 7, 13]);
```

$$[0, 3, 5, 7, 13]$$

```
[ Sol1:=ode::solve({u"(t)=Omega[1]^2*u(t)-sinh(t),
u(0) = 0, u(1) = 0}, u(t));
```

$$\left\{ t \sinh(1) - \sinh(t) \right\}$$

```
[ Sol2:=ode::solve({u"(t)=Omega[2]^2*u(t)-sinh(t),
u(0) = 0, u(1) = 0}, u(t));
```

$$\left\{ \frac{e^{-t}\left(e^{2t}-2\right)}{48} + \frac{e^{-t}\left(2\,e^{2t}-1\right)}{48} + \frac{e^{-3t}\,e^{2}}{16\left(e^{2}+e^{4}+1\right)} - \frac{e^{3t}\,e^{2}}{16\left(e^{2}+e^{4}+1\right)} \right\}$$

```
[ Sol3:=ode::solve({u"(t)=Omega[3]^2*u(t)-sinh(t),
 u(0) = 0, u(1) = 0}, u(t));
```

$$\left\{ \frac{e^{-t}\left(2\,e^{2t}-3\right)}{240} + \frac{e^{-t}\left(3\,e^{2t}-2\right)}{240} + \frac{e^{-5t}\,e^{4}}{48\left(e^{2}+e^{4}+e^{6}+e^{8}+1\right)} \right.$$
$$\left. - \frac{e^{5t}\,e^{4}}{48\left(e^{2}+e^{4}+e^{6}+e^{8}+1\right)} \right\}$$

```
[ Sol4:=ode::solve({u"(t)=Omega[4]^2*u(t)-sinh(t),
 u(0) = 0, u(1) = 0}, u(t));
```

$$\left\{ \frac{e^{-t}\left(3\,e^{2t}-4\right)}{672} + \frac{e^{-t}\left(4\,e^{2t}-3\right)}{672} + \frac{e^{-7t}\,e^{6}}{\sigma_1} - \frac{e^{7t}\,e^{6}}{\sigma_1} \right\}$$

where

$$\sigma_1 = 96\left(e^{2}+e^{4}+e^{6}+e^{8}+e^{10}+e^{12}+1\right)$$

```
[ Sol5:=ode::solve({u"(t)=Omega[5]^2*u(t)-sinh(t),
u(0) = 0, u(1) = 0}, u(t));
```

$$\left\{ \frac{e^{-t}\left(6\,e^{2\,t}-7\right)}{4368} + \frac{e^{-t}\left(7\,e^{2\,t}-6\right)}{4368} + \frac{e^{-13\,t}\,e^{12}}{\sigma_1} - \frac{e^{13\,t}\,e^{12}}{\sigma_1} \right\}$$

where

$$\sigma_1 = 336\left(e^2 + e^4 + e^6 + e^8 + e^{10} + e^{12} + e^{14} + e^{16} + e^{18} + e^{20} + e^{22} + e^{24} + 1\right)$$

```
[ plot(Sol1, Sol2, Sol3, Sol4, Sol5, t=0..1, #G, #L)
```

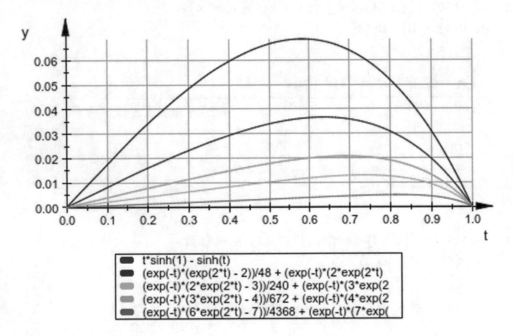

```
# An alternative and more compact solution:
[for omega in [0, 3, 5, 7, 13] do
   Sol:=ode::solve({u"(t) =omega^2*u(t)-sinh(t),
   u(0) = 0, u(1) = 0}, u(t));
   plot(hold(Sol, t=0..1, #G, #L));
   end_for
# Results are not displayed.
```

From the simulation results of the analytical solution from the MuPAD note BVP_EX4_MuPAD.mn and the numerical solutions (Figure 8-6) obtained from BVP_EX4.m for different values of ω, all solutions converge very well. Moreover, with the increase of ω, the magnitude of solution $y(t)$ decreases proportionally. The simulation results show that initial guess functions, such as sine and cosine, converge perfectly with numerical guess values.

Stiff Boundary Value Problem

There are many stiff BVPs similar to IVPs of ODEs, and there are a few methods and techniques that have been developed to solve such problems. In this section, we look at one example for stiff BVPs. Solving them in an efficient way is often a challenge. Let's consider the next example.

Example 5

Here's the example stiff BVP: $\ddot{y} - (\alpha + \beta)\dot{y} + \alpha\beta y = 0$ with BCs of $y(0) = y(1) = 1$ and $\alpha = 100, \beta = -100$.

This problem [2] is very stiff and depends on the values of α, β. The given problem is solved like the previously solved problems by converting the given second-order equation into two first-order differential equations. This is by introducing new variables and then implementing the newly derived system of equations into the script called BVP_EX5.m. The simulation results are shown in Figure 8-5 and Figure 8-6.

```
function BVP_EX5
Solutions_initial = bvpinit(0:.0025:1,[10 10]);
tol=1e-16; % Tolerance
OPT = bvpset('stats','on','abstol',tol,'reltol',tol,'Nmax',1000);
solutions_final=bvp4c(@ODE_fun1,@BC_fun1,Solutions_initial, OPT);
x = 0:.0025:1; y = deval(solutions_final, x);
close all
figure(1), plot(x, y(1,:), 'b', 'linewidth', 2);
title('\it Solutions of: $\frac{d^2y}{dt^2}-(\alpha+\beta)\frac{dy}{dt}+\
alpha+\beta*y(t)=0$', 'interpreter', 'latex'); grid on
xlabel('\it t')
ylabel('\it Solution, y(t)'),
xlim([-.05 1.05])
figure(2)
yyaxis left
```

```
plot(x, y(1,:),'b', 'linewidth', 2)
title('\it Solutions of: $\frac{d^2y}{dt^2}-(\alpha+\beta)\frac{dy}{dt}+\
alpha+\beta*y(t)=0$', 'interpreter', 'latex'); grid on
grid on
ylabel('\it Solution: y(t)')
xlabel('\it t')
yyaxis right
plot(x,y(2,:),'m--','linewidth', 2)
ylabel('\it $\frac{dy}{dt}$', 'Interpreter', 'latex')
axis([-0.05 1.05 -5  5]), grid on
legend('\it Solution: y(t)', '\it Derivative: $\frac{dy}{dt}$',
'Interpreter', 'latex', 'location', 'southwest');
function dydx = ODE_fun1(~,y)
        alpha=100; beta=-100;
    dydx = [ y(2),      (alpha+beta)*y(2)-alpha*beta*y(1)];
end
function res = BC_fun1(y1,yend)
res = [ 1-y1(1) ,    1-yend(1)];
end
end
```

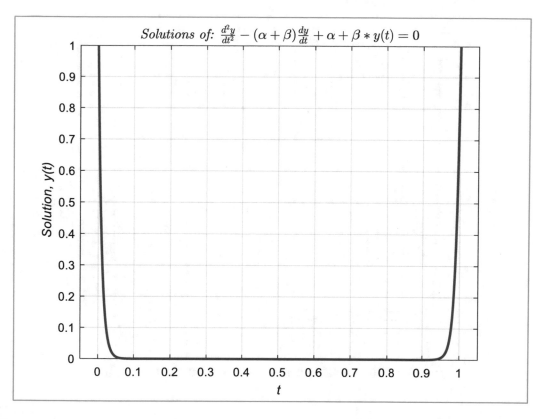

Figure 8-5. *Simulation results of* $\ddot{y} - (\alpha + \beta)\dot{y} + \alpha\beta y = 0$ *with BCs:* $y(0) = y(1) = 1$
and $\alpha = 100, \beta = -100$

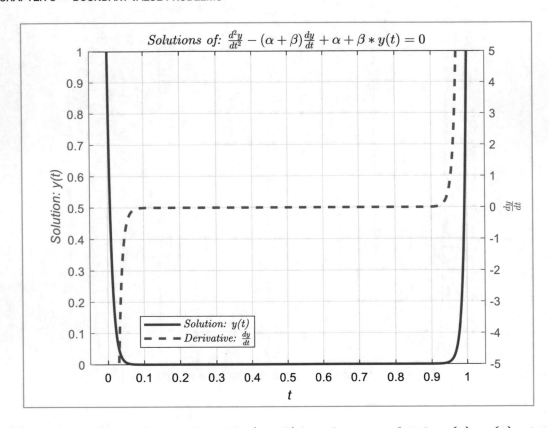

Figure 8-6. *Simulation results of $\ddot{y}-(\alpha+\beta)\dot{y}+\alpha\beta y=0$ with BCs: $y(0)=y(1)=1$ and $\alpha=100$, $\beta=-100$*

An alternative solution is to use the Symbolic Math Toolbox commands in the MuPAD note that solve the problem analytically. Here is a solution script, called BVP_EX5_MuPAD.mn:

```
# BVP_EX5_MuPAD.mn
[ A:=100; B:=-100;

[ Sol:=ode::solve({y"(t) =-(A*B)*y(t)+(A+B)* y'(t),
y(0) = 1, y(1) = 1}, y(t))
```

$$\left\{ \frac{e^{100\,t}\left(\frac{1}{e^{100}}-1\right)}{\frac{1}{e^{100}}-e^{100}} - \frac{e^{100}-1}{e^{100\,t}\left(\frac{1}{e^{100}}-e^{100}\right)} \right\}$$

```
[ plot(Sol, t = 0..1, #G, #L, #3D)
```

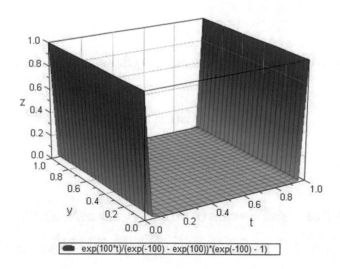

exp(100*t)/(exp(-100) - exp(100))*(exp(-100) - 1)

This concludes the BVP-solving exercises. It should be noted that a solution search procedure of a BVP has to start with a careful study of the problem to identify the BCs (which type) and the stiffness of the given problem. The general procedure of solving BVPs in a MATLAB environment is composed of six steps, as depicted earlier. If it is feasible to compute analytical solutions of a given BVP, a first approach to try is to use the Symbolic Math Toolbox and/or a MuPAD note.

References

[1] Boyce, W. E., Diprima, R.C., *Elementary Differential Equations and Boundary Value Problems* (7th ed.), John Wiley & Son (2003).

[2] Pasic, H., (1999) "Multipoint Boundary-Value Solution of Two-point Boundary-Value Problems," *Journal of Optimization Theory and Applications*: vol 100, No. 2, pp. 397–416, February 1999.

Drill Exercises

The following are some exercises to try on your own.

Exercise 1

Solve the following BVP numerically by using the ODE solvers and analytically with MuPAD:

$$x^2\ddot{y} - 5x\dot{y} + 8y = 0 \text{ with } y(1) = 0 \text{ and } y(2) = 24.$$

Exercise 2

Solve the given BVP of a second-order ODE numerically in MATLAB: $\ddot{y} + 9y = 0$ with $y(0) = 0$ and $y\left(\dfrac{\pi}{2}\right) = 6$. Plot the numerical solutions of the BVP. Compute an analytical solution of the exercise in a MuPAD note and plot the computed analytical solution.

Exercise 3

Solve the given BVP of a second-order ODE numerically in MATLAB: $\ddot{y} + 9y = 0$ with $\dot{y}(0) = 0$ and $\dot{y}\left(\dfrac{\pi}{2}\right) = 6$. Plot the computed numerical solutions. Evaluate the analytical solution of the exercise in a MuPAD note and plot the computed analytical solution.

Exercise 4

Solve the given BVP of a second-order ODE numerically in MATLAB: $x^2\ddot{y} - 5x\dot{y} + 8y = 0$ with $y(1) = 0$ and $\dot{y}(2) = 24$. Plot the numerical solutions of the problem. Compute the analytical solution of the exercise in a MuPAD note and plot the found analytical solution.

Exercise 5

Solve the given BVP of a second-order ODE numerically in MATLAB: $\ddot{y} + y = 0$ with $y(0) = y(2\pi) = 0.5$ and $\dot{y}(0) = \dot{y}(2\pi) = 1$. Plot the numerical solutions of the problem. Compute the analytical solution of the exercise in a MuPAD note and plot the found analytical solution.

Exercise 6

Solve the given BVP of a second-order ODE numerically in MATLAB: $\ddot{y} + 40y = \cos(2t)$ with $y(0) = 0$, $y(\pi) = 0$. Plot the numerical solutions of the problem. Is it possible to compute the given BVP's analytical solution in a MuPAD note?

Exercise 7

Solve the given BVP: $\ddot{y} + 40y = \cos(2t)$ with $\dot{y}(0) = 0, \dot{y}(\pi) = 0$. Plot the numerical solutions of the problem. Plot the numerical solutions of Exercises 6 and 7 in one plot area and compare them.

Exercise 8

Solve the given BVP numerically in MATLAB: $x^2\ddot{y} - 2x\dot{y} + 1000y = 0$ with $y(1) = 0$, $y(e) = 0$. Plot the numerical solutions of the problem. Evaluate the analytical solution of the exercise in a MuPAD note and plot the found analytical solution formulation.

Exercise 9

Solve the given two BVPs numerically in MATLAB: 1) $x^2\ddot{y} + 6x\dot{y} + 200y = 0$ with $y(1) = 0, \dot{y}(\sqrt{e}) = 0$; 2) $x^2\ddot{y} + 6x\dot{y} + 200y = 0$ with $y(1) = 0, y(\sqrt{e}) = 0$.

Plot the numerical solutions of the given two BVPs and compare the computed solutions. Evaluate analytical solutions of the given two BVPs in MuPAD notes and plot the found analytical solutions.

Exercise 10

Solve the given BVP numerically in MATLAB: $\dfrac{d^2f}{dt^2} + \dfrac{2df}{dt} + 40f = 0$ with $f(0) = -0.5$, $f(0.5) = \pi$. Plot the numerical solutions of the problem.

Exercise 11

Solve the given BVP numerically in MATLAB: $\dfrac{d\left[\dfrac{\left(1+t^2\right)dy}{dt}\right]}{dt} + 4y = 0$ with $y(-1) = 0$, $y(1) = 0$. Plot the numerical solutions of the problem. Evaluate their analytical solution in a MuPAD note and plot the found analytical solution.

Exercise 12

Solve the given BVP numerically in MATLAB: $\dfrac{d^2x}{dt^2} + 2x = -1 + |1 - 2t|$ with $x(0) = 0$, $dx(1) = 0$. Plot the numerical solutions of the problem.

Exercise 13

Solve the next BVP numerically in MATLAB: $\ddot{y} + \pi^2 y = \pi^2 t$ with $y(0) = 1$, $y(1) = 0$. Compare your computed numerical solutions with its analytical solution, which is $y = c_1 \sin(\pi t) + \cos(\pi t) + t$. Compute its analytical solution in a MuPAD note and build a 3D plot of the found analytical solution.

PART III

Applications of Ordinary Differential Equations

CHAPTER 9

Spring-Mass-Damper Systems

This chapter covers several essential aspects and approaches how to build simulation models of spring-mass-damper systems in MATLAB and Simulink environments. The equations of motions of one, two, three degree of freedom spring-mass-damper systems are derived and MATLAB/Simulink models are built based on the derived mathematical formulations. Free and forced motions of the spring-mass-damper systems are studied, and linear and non-linear behaviours of the spring-mass-damper systems are considered. The developed simulation models of one, two and three degree of freedom systems can be applied for multipled degree of freedom systems.

Single Degree of Freedom System
Case 1: Free Vibration (Motion)

Let's consider a single degree of freedom (DOF) spring-mass-damper (SMD) system of a nonlinear type formulated by $M\ddot{u} + C|\dot{u}|\dot{u} + Ku = 0$, $u(0) = 0$, and $\dot{u}(0) = 1$, where M is a mass, C is damping, and K is stiffness of the system. Note that the formulation of the system is a second-order homogenous ODE, and there is no external force applied. The system oscillates with respect to its initial conditions. In other words, its response depends on its initial displacement and velocity values.

Likewise, when solving the initial value problems of ODEs, the first step to build a simulation model of the given problem is to rewrite its equation of motion.

$$\ddot{u} = -\frac{C|\dot{u}|\dot{u} + Ku(t)}{M} \tag{9-1}$$

© Sulaymon L. Eshkabilov 2020
S. L. Eshkabilov, *Practical MATLAB Modeling with Simulink*, https://doi.org/10.1007/978-1-4842-5799-9_9

The second-order ODE, shown in Equation (9-1), can be represented by the next two differential equations by introducing a new variable, as explained in Part 1 of the book, to solve the second-order ODEs numerically.

$$\begin{cases} \dot{u}_1 = u_2 \\ \dot{u}_2 = -\dfrac{C|u_2|u_2 + Ku_1}{M} \end{cases} \tag{9-2}$$

The given system model will be of a stiff-type ODE if the magnitude of its mass is much smaller than its stiffness and damping, for instance: $M = 1\,\text{kg}, C = 1001\dfrac{\text{N s}}{\text{m}}, K = 1000\dfrac{N}{m}$.

Let's write a script in a function file (SMDode.m) with three input arguments (M, C, K) based on the first ODEs shown in Equation (9-2). We are using the ODE solver ode15s since the given problem is a stiff ODE type. Note that we also introduce the Jacobian matrix of the given system in Equation (9-2) as the nested function function dfdu = Jacobian(t, u). The expression shown in Equation (9-2) is coded within the nested function function dudt = f(t, u). Having the Jacobian matrix defined speeds up the simulation process of the MATLAB ODE solvers.

```
function  SMDode(Mass,Damping, Stiffness)
%{
SMDode.m. Spring-Mass-Damper system behavior analysis
with the given Mass, Damping and Stiffness values.
Note: an order of input argument data is important.
The system's damper has non-linear properties expressed
with D*|u'|*u' e.g., abs(velocity)*velocity
Solver ode15s is employed; yet, other solvers, viz.
ODE15S, ODE23S, ODE23T, ODE23TB, can be used, as well.
%}

if nargin < 1
%  Some default values for parameters: Mass, Damping & Stiffness
%  in case not specified by the user
    Mass = 1;          % [kg]
    Damping=1001;      % [Ns/m]
    Stiffness=1000;    % [N/m]
  end
```

```matlab
tspan = [0; min(20,5*(Damping/Mass))]; % Several periods
u0 = [0; 1];                           % Initial conditions
% Options for ODESETs can be switched off
options = odeset('Jacobian',@Jacobian);
% [t,u] = ode15s(@f,tspan,u0,[]); % without options
[t,u] = ode15s(@f,tspan,u0,options);

figure;
plot(t,u(:,1),'r-', t, u(:,2), 'b-.', 'MarkerSize', 5,...
    'LineWidth', 1.5);
title(['\it Spring-Mass-Damper System: ',...
'M = ' num2str(Mass), '[kg]' '; D = ' num2str(Damping),...
'[Ns/m]', '; S = ' num2str(Stiffness), '[N/m]']);
xlabel('time t')
grid on
axis tight
axis([tspan(1) tspan(end) -0.1 0.1]); % Axis limits are set
hold on
Acceleration=-(Damping/Mass)*abs(u(:,2)).*u(:,2)-...
(Stiffness/Mass)*u(:,1);
plot(t, Acceleration,'k--', 'MarkerSize', 3, 'LineWidth', 1);
legend('Displacement', 'Velocity', 'Acceleration');
ylabel('Displacement, Velocity, Acceleration')
hold off; xlim([0, 3])
% -----------------------------------------------------------
% Nested functions: Mass, Damping and Stiffness are provided
% by the outer function.
%------------------------------------------------------------
  function dudt = f(t, u)
% Derivative function.  Mass, Damping and Stiffness are provided
% by the outer function or taken default (example values)
  dudt = [            u(2);
        -(Damping/Mass)*abs(u(2))*u(2)-(Stiffness/Mass)*u(1) ];
  end
```

```
% -----------------------------------------------------------
  function dfdu = Jacobian(t, u)
% Jacobian function. Mass, Damping and Stiffness are provided by
% the main function or taken default values for no inputs.
    dfdu = [            0,                        1;
              -(Stiffness/Mass),  -(Damping/Mass)*abs(u(2))-...
              (Damping/Mass)*u(2)*sign(u(2)) ];
  end %
% -----------------------------------------------------------
end  % SMDode.m
```

From the simulation results plotted in Figure 9-1, it is clear that the system is over-damped. In studies of spring-mass-damper systems, it is important to study not only the overall system behavior but also its natural frequency, damping level (lightly damped, under-damped, and over-damped case scenarios), and influence of the system parameter changes. All of these studies can be introduced additionally into a simulation model of the system.

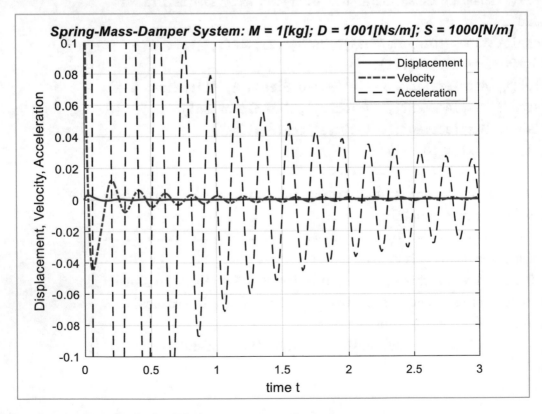

Figure 9-1. *Simulation of the spring-mass-damper system*

Let's reconsider the previous problem of the single degree of freedom SMD system to study its free motion when the system is linear, i.e., $|\dot{u}|$ is not present: $M\ddot{u} + C\dot{u} + Ku = 0$. This can be rewritten by dividing it by M (mass), as shown here:

$$\ddot{u} + \frac{C}{M}\dot{u} + \frac{K}{M}u = 0 \tag{9-3}$$

In this new formulation, the following substitutions can be introduced:

$$\frac{K}{M} = \omega_n^2; \quad \frac{C}{M} = 2\zeta\omega_n \tag{9-4}$$

where ω_n is the natural frequency of the system and ζ is the damping ratio.

The problem formulation can be rewritten again like this:

$$\ddot{u} + 2\zeta\omega_n\dot{u} + \omega_n^2 u = 0 \tag{9-5}$$

That equation has a general solution of the following form:

$$u(t) = Ae^{r_1 t} + Be^{r_2 t} \tag{9-6}$$

where the unknown coefficients A and B are computed from the initial conditions $(u(0) = u_0, \dot{u}(0) = \dot{u}_0)$ using the next general equations:

$$A = \frac{u_0 r_2 - \dot{u}_0}{r_2 - r_1} \tag{9-7}$$

$$B = \frac{\dot{u}_0 - u_0 r_1}{r_2 - r_1} \tag{9-8}$$

By plugging the general solution into the given problem formulation of Equation (9-5), we obtain the following characteristic equation:

$$r^2 + 2\zeta\omega_n r + \omega_n^2 = 0 \tag{9-9}$$

The roots (roots r_1 and r_2) of this quadratic characteristic equation are as follows:

$$r_{1,2} = \left(-2\zeta\omega_n \pm \sqrt{\left(2\zeta\omega_n\right)^2 - 4\omega_n^2}\right) / 4\omega_n^2 \tag{9-10}$$

The roots (r_1 and r_2) depend on the parameters, namely, ω_n and ζ. Let's look at four case scenarios [1, 2] to compute them.

Case Scenario I. Undamped. $\zeta = 0$. The problem formulation is as follows:

$$\ddot{u} + \omega_n^2\, u = 0\,,\; u(0) = u_0\,,\; \dot{u}(0) = \dot{u}_0 \tag{9-11}$$

The solutions of the differential equation in (9-11) are $r_1 = i\,\omega_n$; $r_2 = -i\,\omega_n$, and the solutions for real system cases are available for: $M > 0$ and $K > 0$.

The solution is as follows:

$$u(t) = u_0 \cos\left(\omega_n t\right) + \frac{\dot{u}_0}{\omega_n}\sin\left(\omega_n t\right) \tag{9-12}$$

Note that this solution is obtained using some substitutions, such as the Euler formula we've used in previous chapters ($e^{i\theta} = \cos\theta + i\sin\theta$) and the general solution formulation shown earlier in Equation (9-6).

The following is another formulation of the solution:

$$u(t) = \left(\sqrt{u_0^2 + \frac{\dot{u}_0^2}{\omega_n^2}}\right)\sin\left(\omega_n t + \varphi\right) \tag{9-13}$$

$$\varphi = \tan^{-1}\left(\frac{\omega_n u_0}{\dot{u}_0}\right) \tag{9-14}$$

where φ is the phase shift angle. Note that both solution formulations, shown in Equations (9-12) and (9-13), are equivalent and can be derived from one another.

Case Scenarios II. Underdamped. $0 < c^2 < 4\,MK$ or $0 < \zeta < 1$. The roots of the general solution in Equation (9-6) are as follows:

$$r_1 = \left[-\zeta + i\sqrt{\left(1 - \zeta^2\right)}\right]\omega_n \tag{9-15}$$

$$r_2 = \left[-\zeta - i\sqrt{\left(1 - \zeta^2\right)}\right]\omega_n \tag{9-16}$$

Damping is present, and the damped natural frequency of the system will be as follows:

$$\omega_d = \omega_n \sqrt{\left(1-\zeta^2\right)} \tag{9-17}$$

Thus, the solution is as follows:

$$u(t) = e^{-\zeta\omega_n t} \left(u_o \cos(\omega_d t) + \frac{u_0\zeta\omega_n + \dot{u}_0}{\omega_d} \sin(\omega_d t) \right) \tag{9-18}$$

Case Scenario III. Overdamped. $c^2 > 4\,MK$ or $\zeta > 1$. The roots of the general solution equation shown in Equation (9-6) are as follows:

$$r_1 = \left[-\zeta - \sqrt{\left(\zeta^2 - 1\right)} \right] \omega_n \tag{9-19}$$

$$r_2 = \left[-\zeta + \sqrt{\left(\zeta^2 - 1\right)} \right] \omega_n \tag{9-20}$$

The coefficients of the general solution equation shown in Equation (9-6) are as follows:

$$A = u_0 \left[\frac{\zeta}{2\sqrt{\zeta^2 - 1}} \right] - \frac{\dot{u}_0}{2\omega_n \sqrt{\zeta^2 - 1}} \tag{9-21}$$

$$B = u_0 \left[\frac{\zeta}{2\sqrt{\zeta^2 - 1}} \right] + \frac{\dot{u}_0}{2\omega_n \sqrt{\zeta^2 - 1}} \tag{9-22}$$

Finally, the solution equation is as follows:

$$u(t) = Ae^{r_1 t} + Be^{r_2 t} = \left[u_0 \left[\frac{\zeta}{2\sqrt{\zeta^2 - 1}} \right] + \frac{\dot{u}_0}{2\omega_n \sqrt{\zeta^2 - 1}} \right] e^{\left[-\zeta - \sqrt{\left(\zeta^2 - 1\right)} \right]\omega_n} +$$
$$\left[-u_0 \left[\frac{\zeta}{2\sqrt{\zeta^2 - 1}} \right] + \frac{\dot{u}_0}{2\omega_n \sqrt{\zeta^2 - 1}} \right] e^{\left[-\zeta + \sqrt{\left(\zeta^2 - 1\right)} \right]\omega_n} \tag{9-23}$$

Case Scenario IV. Critically damped. $\zeta = 1$ (*critical damping*), $c^2 = 4MK$. The roots of the general solution equation shown in Equation (9-6) are as follows:

$$r_1 = r_2 = -\omega_n \tag{9-24}$$

The coefficients of the general solution equation shown in Equation (9-6) are as follows:

$$A = u_0, B = \dot{u}_0 + \omega_n u_0 \tag{9-25}$$

The solution equation is composed of two linearly independent solutions.

$$u(t) = (A + Bt)e^{-\omega_n t} \tag{9-26}$$

Here we have skipped some of the details of deriving solutions in four case scenarios. A few essential details of the derived formulations shown in Equations (9-7) to (9-26) can be found in [3].

A complete simulation model of the previous case scenarios can be programmed with a nested function. Out=SMDsysSim(M, D, S, ICs, t) embedded within the m-file SMDsysSIM_compare.m that simulates all four scenarios based on five input entries, namely, M, D, S, ICs, t. The input entries are mass (M), damping (D), stiffness (S), initial conditions (ICs), and time vector (t).

```
clearvars; close all
% Input Data for Scenarios I - IV.
M1=1; D1=0; S1=25;    % Scenario I. No damping
M2=1; D2=2; S2=25;    % Scenario II. Underdamped
M3=1; D3=13; S3=25;  % Scenario III. Over-damped
M4=1; D4=10; S4=25;  % Scenario IV. Critically damped
ICs=[0.1, 0]; t=0:.05:5;
% Displacement:
ut1=SMDsysSim(M1, D1, S1, ICs, t);
ut2=SMDsysSim(M2, D2, S2, ICs, t);
ut3=SMDsysSim(M3, D3, S3, ICs, t);
ut4=SMDsysSim(M4, D4, S4, ICs, t);
% Velocity:
```

```
dut1=diff(ut1);
dut2=diff(ut2);
dut3=diff(ut3);
dut4=diff(ut4);
figure(1)
plot(t, ut1, 'b-',  'linewidth', 1.5), hold on
plot(t, ut2, 'r-.', 'linewidth', 1.5)
plot(t, ut3, 'k:',  'linewidth', 2)
plot(t, ut4, 'g--', 'linewidth', 1.5)
xlabel('time, [s]'); ylabel('System response, [m]'), grid on,
title('1-DOF Spring-Mass-Damper System simulation - 4 Scenarios'),
legend('No damping: \zeta=0',' Underdamped: 0<\zeta<1',...
 'Overdamped: \zeta>1','Critically damped: \zeta=1'), hold off
figure(2)
plot(ut1(2:end), dut1, 'b-',  'linewidth', 1.5),
hold on
plot(ut2(2:end), dut2, 'r-.','linewidth', 1.5)
plot(ut3(2:end), dut3, 'k:', 'linewidth', 2)
plot(ut4(2:end), dut4, 'g--',  'linewidth', 1.5)
xlabel('time, [s]');
ylabel('System response, [m]')
grid on, title('Phase plot: Displacement vs. velocity')
legend('No damping: \zeta=0',' Underdamped: 0<\zeta<1',...
    'Overdamped: \zeta>1','Critically damped: \zeta=1')
xlim([-.1 .1]), ylim([-.025 .025]), axis square
hold off; shg


function Out=SMDsysSim(M, D, S, ICs, t)
% HELP. Simulation model of 1-DOF SMD system.
    omegaN = sqrt(S/M);
    ksi = 0.5*(D/M)*(1/omegaN);
    u0=ICs(1);
    du0=ICs(2);
if ksi==0                % Scenario I. Undamped
    Out=u0*cos(omegaN*t)+(du0/omegaN)*sin(omegaN*t);
```

```
elseif ksi>0 && ksi<1    % Scenario II. Underdamped
 omegaD=omegaN*sqrt(1-ksi^2);
 Out=exp(-ksi*omegaN*t).*(u0*cos(omegaD*t)+(u0*ksi*omegaN+du0)...
*sin(omegaD*t)/omegaD);
elseif ksi>1              % Scenario III. Overdamped
r1=(-ksi-sqrt(ksi^2-1))*omegaN;
r2=(-ksi+sqrt(ksi^2-1))*omegaN;
A=(-du0+(-ksi+(sqrt(ksi^2-1)))*u0*omegaN)/(2*omegaN*sqrt(ksi^2-1));
B=(du0+(ksi+(sqrt(ksi^2-1)))*u0*omegaN)/(2*omegaN*sqrt(ksi^2-1));
Out=(A*exp(r1*t)+B*exp(r2*t));
else  %ksi==1             % Scenario IV. Critically damped
  r1=-omegaN;
  A=u0;
  B=du0+omegaN*u0;
  Out=(A+B*t).*exp(r1*t);
end
end
```

The simulation results of this simulation model demonstrate the influence of the damping ratio values, ζ (Figure 9-2 and Figure 9-3) on the system behaviours. For instance, having ζ under a critical damping (under-damped case) value leads to decay (damping) of the system response, but the decay rate will be exponential depending on the value of ζ. If there is no damping present, then the system response oscillates at its natural frequency value of ω_n. Conversely, in over-damped and critically damped cases, the system response does not oscillate.

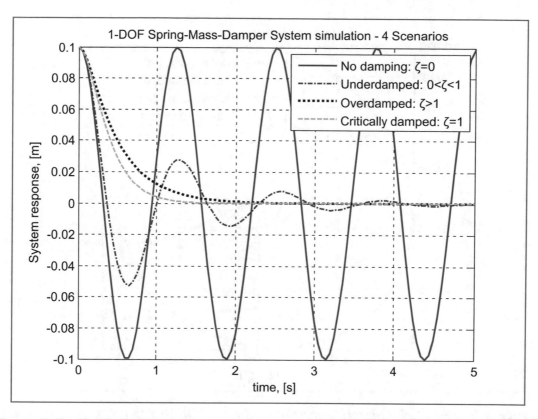

Figure 9-2. *Single-DOF spring-mass-damper system simulation for four different scenarios*

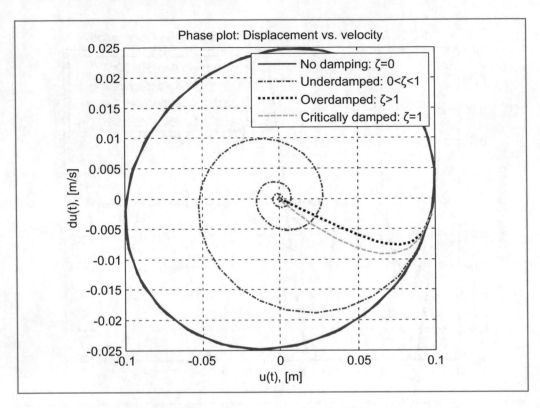

Figure 9-3. *Phase plot of four scenarios of single-DOF spring-mass-damper system*

Moreover, we can study the influence of initial conditions on the system response $u(t)$ to see how much time is required for the system to come to a complete stop. Now, by modifying the previous script, SMDsysSIM_compare.m, we can write another simulation model, SMDsysSIM_compare_ICs.m, that simulates three case scenarios: under-damped, critically damped, and over-damped with three different initial conditions.

```
%%  Study influence of Initial Conditions
clearvars; close all
% Input Data:
M=1; D=5; S=25; t=0:.05:2.5;
% Initial Conditions for three different cases:
ICs1=[0.1, 0];
ICs2=[0.1, 1];
ICs3=[0.1, -1];
% Displacement:
```

```
ut1=SMDsysSim(M, D, S, ICs1, t);
ut2=SMDsysSim(M, D, S, ICs2, t);
ut3=SMDsysSim(M, D, S, ICs3, t);
figure, subplot(311)
plot(t, ut1, 'b-', 'linewidth', 1.5), hold on
plot(t, ut2, 'r--', 'linewidth', 2),
plot(t, ut3, 'k:', 'linewidth', 2), grid on
legend('u_0=0.1, du_0=0','u_0=0.1, du_0=1','u_0= 0.1, du_0= -1')
ylabel(' u(t), [m]')
gtext({' \zeta < 1'}, 'fontsize', 13), hold off ,axis tight
% Input Data for Scenarios III. Critically damped
M2=1; D2=10; S2=25; t=0:.05:2.5;
ICs1=[0.1, 0]; ICs2=[0.1, 1]; ICs3=[0.1,-1];
% Displacement:
ut1=SMDsysSim(M2, D2, S2, ICs1, t);
ut2=SMDsysSim(M2, D2, S2, ICs2, t);
ut3=SMDsysSim(M2, D2, S2, ICs3, t);
subplot(312)
plot(t, ut1, 'b-', 'linewidth', 1.5), hold on
plot(t, ut2, 'r--', 'linewidth', 2),
plot(t, ut3, 'k:', 'linewidth', 2), grid on
legend('u_0=0.1, du_0=0','u_0=0.1, du_0=1','u_0=0.1, du_0= -1')
ylabel(' u(t), [m]')
gtext({' \zeta = 1'}, 'fontsize', 13), axis tight; hold off
% Input Data for Scenario IV. Over-damped case
M3=1; D3=13; S3=25; t=0:.05:2.5;
ICs1=[0.1, 0]; ICs2=[0, 0.5]; ICs3=[-0.1, 0];
% Displacement:
ut1=SMDsysSim(M3, D3, S3, ICs1, t);
ut2=SMDsysSim(M3, D3, S3, ICs2, t);
ut3=SMDsysSim(M3, D3, S3, ICs3, t);
subplot(313), plot(t, ut1, 'b-', 'linewidth', 1.5), hold on
plot(t, ut2, 'r--', 'linewidth', 2),
plot(t, ut3, 'k:', 'linewidth', 2), grid on
legend('u_0=0.1, du_0=0','u_0=0, du_0=0.5','u_0= -0.1, du_0=0')
```

```
xlabel('t, [s]'); ylabel(' u(t), [m]')
gtext({'\zeta >1'}, 'fontsize', 13), hold off, axis tight
function Out=SMDsysSim(M, D, S, ICs, t)
% HELP. Simulation model of 1-DOF SMD system.
% Out=SMDsysSim(M, D, S, ICs, t)
    omegaN = sqrt(S/M);
    ksi = 0.5*(D/M)*(1/omegaN);
    u0=ICs(1);
    du0=ICs(2);
if ksi==0                    % Scenario I. Undamped
    Out=u0*cos(omegaN*t)+(du0/omegaN)*sin(omegaN*t);
elseif ksi>0 && ksi<1    % Scenario II. Underdamped
  omegaD=omegaN*sqrt(1-ksi^2);
  Out=exp(-ksi*omegaN*t).*(u0*cos(omegaD*t)+(u0*ksi*omegaN+du0)...
*sin(omegaD*t)/omegaD);
elseif ksi>1                 % Scenario III. Overdamped
r1=(-ksi-sqrt(ksi^2-1))*omegaN;
r2=(-ksi+sqrt(ksi^2-1))*omegaN;
A=(-du0+(-ksi+(sqrt(ksi^2-1)))*u0*omegaN)/(2*omegaN*sqrt(ksi^2-1));
B=(du0+(ksi+(sqrt(ksi^2-1)))*u0*omegaN)/(2*omegaN*sqrt(ksi^2-1));
Out=(A*exp(r1*t)+B*exp(r2*t));
else  %ksi==1             % Scenario IV. Critically damped
  r1=-omegaN;
  A=u0;
  B=du0+omegaN*u0;
  Out=(A+B*t).*exp(r1*t);
end
end
```

As demonstrated in Figure 9-4, the simulation results for three different values of ζ show that if there is considerable damping present in the system, free motion of a system decays over time. Nevertheless, the whole decay (die) time of free motion depends on the initial conditions, i.e., initial excitation values (displacement and velocity of the system). The previously discussed simulation model approaches for free vibration (motion) of a single DOF system can be also applied with some changes to free vibration of multidegree of freedom (linear) system simulations.

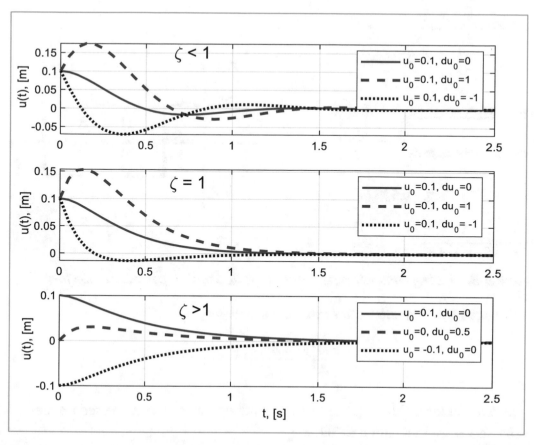

Figure 9-4. *Study the influence of initial conditions of the system on its responses*

Case 2: Forced Vibration (Motion)

Let's consider the case of the single-DOF SMD system excited with an external force that is a rectangular type of periodic wave force, as shown in Figure 9-5.

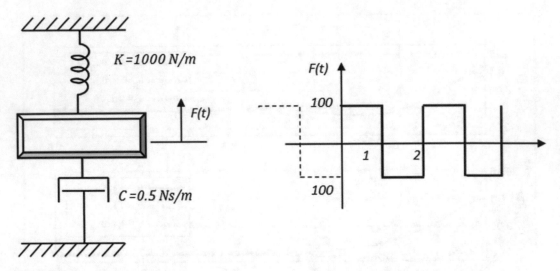

Figure 9-5. *Spring-mass-damper system excited with a periodic wave force*

The equation of motion of the system is as follows:

$$\frac{md^2 y}{dt^2} + \frac{Cdy}{dt} + Ky = F(t) \tag{9-27}$$

where external force $F(t)$ is a periodic wave function that can be expressed via the Fourier series approximation as follows:

$$F(t) = \sum_{n=1}^{\infty}(1 - \cos n\pi)\sin(n\pi t) = \frac{400}{\pi}\sin(\pi t) - \frac{400}{3\pi}\sin(3\pi t) + \frac{80}{\pi}\sin(5\pi t) - \cdots \tag{9-28}$$

Now, the final formulation of the system model in Equation (9-27) will be as follows:

$$\frac{md^2 y}{dt^2} + \frac{Cdy}{dt} + Ky = \frac{400}{\pi}\sin(\pi t) - \frac{400}{3\pi}\sin(3\pi t) + \frac{80}{\pi}\sin(5\pi t) - \cdots \tag{9-29}$$

First simulation model. Using Equation (9-29), we create a simulation model with MATLAB scripts. The periodic wave function formulation in Equation (9-28) can be expressed as a function called F=func_pulse(t, n).

```
function F=func_pulse(t, n)
%HELP. t is time vector. n is number of terms in pulse approximation.
F(1,:)=(200/pi)*(1-cos(pi))*sin(pi*t);
for ii=2:n
    F(:,:)=F(:,:)+(200/(ii*pi))*(1-cos(ii*pi))*sin(ii*pi*t);
end
```

We express the second-order ODE, which is the left side of Equation (9-29), via the anonymous function handle DDx and embed the force function F=func_pulse(t, n) within an m-file called Sim_func_pulse.m with the ode45 solver. This simulates the system behavior and plots the simulation results. Note that in this simulation model, zero initial conditions are taken. The simulation results of the model aer shown in Figure 9-6 and Figure 9-7.

```
% F = func_pulse(t, n) produces rectangular periodic pulses.
% DDx = @(t, x) anonymous function handle.
clearvars; close all
M = 10; C = 0.5; K = 1000; % System parameters
t=0:pi/1000:13;            % Time
n=113;                     % Terms for Fourier Series
ICs=[0,0];                 % Initial Conditions
F=func_pulse(t, n);        % Fourier Series values
% System model equation is expressed by two 1st order ODEs
DDx=@(t,x)([x(2); (1/10)*(func_pulse(t,n)-C*x(2)-K*x(1))]);
opts_ODE=odeset('RelTol', 1e-6, 'AbsTol', 1e-8);
[time, Xt]=ode45(DDx, t, ICs, opts_ODE);
subplot(211)
yyaxis left
plot(t, F, 'k-');
ylabel('Input F(t), [N]')
yyaxis right
plot(t, Xt(:,1), 'r--')
ylabel('Output x(t), [m]')
legend('F(t) Input', 'x(t) Output'), grid on
title(['\it SMD system excited ',...
'by periodic pulse force'])
axis tight
```

```
subplot(212)
yyaxis left
plot(t,Xt(:,1),'k-')
ylabel('Displacement x(t), [m]')
yyaxis right
plot(t, Xt(:,2),'b-.','linewidth', 1.5)
legend('x(t)','dx(t)')
ylabel('Velocity dx(t), [m/s]')
xlabel('time [sec]'), grid on, axis tight
figure
plot(Xt(:,1), Xt(:,2)),
xlabel('x(t)'), ylabel('dx(t)')
title('Phase plot; x vs. dx'), grid on
axis square
function F=func_pulse(t,n)
%HELP. t is time vector. n is number of terms in pulse approximation.
F(1,:)=(200/pi)*(1-cos(pi))*sin(pi*t);
for ii=2:n
    F(:,:)=F(:,:)+(200/(ii*pi))*(1-cos(ii*pi))*sin(ii*pi*t);
end
end
```

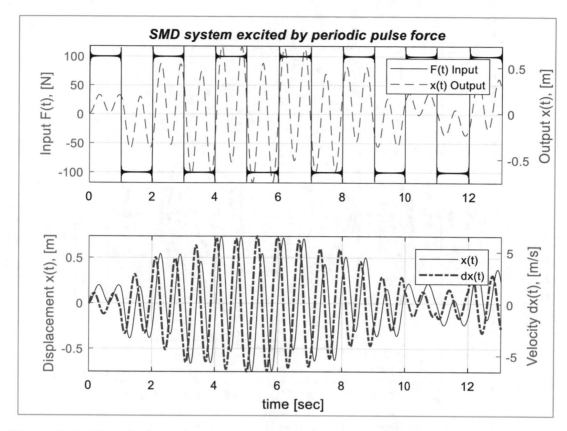

Figure 9-6. *Simulation of spring-mass-damper system excited with a periodic pulse force*

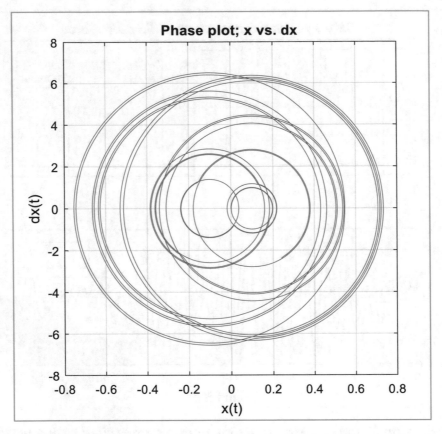

Figure 9-7. *Phase plot: displacement versus velocity of the SMD system excited with a periodic rectangular wave force*

Now let's employ the Simulink modeling approach to simulate the system behavior.

The second simulation model is built in Simulink. In Simulink, there are a few blocks with which periodic pulse signals can be generated (see [4] for more details about how to build Simulink models). They are signal generator, pulse generator, series of step function, signal builder, and repeating sequence. Pulses__Input.mdl (shown in Figure 9-8) is the final model of the spring-mass-damper system subject to external force, namely, repeated pulses.

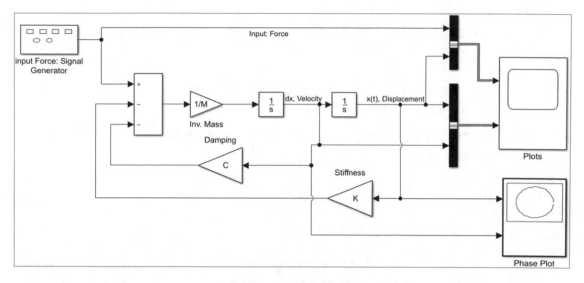

Figure 9-8. *Simulink model (`Pulses__Input.mdl`) of the spring-mass-system subject to a pulse external force*

The simulation results (input signal versus system response, displacement versus velocity in the Plots–Scope block, displacement versus velocity in the Phase Plot block) of the Simulink model (`Pulses__Input.mdl`) shown in Figure 9-8 are well converged with the results in Figures 9-6 and 9-7 from the m-file `Sim_func_pulse.m`. The simulation results obtained from the Scope block called [Plots] and the XY Scope block called [Phase Plot] are not shown here. The x-y axis limits of the XY Scope block, shown in [Phase Plot], are set to [-0.8 0.8 -8 8]. Note that the parameters of the [Signal Generator] block are as follows:

Output various wave forms: Y(t) = Amp∗Waveform(Freq, t)

Wave form: square

Time (t): Use simulation time

Amplitude: -100

Frequency: 0.5

Units: Hertz

The values of the other gain parameters *M, C,* and *K,* of the model (`Pulses__Input.mdl`), are 10, 0.5, and 1000, respectively.

Let's consider another case with a sinusoidal periodic forced motion of a single-DOF SMD system with $F(t) = F_0 \cos(\omega t)$.

315

Equation (9-27), which is $\dfrac{md^2y}{dt^2} + \dfrac{Cdy}{dt} + Ky = F(t)$, has a general solution, shown here:

$$x(t) = Ae^{-\zeta\omega_n t}\sin(\omega_d t + \varphi) + X\cos(\omega t - \theta) \tag{9-30}$$

where ω is the applied force's frequency and φ is the phase shift. The constants A and X are computed from the initial conditions [1, 2, 3].

$$A = \frac{x_0 - X\cos\theta}{\sin\varphi} \tag{9-31}$$

$$X = \frac{f_0}{\sqrt{\left(\omega_n^2 - \omega^2\right)^2 + \left(2\zeta\omega_n\omega\right)^2}} \tag{9-32}$$

where $f_0 = \dfrac{F_0}{m}$ is normalized force magnitude. The phase θ is computed from the following formulation:

$$\theta = \tan^{-1}\frac{2\zeta\omega_n\omega}{\omega_n^2 - \omega^2} \tag{9-33}$$

The previous formulations of magnitude X and phase θ are of great importance in general practice. The transient response $x_t(t) = Ae^{-\zeta\omega_n t}\sin(\omega_d t + \varphi)$ part of the general solution is relatively small in amplitude, and its influence on the response is considerably smaller than the steady-state response magnitude. Therefore, in practice, we deal with the steady-state response part of the solution, which is $x(t) = X\cos(\omega t - \theta)$.

From the formulations of X and θ, by factoring out ω^2, dividing the magnitude by $\dfrac{F_0}{m}$ or f_0, and substituting $r = \dfrac{\omega}{\omega_n}$, the following expressions of the normalized magnitude of the system response and phase are obtained:

$$\frac{XK}{F_0} = \frac{X\omega_n^2}{f_0} = \frac{1}{\sqrt{\left(1 - r^2\right)^2 + \left(2\zeta r\right)^2}} \tag{9-34}$$

$$\theta = \tan^{-1}\frac{2\zeta r}{1 - r^2} \tag{9-35}$$

These formulations are simulated with the following script (zeta_1DOF_SMD.m):

```
function zeta_1DOF_SMD(M, zeta, S, omega)
% HELP. Study damping ratio (zeta) effects on Amplitude change
% M - Mass of the system
% zeta - Damping ratio
% S - Stiffness of the system
% omega - frequency
if nargin<1
M=2.5; S=25;
omega=0:.2:10;
zeta=0:0.2:1;
end
omegaN=sqrt(S/M);
r=omega./omegaN;
Mag=zeros(length(zeta), length(r));
Theta=ones(length(zeta), length(r));
for ii=1:length(zeta)
    for k=1:length(r)
        Mag(ii,k)=1/sqrt((1-r(k)^2)^2+(2*zeta(ii)*r(k))^2);
        Theta(ii,k)=atan2(2*zeta(ii)*r(k), (1-r(k)^2));
    end
end
Labelit = {};
Colorit = 'bgrmkgbckmbgrygr';
Lineit  = '--:-:--:-:--:----:----:--';
Markit  = 'oxs+*^v<p>.xsh+od+*^v';
for m=1:length(zeta)
    Stylo    = [Colorit(m) Lineit(m) Markit(m)];
    Labelit{m} = ['\zeta= ' num2str(zeta(m))];
    semilogy(r, Mag(m,:), Stylo, 'linewidth', 1.5), hold on
end
```

```
legend(Labelit{:}); axis tight; grid on; ylim([0, 5])
title('\zeta influence on Response Amplitude change')
xlabel('r = \omega/\omega_n'), ylabel('\it Normalized Amplitude'),
hold off
figure
for m=1:length(zeta)
    Stylo     = [Colorit(m) Lineit(m) Markit(m)];
    Labelit{m} = ['\zeta= ' num2str(zeta(m))];
    plot(r, Theta(m,:), Stylo, 'linewidth', 1.5)
    hold on
end
legend(Labelit{:}, 'location', 'northwest')
axis tight, grid on;
axis([0, 2 -.2 3.35])
title('\zeta  effects on Phase changes')
xlabel('r = \omega/\omega_n ')
ylabel('Phase, \theta(r)')
```

Again, in our simulation studies of damping ratio (ζ) with the normalized system amplitude (Figure 9-9) and the phase shift (Figure 9-10), we can conclude that the resonance peak occurs when the system's natural frequency equals the periodic excitation force's excitation frequency as expected in theoretical evaluations, and the system behavior will become uncontrollable. To avoid such circumstances, we need to apply damping. Note that in the script, the function atan2 is used to evaluate the phase angle $\theta(r)$ values since it computes the values of $\theta(r)$ in the range of $[0, \pi]$, whereas the tan function computes in the range of $\left[-\dfrac{\pi}{2}, \dfrac{\pi}{2} \right]$, which is not appropriate in this case.

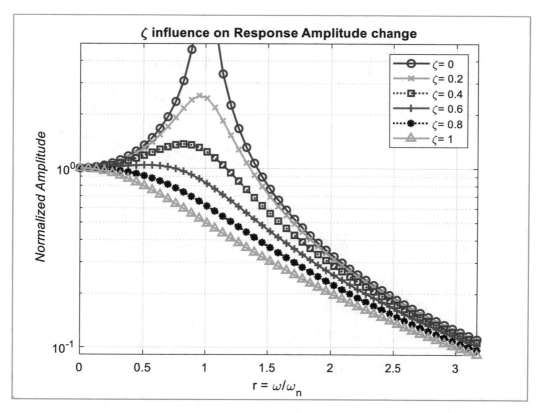

Figure 9-9. *Influence of the damping ratio (ζ) on the normalized system response amplitude*

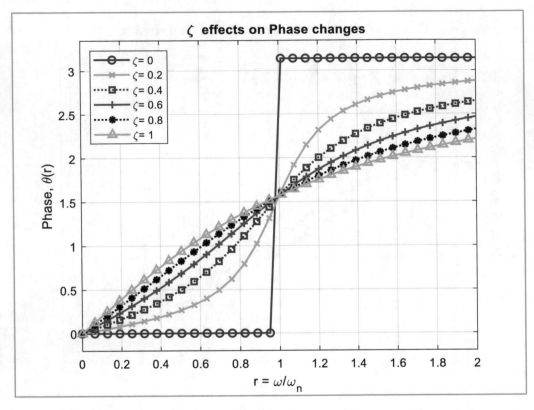

Figure 9-10. *Influence of the damping ratio (ζ) on the system phase θ change*

Two Degrees of Freedom System

The motions of many mechanical systems are studied by modeling them as SMD systems. Let's consider the two-DOF SMD system subject to external force shown in Figure 9-11. Note that the friction under mass 1 and 2 is taken to be linear.

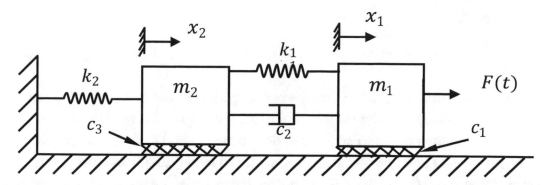

Figure 9-11. *Two mass spring-mass-damper system subject to external impulse force F(t)*

Before we start to derive the system equations, Figure 9-12 shows an alternative equivalent physical model of the given system by substituting the friction damping under body mass 1 and 2 with dampers applied to each mass.

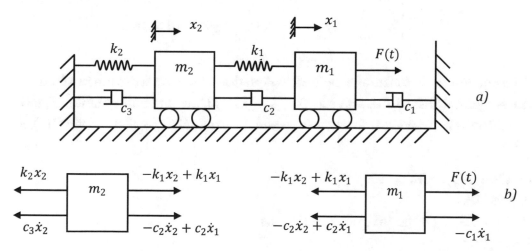

Figure 9-12. *a) Alternative physical model of the system; b) free-body diagrams of mass 1, 2*

Using the free-body diagram of the system and applying D'Alembert's principle for acting forces on each mass, we can write the equations of motion of the system.

For mass 1:

$$F(t) = m_1 \ddot{x}_1 + (c_1 + c_2)\dot{x}_1 - c_2 \dot{x}_2 + k_1 x_1 - k_1 x_2 \tag{9-36}$$

For mass 2:

$$0 = m_2\ddot{x}_2 + (c_2 + c_3)\dot{x}_2 + (k_1 + k_2)x_2 - c_2\dot{x}_1 - k_1x_1 \tag{9-37}$$

Note that m_1, m_2, c_1, c_2, c_3, k_1, k_2 are the system parameters of mass, damping, and stiffness, respectively. x_1 and x_2 are displacements of mass 1 and 2, respectively, and $\dot{x}_1, \dot{x}_2, \ddot{x}_1, \ddot{x}_2$ are the velocities and accelerations of mass 1 and mass 2, respectively.

By introducing new state variables, we rewrite Equations (9-36) and (9-37) as a system of first-order ODEs. To avoid confusion, the variables are renamed as $x_1 = y_1$ and $x_2 = y_2$.

$$\begin{cases} \dot{x}_1 = y_2 \\ \dot{y}_2 = \dfrac{1}{m_1}\Big[F(t) - (c_1 + c_2)y_2 + c_2y_4 - k_1y_1 + k_1y_3\Big] \\ \dot{x}_2 = y_4 \\ \dot{y}_4 = \dfrac{1}{m_2}\Big[c_2y_2 - (c_2 + c_3)y_4 + k_1y_1 - (k_1 + k_2)y_3\Big] \end{cases} \tag{9-38}$$

Based on the system model defined in Equation (9-38), we write the MATLAB models, named SMsysSIM.m and two_DOF_sys_sim_ALL.m, and build a Simulink model named Two_DOF_SMsys.mdl for the case when the external force, $F(t) = F_0 \sin(10t)$, is applied.

```
function SMsysSIM(t, ICs)
% SMsysSIM.m
% t is time; ICs is initial conditions.
if nargin<1
% If no input arguments and Initial conditions are given, then
t=[0,5]; ICs=[1; 1.5; -1; 1];
OPTs=odeset('OutputFcn', @odeplot,'reltol', 1e-4,'abstol',1e-6);
ode45(@SMsys1,t, ICs, OPTs)
else
Opts=odeset('OutputFcn', @odeplot,'reltol', 1e-4,'abstol',1e-6);
ode45(@SMsys1,t, ICs, Opts)
end
```

```
function DX=SMsys1(t, y)
k1=90; k2=85; c1=2; c2=1; c3=3; m1=5; m2=3; F0=3.3;
Dx(1)= y(2);
Dx(2)=(1/m1)*(F0*sin(10*t)-(c1+c2)*y(2)+c2*y(4)-k1*y(1)+k1*y(3));
Dx(3)= y(4);
Dx(4)= (1/m2)*(c2*y(2)-(c2+c3)*y(4)+k1*y(1)-(k1+k2)*y(3));
DX=[Dx(1); Dx(2); Dx(3); Dx(4)];
end
end
```

This function file (SMsysSIM.m) computes the displacement and velocity values of mass 1 and 2 and plots them in a single plot without figure attributes, such as plot title, legends, etc.

The next m-file (two_DOF_sys_sim_ALL.m) computes the displacement and velocity values of mass 1 and 2 with the ODE solvers (ode23, ode23s, ode23tb, and ode113) and the Simulink model (Two_DOF_SMsys.mdl) and plots them separately with all the plot attributes, as shown in Figure 9-14, Figure 9-15, and Figure 9-16.

```
% two_DOF_sys_sim_ALL.m
clearvars; close all
ICs=[1; 1.5; -1; 1];
k1=90; k2=85; c1=2; c2=1; c3=3; m1=5; m2=3; F0=3.3;
F=@(t, y)[y(2);(1/m1)*(F0*sin(10*t)-(c1+c2)*y(2)+c2*y(4)- ...
    k1*y(1)+k1*y(3)); y(4); ...
(1/m2)*(c2*y(2)-(c2+c3)*y(4)+k1*y(1)-(k1+k2)*y(3))];
t=0:.005:5;
Opts=odeset('OutputFcn',@odeplot,'reltol', 1e-4, 'abstol', 1e-6);
ode45(F, t, ICs, Opts)
xlabel('\it time'), ylabel('x_1, dx_1, x_2,  dx_2'), grid on
legend('x_1(t)', 'dx_1(t)', 'x_2(t)', 'dx_2(t)')
title('Simulation of 2-DOF system')
%% Testing different solvers with different ODE settings
% ODE45 with fixed time step of 0.005
ts=0:.005:5; % fixed time step of 0.005
[t1, y1]=ode45(F, ts, ICs, []);
%% ODE23tb with variable time step
```

```
t=[0, 5];
[t2, y2]=ode23tb(F, t, ICs, []);
%% ODE23 calls an embedded function SMsys. A fixed time step of 0.005
[t3, y3]=ode23(@SMsys, ts, ICs);
%% ODE113 with the varying step size and function handle
[t4, y4]=ode113(@(t, y)[y(2); (1/m1)*(F0*sin(10*t)- ...
    (c1+c2)*y(2)+c2*y(4)-k1*y(1)+k1*y(3)); y(4);
(1/m2)*(c2*y(2)-(c2+c3)*y(4)+k1*y(1)-(k1+k2)*y(3))],t, ICs);
%% SIMULINK model: Two_DOF_SMsys.slx with ODE23S & rel. error tol. 1e-6
OPTs=simset('reltol', 1e-6, 'solver', 'ode23s');
XOUT=sim('Two_DOF_SMsys', t, OPTs); % Simulates model: Two_DOF_SMsys.slx
figure('name', 'Displacement: x1')
plot(t1,y1(:,1), 'b-o'), hold on
plot(t2,y2(:,1), 'r--s')
plot(t3,y3(:,1), 'm-.x'),
plot(t4,y4(:,1), 'g:+')
plot(XOUT.tout, XOUT.yout{1}.Values.Data, 'k-'), grid on
legend('ode45','ode23tb','ode23',...
    'ode113','Simulink(ode23s ~var)')
title('Displacement of body mass: m_1')
xlabel('time, [sec]'), ylabel('x_1(t)')
figure('name', 'Displacement: x2')
plot(t1,y1(:,3), 'b-o'), hold on
plot(t2,y2(:,3), 'r--s')
plot(t3,y3(:,3), 'm-.x'),
plot(t4,y4(:,3), 'g:+')
plot(XOUT.tout, XOUT.yout{2}.Values.Data, 'k-'), grid on
legend('ode45','ode23tb','ode23',...
    'ode113','Simulink(ode23s ~var)')
title('Displacement of body mass: m_2')
xlabel('time, [sec]'), ylabel('x_2(t)')
% Function called by ODE23TB
function DX=SMsys(t,y)
k1=90; k2=85; c1=2; c2=1; c3=3; m1=5; m2=3; F0=3.3;
```

```
Dx(1)= y(2);
Dx(2)=(1/m1)*(F0*sin(10*t)-(c1+c2)*y(2)+c2*y(4)-k1*y(1)+k1*y(3));
Dx(3)= y(4);
Dx(4)= (1/m2)*(c2*y(2)-(c2+c3)*y(4)+k1*y(1)-(k1+k2)*y(3));
DX=[Dx(1); Dx(2); Dx(3); Dx(4)];
end
```

Note that the Simulink model (Two_DOF_SMsys.mdl) takes all the parameter (m1, m2, k1, k2, c1, c2, c3, F0) values and initial condition values (defined in the integrator blocks with these variable names: ICs(1), ICs(2), ICs(3), ICs(4)) from the simulation m-file called two_DOF_sys_sim_ALL.m. Note that the input plot (the sine wave in the Simulink model Two_DOF_SMsys.mdl) has Amplitude set to F0 and Frequency (rad/sec) set to 10.

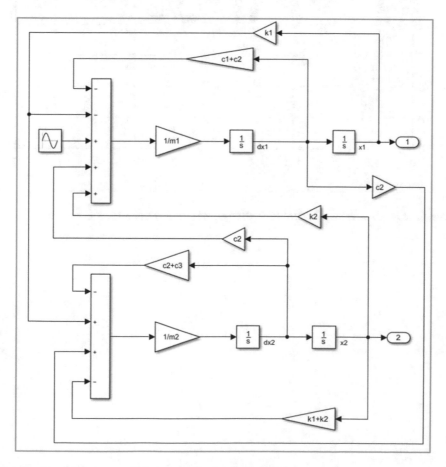

Figure 9-13. *Simulink model, Two_DOF_SMsys.slx*

We run the model two_DOF_sys_sim_ALL.m that automatically calls/simulates the Simulink model (Two_DOF_SMsys.mdl) and function file. Subsequently, we obtain the next plot figures (Figure 9-14, Figure 9-15, and Figure 9-16) displaying the simulation results.

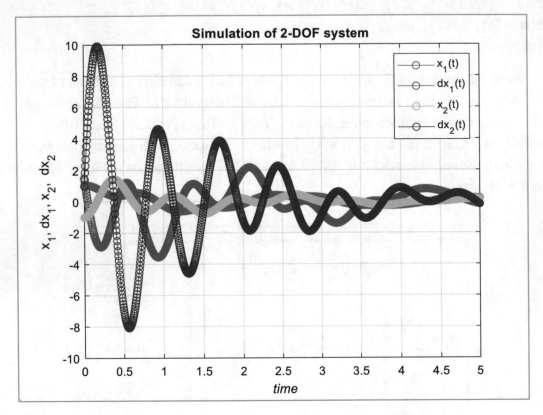

Figure 9-14. *Displacement and velocity of mass 1 and mass 2 bodies*

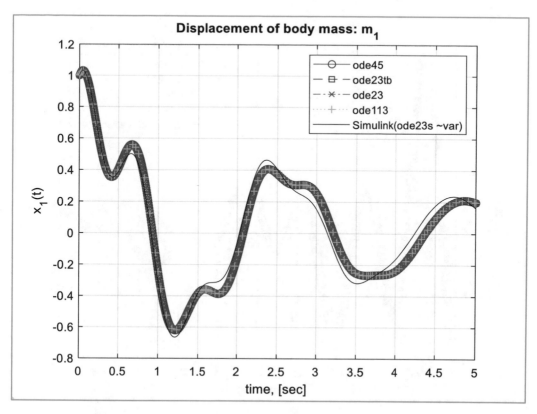

Figure 9-15. *Displacement of body mass 1 computed with different ODE solvers of MATLAB/Simulink*

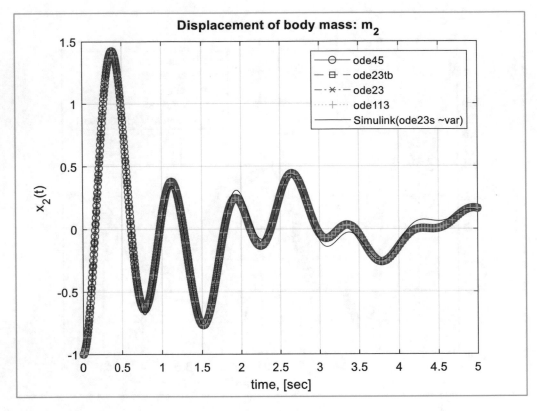

Figure 9-16. *Displacement of body mass 2 computed with different ODE solvers of MATLAB/Simulink*

The simulation results in Figure 9-14 and 9-15 show that all of the employed ODE solvers (ode23, ode23tb, ode23s, ode45, ode113) have resulted in well-converged solutions.

Three Degrees of Freedom System

Let's consider the three-DOF spring-mass-damper system shown in Figure 9-17 that have three masses interconnected with springs and dampers. It is similar to the previous two DOF systems. We will derive system equations of motion and build simulation models.

The equations of motion of the system with respect to its free-body diagrams (Figure 9-17) can be expressed with the system of second-order differential equations shown in Equation (9-39):

$$\begin{cases} m_1\ddot{x}_1 + k_1 x_1 + k_2\left(x_1 - x_2\right) + c_1\dot{x}_1 + c_2\left(\dot{x}_1 - \dot{x}_2\right) = F_1(t) \\ m_2\ddot{x}_2 + k_2\left(x_2 - x_1\right) + k_3\left(x_2 - x_3\right) + c_2\left(\dot{x}_2 - \dot{x}_1\right) + c_3\left(\dot{x}_2 - \dot{x}_3\right) = F_2(t) \\ m_3\ddot{x}_3 + k_3\left(x_3 - x_2\right) + 2k_4 x_3 + c_3\left(\dot{x}_3 - \dot{x}_2\right) = F_3(t) \end{cases} \quad (9\text{-}39)$$

Equation (9-39) can be rewritten as follows:

$$\begin{cases} m_1\ddot{x}_1(t) + \left(k_1 + k_2\right)x_1(t) - k_2 x_2(t) + \left(c_1 + c_2\right)\dot{x}_1(t) - c_2\dot{x}_2(t) = F_1(t) \\ m_2\ddot{x}_2(t) - k_2 x_1(t) + \left(k_2 + k_3\right)x_2(t) - k_3 x_3(t) \\ \qquad + \left(c_2 + c_3\right)\dot{x}_2(t) - c_2\dot{x}_1(t) - c_3\dot{x}_3(t) = F_2(t) \\ m_3\ddot{x}_3(t) - k_3 x_2(t) + \left(k_3 + 2k_4\right)x_3(t) - c_3\dot{x}_2(t) + c_3\dot{x}_3(t) = F_3(t) \end{cases} \quad (9\text{-}40)$$

Note that m_1, m_2, m_3, c_1, c_2, c_3, k_1, k_2, and k_3 are the system parameters of mass, damping and stiffness, respectively. x_1, x_2, and x_3 are the displacements of mass 1, 2, and 3, respectively. $\dot{x}_1$, $\dot{x}_2$, and $\dot{x}_3$ are the velocities, and $\ddot{x}_1$, $\ddot{x}_2$, and $\ddot{x}_3$ are the accelerations of mass 1, 2, 3, respectively. $F_1(t)$, $F_2(t)$, and $F_3(t)$ are applied external forces on mass 1, 2, 3, respectively.

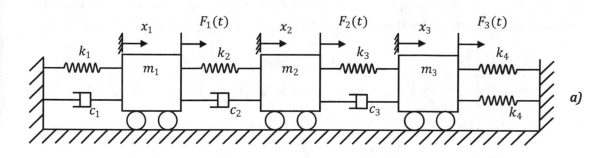

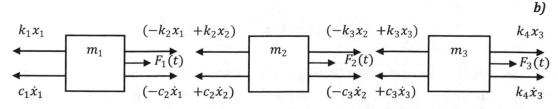

Figure 9-17. *Three degrees of freedom system: a) a physical model of the system with three masses rolling on a flat surface without any friction; b) free body diagram of the system*

Note that the system equations in Equation (9-39) and Equation (9-40) are second-order ODEs, which are linear and time invariant. They are coupled by the coordinate systems of $x_1(t)$, $x_2(t)$, and $x_3(t)$. Therefore, these equations must be solved simultaneously. For numerical simulations, the following values for the system parameters are taken: $m_1 = 2.5$ [kg]; $m_2 = 2.0$ [kg]; $m_3 = 3.0$ [kg];

$$k_1 = 25\left[\frac{N}{m}\right]; k_2 = 5\left[\frac{N}{m}\right]; k_3 = k_4 = 30\left[\frac{N}{m}\right]; c_1 = 2.5\left[\frac{Ns}{m}\right]; c_2 = 0.5\left[\frac{Ns}{m}\right]; c_3 = 3.5\left[\frac{Ns}{m}\right].$$

We simulate this system with forced and free motion.

- **Forced motion** (vibration) case with zero initial conditions:

  e.g.: $F_1(t) = 20 \sin(120t)$; $F_2(t) = 0$; $F_3(t) = 20\Phi(t - t_0)$;

  Note that $F_3(t)$ is a Heaviside step function $\Phi(t - t_0)$ applied at $t_0 = 0$.

- **Free motion** (vibration) case. No external forces are applied: $F_1(t) = F_2(t) = F_3(t) = 0$;. There is an initial (excitation) displacement applied on mass 1 only:

  $x_1(0) = 0.5$, $x_2(0) = 0$, $x_3(0) = 0$; $\dot{x}_1(0) = 0$, $\dot{x}_2(0) = 0$, $\dot{x}_3(0) = 0$; .

Note that before writing a simulation model, we rewrite the system equations in Equation (9-39) via first-order ODEs. The system has three second-order (acceleration) variables, $\ddot{x}_1$, $\ddot{x}_2$, and $\ddot{x}_3$, which will be substituted with new variables, i.e., $x_1 = y_1$, $\dot{x}_1 = y_2$, $\ddot{x}_1 = \dot{y}_2$; $x_2 = y_3$, $\dot{x}_2 = y_4$, $\ddot{x}_2 = \dot{y}_4$; $x_3 = y_5$, $\dot{x}_3 = y_6$, and $\ddot{x}_3 = \dot{y}_6$. Using these new variables, we can write the next system of first-order coupled ODE equations.

$$\begin{cases} \dot{x}_1 = y_2 \\ \dot{y}_2 = \left(\frac{1}{m_1}\right)\left(F_1(t) - (k_1 + k_2)y_1 + k_2 y_3 - (c_1 + c_2)y_2 + c_2 y_4\right) \\ \dot{x}_2 = y_4 \\ \dot{y}_4 = (1/m_2)\left(F_2(t) + k_2 y_1 - (k_2 + k_3)x_3 + k_3 y_5 - (c_2 + c_3)y_4\right) \\ \dot{x}_3 = y_6 \\ \dot{y}_6 = \left(\frac{1}{m_3}\right)\left(F_3(t) + k_3 y_3 - (k_3 + 2k_4)y_5 + c_3 y_4 - c_3 y_6\right) \end{cases} \quad (9\text{-}41)$$

Based on the formulations in Equation (9-41), we write a function file called three_MASS.m with the ode45 solver. Note that the function file contains a nested function called third_DOF(t, y) that defines the system equations given in Equation (9-41). The system parameter values and external forces are also defined within this nested function. The Heaviside step function is expressed via MATLAB's built-in function step().

330

```
% three_MASS.m
% HELP. Simulation of highly coupled three degree of freedom system.
clearvars; clc
t=0:.01:120;  % Simulation time span
IC=[0;0;0;0;0;0];            % Case # 1
% IC=[0.5; 0; 0; 0; 0; 0];  % Case # 2

F1=20*sin(120*t);F2=0; t0=0; F3=20*stepfun(t,t0); % Case # 1
% F1=0; F2=F1; F3=F1;                             % Case # 2
k1=25; k2=5; k3=30; k4=k3; c1=2.5; c2=.5; c3=3.5;
m1=2.5; m2=2; m3=3;
[t, xyz]=ode45(@third_DOF, t, IC, []);
% SIMULINK model: ThreeDOFsys.slx with ODE113 & relative error tol. 1e-6
OPTs=simset('reltol', 1e-6, 'solver', 'ode113');
[tt, YY, X123]=sim('ThreeDOFsys', [0, 20], OPTs);% Run: ThreeDOFsys.slx

figure('name', 'Matlab vs. Simulink')
plot(t(1:2000),xyz(1:2000,1),'bo',t(1:2000),xyz(1:2000,3), ...
    'rx', t(1:2000), xyz(1:2000,5), 'ks', ...
    tt(1:5000),X123(1:5000,1),'r-',tt(1:5000),X123(1:5000,2), ...
    'b--', tt(1:5000), X123(1:5000,3), 'g-.','linewidth',1.5)
legend('Mass 1: x_1(t) (Matlab)','Mass 2: x_2(t) (Matlab)', ...
'Mass 3: x_3(t) (Matlab)', ...
'Mass 1: x_1(t) (Simulink)', 'Mass 2: x_2(t) (Simulink)', ...
'Mass 3: x_3(t) (Simulink)')
title('\it Three DOF SMD (coupled) system')
xlabel('\it time, [sec]')
ylabel('\it Displacement, x_1(t), x_2(t), x_3(t)'), grid on
figure('name', 'Animation')

for k=1:t(end)-20
% Note: Displacement magnitudes of x1, x2, x3 increased
% by the factor of 10 for better visualization
plot(-20,0,'*', 20,0, '*', -10+10*xyz(k+20,1), 0, 'r-o',...
  -5+10*xyz(k+20,3),0,'g-o',...
10+10*xyz(k+20,5),0,'b-o','markersize', 35, ...
  'markerfacecolor', 'c');
```

```
grid on; Motion(k)=getframe;
end
movie(Motion)
function  dx = third_DOF(t, y)
% HELP: three degree of freedom system
k1=25; k2=5; k3=30; k4=k3; c1=2.5; c2=.5; c3=3.5;
m1=2.5; m2=2; m3=3;
F1=20*sin(120*t);F2=0; t0=0; F3=20*stepfun(t,t0);      % Case # 1
% F1=0; F2=0; F3=0;                                    % Case # 2
dx=zeros(6,1);
dx(1)=y(2);
dx(2)=(1/m1)*(F1-(k1+k2)*y(1)+k2*y(3)-(c1+c2)*y(2)+c2*y(4));
dx(3)=y(4);
dx(4)=(1/m2)*(F2+k2*y(1)-(k2+k3)*y(3)+k3*y(5)-...
    (c2+c3)*y(4)+c2*y(2)+c3*y(6));
dx(5)=y(6);
dx(6)=(1/m3)*(F3+k3*y(3)-(k3+2*k4)*y(5)+c3*y(4)-c3*y(6));
end
end
```

This simulation model includes forced (Case #1) and free (Case #2) motion inputs, such as input forces and nonzero initial conditions (for mass 1) for forced and free motion cases, respectively.

Moreover, it incorporates the commands to call and execute the Simulink model of the system, as shown in Figure 9-18. Note that the Simulink model, ThreeDOFsys.slx, takes the numerical values of all the parameters, and the initial conditions are set inside the integrator blocks from the simulation function file, three_MASS.m. The external force magnitudes of the Simulink model are initiated by selecting Model Properties ➤ Callbacks ➤ InitFcn.

F1 = 30; F2=0; F3=30; % for Case #1

F1 = 0; F2=0; F3=0; % for Case #2

Note that in the Simulink model ThreeDOFsys.slx, the [Sine Wave] block is used for an input force signal $F_1(t)$ applied on mass 1 with the following parameters: *Amplitude =1, Frequency=120, Phase=0, Bias =0, Sample time=0.* It has the following output formulation: $O(t)$ = *Amplitude* $*$ sin (*Frequency* $* t +$ *Phase*) + *Bias.* The input force signal is $F_1(t) = O(t) * F1$, where F1 is a constant gain value. The force applied on mass 3 has a constant value equal to F3. The force applied on mass 2 is 0. Thus, F2 = 0, and it is ignored in the model.

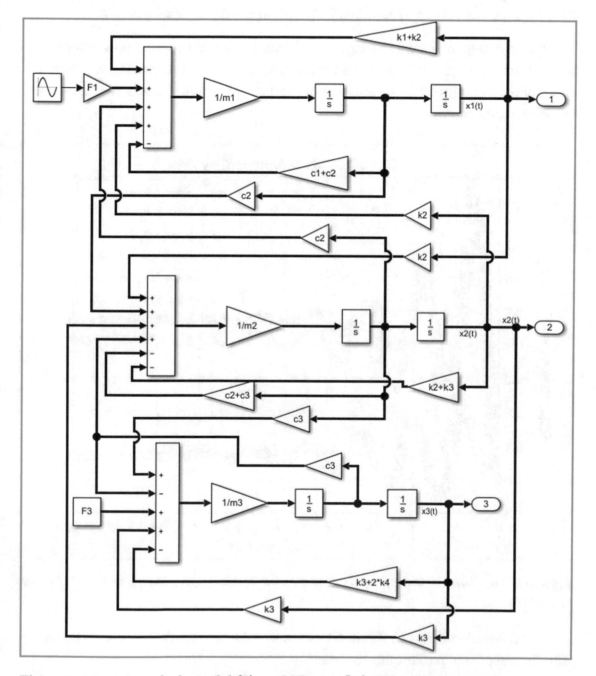

Figure 9-18. *A Simulink model (ThreeDOFsys.slx) of the three-DOF system*

Forced motion case. In this case, we take all zero initial conditions and two nonzero applied external forces:

```
IC=[0;0;0;0;0;0];                                    % Case # 1
F1=20*sin(120*t);F2=0; t0=0; F3=20*stepfun(t,t0); % Case # 1
```

By simulating the model (`three_MASS.m`) that recalls and executes the Simulink model (`threeDOFsys.slx`), we obtain the simulation results shown in Figure 9-19 and an animation plot (not shown here) demonstrating the motion of the three masses with respect to each other.

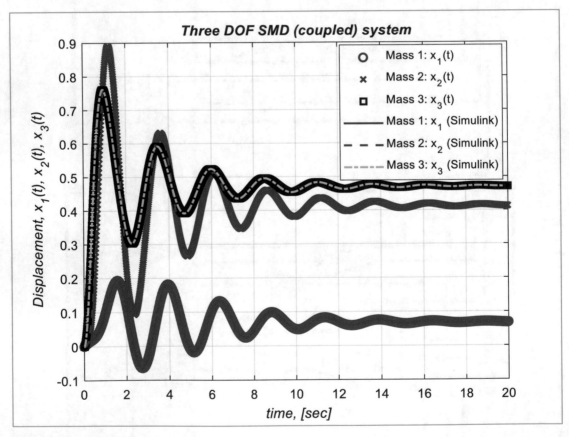

Figure 9-19. *Simulation of the three degree of freedom coupled system with forced vibration*

The simulation results in Figure 9-19 show that the MATLAB model results match with the ones from the Simulink model.

Free motion case. In this case, we take all zero initial conditions and two nonzero applied external forces.

```
IC=[0.5;0;0;0;0;0];      % Case # 2
F1=0; F2=0; F3=0;        % Case # 2
```

The simulation results of Case #2 (free motion, shown in Figure 9-20) show that the small initial displacement (nonzero) on mass 1 result in displacements in the other two masses as well. The computed results by the MATLAB and Simulink models converge perfectly even though different solvers and step sizes are used.

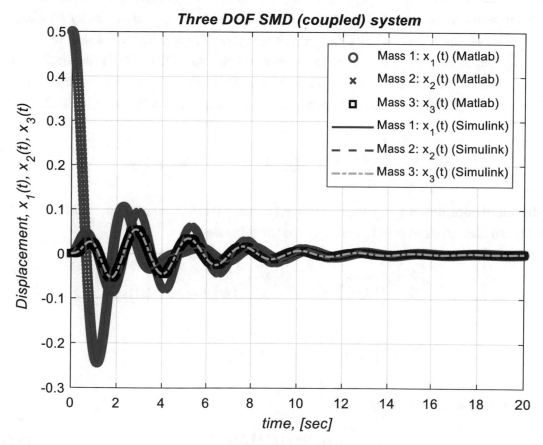

Figure 9-20. *Simulation results of free motion of the three-DOF SMD system*

Matrix Approach for *n*-Degree of Freedom System

When a given n-DOF system is linear, then there is another more efficient way, which is a matrix approach, to build a simulation model of such systems. Let's consider the forced response of a damped linear n-degree of freedom spring-mass-damper system. The most general formulation of such systems can be expressed in the following way:

$$[M]\ddot{x}+[C]\dot{x}+[K]x=[B]F(t) \tag{9-42}$$

with the initial conditions $x(0)=x_0$ and $\dot{x}(0)=\dot{x}_0$, where $[M]$, $[C]$, $[K]$ are mass, damper, and spring matrices, and $[B]$ is a column force matrix of $F(t)$ applied force. By dividing both sides of Equation (9-1) by $[M]$ or multiplying by $[M^{-1}]$ and introducing new variables, second-order ODEs are converted into first-order ODEs. The introduction of new variables, for instance, for a second-order system will be $x=y_1$, $\dot{x}=y_2$, and $\ddot{x}=\dot{y}_2$. With these substitutions, the next system of two differential equations is obtained.

$$\begin{cases} \dot{y}_1 = y_2 \\ \dot{y}_2 = \left[M^{-1}\right]\left[B\right]F(t)-\left[M^{-1}\right]\left[K\right]y_1 -\left[M^{-1}\right]\left[C\right]y_2 \end{cases} \tag{9-43}$$

with initial conditions $y_1(0)=x_0$, $y_2(0)=\dot{y}_1(0)=\dot{x}_0$. The formulation in Equation (9-43) can be written in a more explicit form with the following:

$$\dot{y}=\begin{bmatrix} \dot{y}_1 \\ \dot{y}_2 \end{bmatrix}=\begin{bmatrix} [0]y_1 +[I]y_2 \\ \left[M^{-1}\right][B]F(t)-\left[M^{-1}\right][K]y_1 -\left[M^{-1}\right][C]y_2 \end{bmatrix}=$$

$$\begin{bmatrix} [0] \\ \left[M^{-1}\right][B]F(t) \end{bmatrix}+\begin{bmatrix} [0]y_1 +[I]y_2 \\ -\left[M^{-1}\right][K]y_1 -\left[M^{-1}\right][C]y_2 \end{bmatrix} \tag{9-44}$$

This is in a more general form:

$$\dot{y}(t)=f(t)+[A]y(t) \tag{9-45}$$

where A is the state matrix defined by the following equation:

$$A = \begin{bmatrix} 0 & I \\ -[M^{-1}][K] & -[M^{-1}][C] \end{bmatrix}$$

(9-46)

Moreover, we can write the following:

$$y(t) = \begin{bmatrix} y_1(t) \\ y_2(t) \end{bmatrix}, f(t) = \begin{bmatrix} 0 \\ M^{-1}BF(t) \end{bmatrix}, y_0 = \begin{bmatrix} y_1(0) \\ y_2(0) \end{bmatrix} = \begin{bmatrix} x_0 \\ \dot{x}_0 \end{bmatrix}$$

(9-47)

To summarize, after evaluating matrix [A] and a force vector $f(t)$, if the given system is subject to external force, we can solve the equation $\dot{y}(t) = f(t) + [A]y(t)$ numerically.

Example. Let's consider previously studied three-DOF SMD system.

$$[M] = \begin{bmatrix} 2.5 & 0 & 0 \\ 0 & 2 & 0 \\ 0 & 0 & 3 \end{bmatrix}, [C] = \begin{bmatrix} c_1 + c_2 & -c_2 & 0 \\ -c_2 & c_2 + c_3 & -c_3 \\ 0 & -c_3 & c_3 \end{bmatrix} = \begin{bmatrix} 3 & -0.5 & 0 \\ -0.5 & 4 & -3.5 \\ 0 & -3.5 & 3.5 \end{bmatrix},$$

$$[K] = \begin{bmatrix} k_1 + k_2 & -k_2 & 0 \\ -k_2 & k_2 + k_3 & -k_3 \\ 0 & -k_3 & k_3 + 2k_4 \end{bmatrix} = \begin{bmatrix} 30 & -5 & 0 \\ -5 & 35 & -30 \\ 0 & -30 & 90 \end{bmatrix}$$

Again, we consider two cases: forced and free motions.

Forced motion (vibration) case. This case, with zero initial conditions and external forces applied on mass 2 and mass 1 and 3 are set free:

$$F_1(t) = 0; F_2(t) = 20 \, \Phi(t - t_0); F_3(t) = 0;$$

Note that $F_2(t)$ is the Heaviside step function $\Phi(t - t_0)$ of maginutude 20 applied at $t_0 = 0$.

We write the simulation model script of the problem, MASS3.m, with a nested function, DOF3(t, x).

```
function MASS3
% HELP. Simulation of three-degree-of-freedom system.
t=0:.01:120;            % Simulation time span
 IC=[0;0;0;0;0;0];     % Case # 1. Forced Motion
```

```
% IC=[0.5;0;0;0;0;0]; % Initial conditions
[t, x123]=ode45(@DOF3, t, IC, []);
figure('name', 'Matrix Approach')
H=plot(t, x123(:,1), t, x123(:,3), t, x123(:,5));
H(1).Color = [1 0 0];
H(2).Color = [0 1 0];
H(3).Color = [0 0 1];
H(1).LineStyle = '-';
H(2).LineStyle = '-.';
H(3).LineStyle = '--';
H(1).LineWidth = 1.5;
H(2).LineWidth = 2.0;
H(3).LineWidth = 2.0;
legend('Mass 1: x_1(t)', 'Mass 2: x_2(t)', 'Mass 3: x_3(t)')
ylabel('\it x_1(t), x_2(t), x_3(t)')
xlabel('\it time')
title('\it Forced Motion: Displacement'), grid on
% title('\it Free Motion: Displacement'), grid on
xlim([0, 20])
figure(2)
G=plot(t, x123(:,2), t, x123(:,4), t, x123(:,6));
G(1).Color = [1 0 0];
G(2).Color = [0 1 0];
G(3).Color = [0 0 1];
G(1).LineStyle = '-';
G(2).LineStyle = '-.';
G(3).LineStyle = '--';
G(1).LineWidth = 1.5;
G(2).LineWidth = 2.0;
G(3).LineWidth = 2.0;
legend('Mass 1: dx_1(t)', 'Mass 2: dx_2(t)', 'Mass 3: dx_3(t)')
title('\it Forced Motion: Velocity'), grid on
% title('\it Free Motion: Velocity'), grid on
xlabel('\it time'), ylabel('\it dx_1, dx_2, dx_3')
xlim([0, 20])
```

```
function  dx = DOF3(t, x)
    % HELP: three-degree-of-freedom system simulation
    k1=25; k2=5; k3=30; k4=k3; c1=2.5; c2=.5; c3=3.5;
    m1=2.5; m2=2; m3=3; M=[m1 0 0; 0 m2 0; 0 0 m3];
    K=[k1+k2, -k2, 0; -k2, k2+k3, -k3; 0, -k3, k3+2*k4];
    C=[c1+c2, -c2, 0; -c2, c2+c3, -c3; 0, -c3, c3];
    % Case 1: External forces applied
     F1= 0; t0=0; F2=20*stepfun(t,t0); F3=0;
    % Case 2: Free vibration (No force applied)
    % F1=0;F2=0;F3=0;
    B=[F1;F2;F3]; A=[zeros(3) eye(3); -inv(M)*K, -inv(M)*C];
    f=(M\eye(3))*B; dx=A*x+[0;0;0;f];
  end
end
```

Note that an inverse matrix computing function is computationally costly for large systems. Thus, a backslash (\) operator is employed instead of `inv()` because it is computationally more efficient and accurate than an inverse matrix computation approach. For example, `M\eye(3)` is equivalent to `inv(M)` but more accurate and efficient than `inv(M)`.

By executing the model `MASS3.m,` we obtain the simulation results of the forced motion, as shown in Figure 9-21 and Figure 9-22.

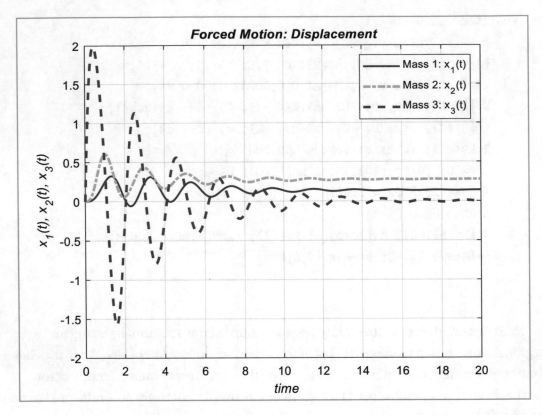

Figure 9-21. *Simulation results of forced motion of the three-DOF SMD system*

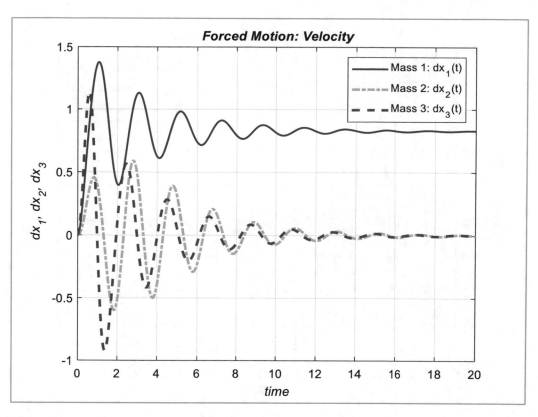

Figure 9-22. *Simulation results of forced motion of the three-DOF SMD system*

Free motion (vibration) case. Initially, a small displacement is applied on mass 2 alone, and zero initial conditions are given to other two masses: $x_1(0) = 0.5$, $x_2(0) = 0$, $x_3(0) = 0$; $\dot{x}_1(0) = 0$, $\dot{x}_2(0) = 0$, $\dot{x}_3(0) = 0$, $F_1(t) = F_2(t) = F_3(t) = 0$. The simulation results are shown in Figure 9-23 and Figure 9-24.

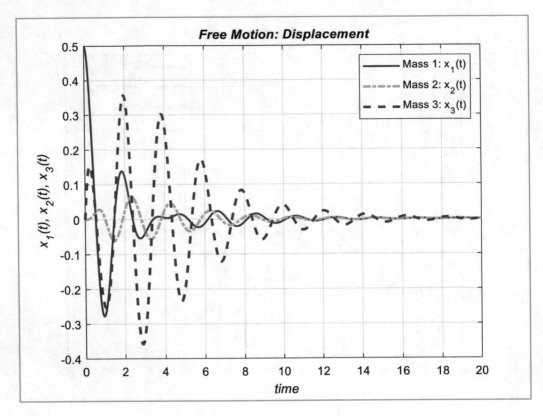

Figure 9-23. *Free motion: displacement*

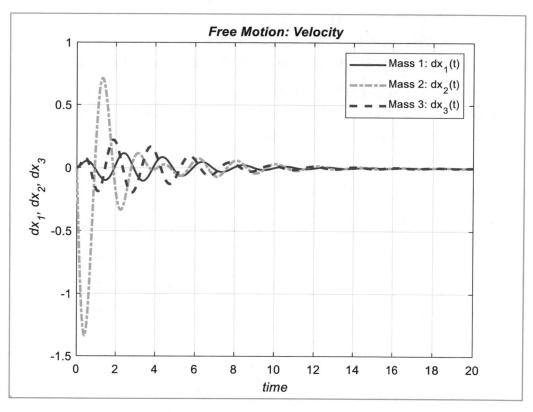

Figure 9-24. *Free motion: velocity*

The developed simulation models of the three-DOF SMD system with free and forced motion can be applied for n-DOF systems with minor adjustments to the system parameters. The matrix approach is relatively simple and computationally efficient.

References

[1] Inman, D. J., 2008, *Engineering Vibration* (3rd ed.), Pearson Prentice Hall, pp. 15–28, 282–288.

[2] Dimarogonas, A., 1996, *Vibration for Engineers* (2nd ed.), Prentice Hall, pp. 84–89.

[3] Weisstein, Eric W. "Critically Damped Simple Harmonic Motion." From MathWorld: A Wolfram Web Resource. `http://mathworld.wolfram.com/ CriticallyDampedSimpleHarmonicMotion.html`, viewed on September 20, 2019.

[4] Sulaymon Eshkabilov, *Beginning MATLAB and Simulink*, Apress, New York: ISBN (pbk) 978-1-4842-5060-0, ISBN (electronic) 978-1-4842-5061-7 (2019). DOI: `https://doi.org/10.1007/978-1-4842-5061-7`.

CHAPTER 10

Electromechanical and Mechanical Systems

This chapter covers electromechanial and mechanical systems. We discuss the issues how to build simulation models of electro-mechanical and mechanical systems in the examples of DC motor with input voltage and flexible load, microphone, motor-pump gear box, and double pendulum. We demonstrate linear and non-linear behaviors of the elctromechanical and mechanical systems by developing and simulating the models in MATLAB and Simulink environments.

Modeling a DC Motor

DC motors are the most widely used electric motors in electronics and various large-scale and small-scale machines to generate torque from electric power sources. Figure 10-1 shows a simplified physical model of a DC motor with input voltage and output torque.

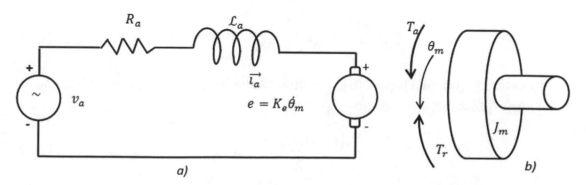

Figure 10-1. *DC motor: a) electric circuit of the armature; b) free-body diagram of the rotor*

We can assume that the rotor has an inertia mass, J_m, and a viscous friction constant, c. The free-body diagram of the rotor disc shows positive directions. There are two torques: the applied torque, $T_a = i_a K_t$, and the resistance torque, $T_r = C\dot{\theta}_m = \dfrac{Cd\theta}{dt}$. The resistance torque is applied in the opposite direction of the applied torque (T_a). By applying Newton's law from the free body diagram, we obtain the differential equation shown in Equation (10-1).

$$\frac{J_m d^2\theta}{dt^2} + T_r = T_a \Leftrightarrow \frac{J_m d^2\theta}{dt^2} + \frac{Cd\theta}{dt} = i_a K_t \tag{10-1}$$

where K_t is a torque constant, θ_m is an angular displacement of the motor shaft, and i_a is the flowing armature current. From the electrical circuit, using Kirchhoff's law, the next differential equation, as shown in Equation (10-2), of the circuit voltage is obtained. It takes into account a back electromotive-force (EMF) voltage, $v_b = K_e \dfrac{d\theta}{dt}$.

$$\frac{\mathcal{L}_a di_a}{dt} + R_a i_a = v_a - v_b \Leftrightarrow \frac{\mathcal{L}_a di_a}{dt} + R_a i_a = v_a - K_e \frac{d\theta}{dt} \tag{10-2}$$

where $\mathcal{L}_a$ is the inductance of the circuit, R_a is the resistance of the circuit, K_e is an EMF constant, and v_a is a voltage of the circuit. The two differential equations in Equations (10-1) and (10-2) can be rewritten in the s domain using the Laplace transformations by considering zero initial conditions.

$$J_m s^2 \theta_m(s) + Cs\theta_m(s) = I_a(s)K_t \tag{10-3}$$

$$\mathcal{L}_a s I_a(s) + R_a I_a(s) = V_a(s) - K_e s\theta_m(s) \tag{10-4}$$

By defining $I_a(s)$ from Equation (10-4) and substituting it into Equation (10-3), the following expressions are derived:

$$\frac{V_a(s) - K_e \theta_m(s)}{\mathcal{L}_a + R_a} = I_a(s) \tag{10-5}$$

$$J_m s^2 \theta_m(s) + Cs\theta_m(s) = \frac{V_a(s) - K_e s\theta_m(s)}{\mathcal{L}_a s + R_a} K_t \tag{10-6}$$

$$\left[J_m s^2 + Cs + \frac{sK_t K_e}{\mathcal{L}_a s + R_a} \right] \theta_m(s) = \frac{V_a(s)K_t}{\mathcal{L}_a s + R_a} \tag{10-7}$$

From the expression in Equation (10-7), the transfer function, i.e., the output $\theta_m(s)$ divided by the input $V_a(s)$, can be evaluated from this:

$$T(s) = \frac{\theta_m(s)}{V_a(s)} = \frac{K_t}{J_m \mathcal{L}_a s^3 + (C\mathcal{L}_a + J_m R_a)s^2 + s(K_t K_e + R_a C)} \tag{10-8}$$

Or an alternative of this transfer function is the angular velocity $\omega(s)$ divided by the Input voltage, $V_a(s)$.

$$T_a(s) = \frac{s\theta_m(s)}{V_a(s)} = \frac{\omega(s)}{V_a(s)} = \frac{K_t}{J_m \mathcal{L}_a s^2 + (C\mathcal{L}_a + J_m R_a)s + (K_t K_e + R_a C)} \tag{10-9}$$

Using the transfer function formulations shown in Equations (10-8) and (10-9), we can build the Simulink model, DC_motor.slx, of the DC motor (see Figure 10-2).

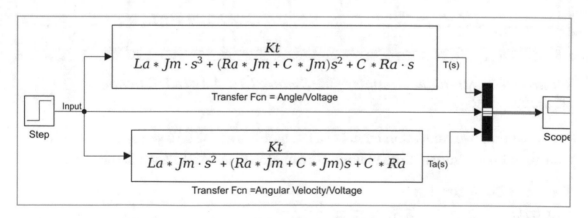

Figure 10-2. *Simulink model of DC motor,* DC_motor.slx

Note that all the parameters of the model are defined by selecting Model Properties: DC_motor ➤ Callbacks ➤ InitFcn.

     Jm=0.01; % [kgm^2]

     C=0.1; % [Nms]

     Ke=0.01; % [Nm/A]

Kt=1; % [Nm/A]

Ra=1.25; % [Ohm]

La=0.5; % [H]

The input signal is a step function representing the constant voltage of 12 V applied at a step time of one second. The simulation results of the model, as shown in Figure 10-3, were obtained from the [Scope] block of the model, DC_motor.slx.

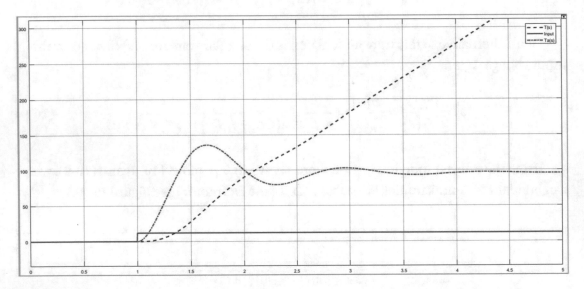

Figure 10-3. *Simulation results of the Simulink model of a DC motor,* DC_motor.slx

An alternative simulation model of a DC motor is a MATLAB script based on the fourth/fifth-step Runge-Kutta method. The model DC_Motor_SIM.m is shown here:

```
function DC_Motor_SIM
h=0.001;                % Time step
t=0:h:5;                % Simulation time
theta0 = [0, 0];        % Initial conditions
ia0    = 0;             % Initial conditions
u=[theta0, ia0];
for m=1:length(t)-1
k1(m,:)=DC_Motor(t(m), u(m,:));
k2(m,:)=DC_Motor(t(m), u(m,:)+h*k1(m,:)./2);
```

```matlab
k3(m,:)=DC_Motor(t(m), u(m,:)+h*k2(m,:)./2);
k4(m,:)=DC_Motor(t(m), u(m,:)+h*k3(m,:));
u(m+1,:)=u(m,:)+h*(k1(m,:)+2*k2(m,:)+2*k3(m,:)+k4(m,:))./6;
end
close all
figure(1)
plot(t,u(:,1),'b',t,u(:,2),'r--',t,u(:,3),'k:','linewidth', 1.5)
title('\it Simulation of DC Motor for V_a = 12 [V] input')
xlabel('\it Time, t [s]'), ylabel('\it \theta_i(t), i_a(t)')
legend('\theta_m(t)','\omega(t)','i_a(t)', 'location', 'northwest')
grid on
ylim([0, 320])
function dudx=DC_Motor(t, u)
% DC_Motor(u) is a nested function.
% Parameters used in this simulation are:
Jm=0.01;    % [kgm^2]
C=0.1;      % [Nms]
Ke=0.01;    % [Nm/A]
Kt=1;       % [Nm/A]
Ra=1.25;    % [Ohm]
La=0.5;     % [H]
Va=12*stepfun(t, 1);    % [V]
dudx=zeros(1,3);
dudx(1)=u(2);
dudx(2)=(1/Jm)*(u(3)*Kt-C*u(2));
dudx(3)=(1/La)*(Va-Ke*u(2)-Ra*u(3));
return
```

Figure 10-4 shows the simulation results of the MATLAB model DC_Motor_SIM.m.

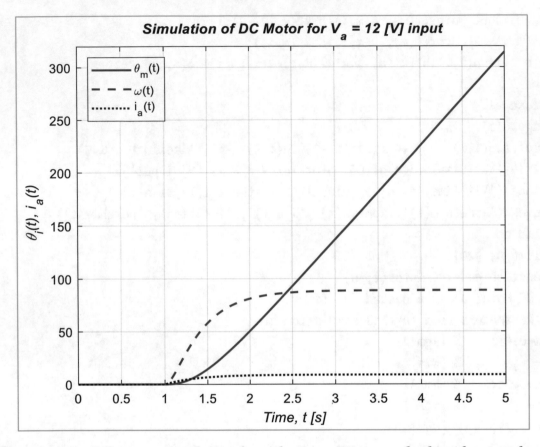

Figure 10-4. *DC motor simulation from the Rung-Kutta method implemented via scripts*

Modeling a DC Motor with Flexible Load

Figure 10-5 shows a physical model of a DC motor with flexible load; it is composed of electrical and mechanical components, such as the voltage source, inductor, resistor, rotational shaft, and load disk. This exercise is similar to the previous one except for one difference, which is that, in this case, there is a flexible load applied.

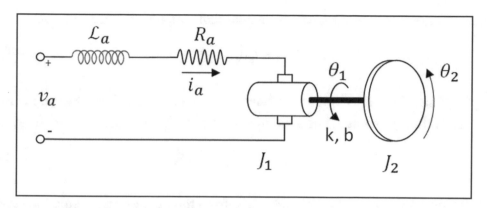

Figure 10-5. *DC motor with a flexible load*

For the electromechanical system shown in Figure 10-5 (this exercise was taken from [1]), the following differential equations of motion and the relationships of the system components can be derived:

$$\mathcal{L}_a \frac{di_a}{d_t} + R_a i_a = v_a - k_e \theta_1 \tag{10-10}$$

Also, the electromechanical torque (τ_m) of the motor and the torque of the flexible shaft (τ_f) can be found from the following:

$$\tau_m = K_t i_a \tag{10-11}$$

$$\tau_f = k(\theta_1 - \theta_2) \tag{10-12}$$

Now, another equation of motion torque balances can be written by applying Newton's second law that states an object's mass represents the proportional relationship between the applied force and the resulting acceleration. Applying this law to a rotating object reveals that every object has corresponding rotational inertia relating to an applied torque to a resulting angular acceleration (e.g., $\tau = J\dot{\omega}$). In other words, rotational mass is represented equivalently by the moment of inertia.

$$\tau_e = J_1 \ddot{\theta}_1 + B_1 \dot{\theta}_1 + \tau_f \tag{10-13}$$

Similarly, the torque equation for a flexible shaft can be written as follows:

$$\tau_f = J_2\ddot{\theta}_2 + B_2\dot{\theta}_2 \tag{10-14}$$

By plugging τ_f from the expression shown in Equation (10-12) into Equation (10-14), we obtain another differential equation of motion for a flexible shaft part of the system.

$$k(\theta_1 - \theta_2) - B_2\dot{\theta}_2 = J_2\ddot{\theta}_2 \tag{10-15}$$

Finally, by setting $\tau_e = \tau_m$ and plugging in τ_f from the expression in Equation (10-12), we derive a third differential equation to build a complete model of the system.

$$K_t i_a = J_1\ddot{\theta}_1 + B_1\dot{\theta}_1 + k(\theta_1 - \theta_2) \tag{10-16}$$

Now using the three differential equations, specifically, Equation (10-10), Equation (10-15), and Equation (10-16), a simulation model of the system is built in Simulink (Figure 10-6).

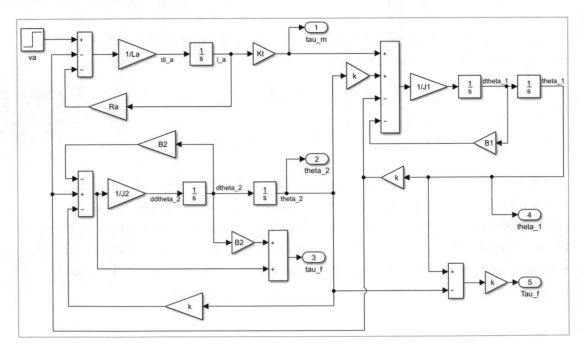

Figure 10-6. *Simulink model DC_Motor_w_Load.slx of the system DC motor with flexible load*

Note that in the Simulink model (DC_Motor_w_Load.slx) of the system DC motor with flexible load shown in Figure 10-6, the signal names (di_a and i_a) represent di_a and i_a, and dtheta_1, theta_1, ddtheta_2, dtheta_2, and theta_2 represent $\dot{\theta}_1, \theta_1, \ddot{\theta}_2, \dot{\theta}_2, \theta_2$. These signal names are added using the right-click options (the signal properties) of signals. Moreover, the output blocks called tau_m, theta_2, tau_f (and Tau_f), and theta_1 represent τ_m, θ_2, τ_f, and θ_1. Note that the numerical results of the two output blocks tau_f and Tau_f are based on the formulations given in Equation (10-14) and Equation (10-12).

For the simulation of the model, the following is the data for a brushed DC motor A-max 32:

Input: step function of $v_a = 12$ [V] applied at a step time of 1 second

Resistance: $R_a = 7.17$ [Ω]

Spring const: $k = 0.01 \left[\dfrac{N}{m} \right]$

Viscous damping: $B_1 = 7.05 * 10^{-5}$[N m sec]

Rotor's inertia: $J_1 = 4.14 * 10^{-5} [kg * m^2]$

Load inertia: $J_2 = 2.12 * 10^{-3} [kg * m^2]$

Viscous friction: $B_2 = 3 * 10^{-5}$ [N m sec]

Electrical const. of motor: $k_e = 0.29$ [V * s]

Torque constant: $K_t = 4.6 * 10^{-4} \left[N * \dfrac{m}{A} \right]$

Armature inductance: $\mathcal{L}_a = 9.53 * 10^{-4} [H]$

Note that these input variables can be defined via MATLAB workspace or (more preferably) via Model Settings (Ctrl+E) [Model Settings ▾ SETUP] ➤ [Model Properties] ➤ Callbacks ➤ InitFcn: Model initialization function. The latter option is more preferable because all of our assigned input variables will be attached to the model.

Torque of the motor: $\tau_m = ?$

Torque of the flexible shaft: $\tau_f = ?$

Output (angular displacement of motor shaft): $\theta_1 = ?$

Output (angular displacement of load disk): $\theta_2 = ?$

The computed outputs $(\tau_m, \tau_f, \theta_1, \theta_2)$ are obtained from the output block port numbers 1, 3, 5, 4, and 2, respectively. The plots of the simulation results are obtained with the next script (Sim_DC_Motor_w_Load.m):

```
clc; close all
figure('name', 'Torque plot')
subplot(211)
plot(out.yout{1}.Values, 'k-'), grid on
ylabel('\it Torque, \tau_m, [Nm]')
title('\it Motor torque, \tau_m')
subplot(212)
plot(out.yout{3}.Values, 'b--x'), grid on
hold on
plot(out.yout{5}.Values, 'k-'),
ylabel('\it Torque, \tau_f, [Nm]')
legend('\tau_f from Eqn. (14)', '\tau_f from Eqn. (12)')
title('\it Torque on the flexible shaft, \tau_f'), hold on
hold off
xlim([0, 13])

figure('name', 'Angular displacement')
subplot(211)
plot(out.yout{4}.Values, 'k-'), grid on
ylabel('\it  \theta_1')
title('\it Angular displacement of the motor shaft, \theta_1')
subplot(212)
plot(out.yout{2}.Values, 'b-'), grid on
ylabel('\it \theta_2')
title('\it Angular displacement of the flexible shaft, \theta_2')
```

To summarize, we demonstrate in the simulation results (Figure 10-7 and Figure 10-8) of the model (Figure 10-6) that there is an instability region in the torque values of a flexible load shaft. This is due to the shaft's flexibility property.

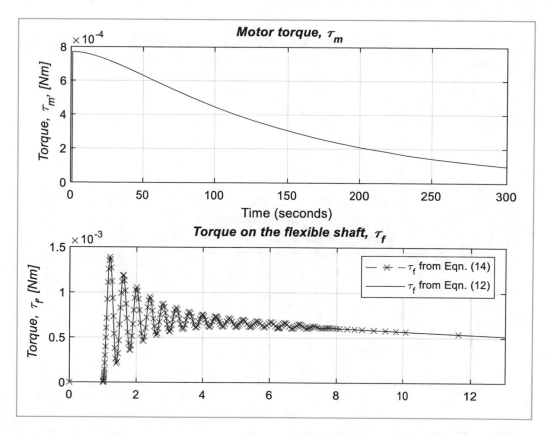

Figure 10-7. *Angular velocities of the motor shaft ($\dot{\theta}_1$) and flexible load shaft ($\dot{\theta}_2$)*

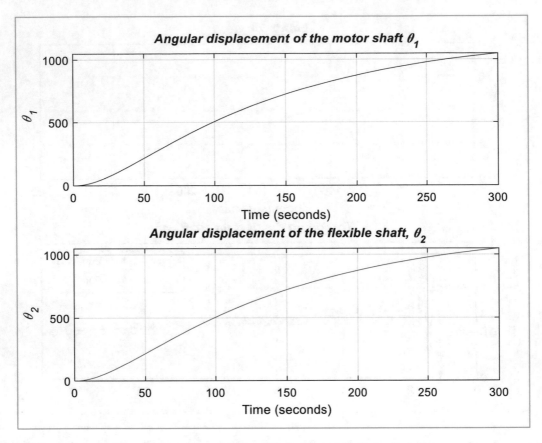

Figure 10-8.   *Torque of the motor (τ_e) and flexible load shaft (τ_f)*

The simulation model (Figure 10-6) of the DC motor with a flexible load in Simulink is one of many possible approaches to simulation model development. This Simulink model (Figure 10-6) can be simplified and optimized for better readability using various blocks. In addition, similar simulation models can be developed by writing scripts.

Modeling a Microphone

A simplified physical model (Figure 10-9) of a microphone consists of electrical and mechanical components. Its electrical components are the capacitor plate (a), which is rigidly fastened to the frame of the microphone; the charge source (v); the resistor (R); and the inductance (L). The mechanical parts of the model are the movable plate (membrane (b)) with a mass of m, connected to the frame with a set of springs of constant K and dampers with a constant of D. Sound waves pass through the mouthpiece and exert a force, $F_s(t)$, on a movable plate (membrane (b)).

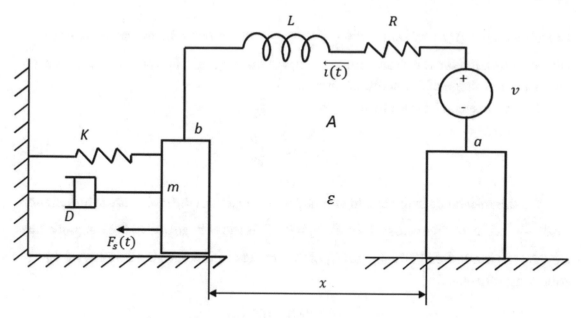

Figure 10-9. *Simplified physical model of a microphone*

The capacitance C is a function of the distance x between the plates, $C(x) = \dfrac{\varepsilon A}{x}$, where ε is a dielectric constant of the material between plates. (The air's dielectric constant at 20^0 C is 1.0059, and A is the surface area of the plates.) The charge q and the voltage v across the plates are related by the formulation of $q = C(x)v$. The electric field in turn produces the following force F_e on the movable plate (b) that opposes its motion: $F_e = \dfrac{q^2}{2\varepsilon A}$. The force F_e is equivalent to the force ($F = Blv$) induced due to magnetic flux density (B) and coil length (l), and thus, we substitute F_e with $F = Blv$. The differential equations that describe the operation of this system consist of two parts, one of which is the mechanical part and the other is the electrical part.

Here's the mechanical part:

$$F_s(t) = m\ddot{x} + D\dot{x} + Kx(t) + F \qquad (10\text{-}17)$$

$F_s(t)$ is the force of the moving air on the movable plate (b) (membrane). m is the mass of the movable plate (b) (membrane). D is the friction or damping constant. K is the stiffness or spring constant of the movable plate (b) (membrane). $x(t)$ is the displacement of the movable plate (b) (membrane). $\dot{x} = \dfrac{dx}{dt}$ is the velocity of the movable plate (*b*) (membrane). $\ddot{x} = \dfrac{d^2 x}{dt^2}$ is the acceleration of the movable plate (*b*) (membrane). $F = Blv$ is the force induced due to the magnetic flux and the current in the coil. *v* is the voltage of the electrical circuit.

Here's the electrical part in voltages:

$$F_v = Ri + L\frac{di}{dt} \tag{10-18}$$

R is the resistance of the coil in Ohms. *i* is the current in the coil. *L* is the inductance, and $\dfrac{Ldi}{dt}$ is the voltage across the coil. $F_v = Bl\dfrac{dx}{dt}$ is the voltage due to the magnetic flux density (*B*) multiplied by the length (*l*) of the coil. For numerical simulations, we take the following values:

$$v = 4.8 * 10^{-3}\ [V];$$

$$i_0 = 3\ mA = 3 * 10^{-3}\ [A];$$

$$m = 50\ g = 50 * 10^{-3}\ [kg];$$

$$K = 0.1 \left[\frac{N}{m}\right];$$

$$D = 3 * 10^{-5}\ [\text{N m sec}];$$

$$Bl = 0.5;$$

$$R = 5\ [\Omega];$$

$$L = 45\ [H],\ F_s = 10^{-4}[N]$$

Finally, summing up the previous two equations for the mechanical and electrical parts of the system (the microphone), we derive the following system of equations:

$$\begin{cases} F_s(t) = \dfrac{md^2 x}{dt^2} + \dfrac{Ddx}{dt} + Kx(t) + Blv \\[2mm] \dfrac{Bldx}{dt} = Ri + L\dfrac{di}{dt} \end{cases} \tag{10-19}$$

That can be also rewritten as follows:

$$\begin{cases} \dfrac{d^2x}{dt^2} = \dfrac{1}{m}\left[F_s(t) - \dfrac{Ddx}{dt} - Kx(t) - Blv \right] \\ \dfrac{di}{dt} = \dfrac{1}{L}\left[\dfrac{Bldx}{dt} - Ri \right] \end{cases} \qquad (10\text{-}20)$$

A simulation model of the system is developed by taking $F_s(t)$ as an excitation force signal treated as an input signal.

Figure 10-10 shows a simulation model (Microphone_Model.mdl) in Simulink. Here is a script (Sim_Microphone_Model.m) to display its simulation results:

```
clc; close all; clearvars
out=sim('Microphone_Model');
figure('name', 'Microphone')
yyaxis left
plot(out.yout{1}.Values, 'b--')
ylabel('\it Displacement, [mm]')
yyaxis right
plot(out.yout{2}.Values, 'k-')
xlabel('\it Time, [s]')
ylabel('\it i(t), [mA]')
legend('Displacement', 'Current', 'location', 'southeast')
title('Microphone simulation')
```

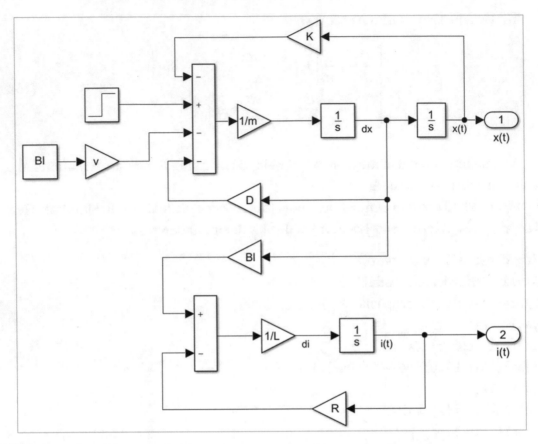

Figure 10-10. *Simulink model (Microphone_Model.slx) of the microphone*

From the simulation model results shown in Figure 10-11, it is clear that the motion of the membrane (movable part) of the microphone is periodic, and the current flow in the electrical components of the system is also periodic when the system is excited with a step reference signal, such as a step 4.8 [mV] input voltage.

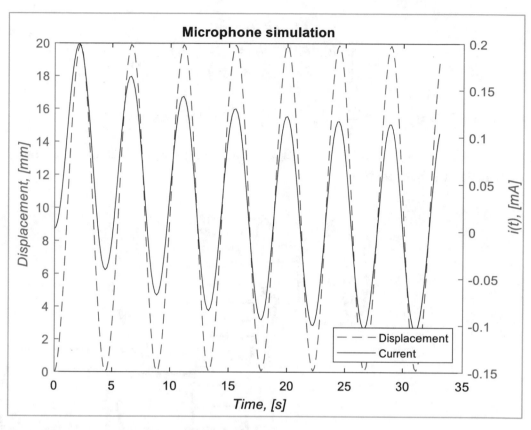

Figure 10-11. *Simulation of the microphone's membrane*

Modeling Motor: Pump Gear Box

A motor is powering a pump through the gear box shown in Figure 10-12. The rotors have moments of inertia ($J_1 = 1.5\ kgm^2$, $J_2 = 0.5\ kgm^2$, $J_3 = 1.0\ kgm^2$, $J_4 = 2.0\ kgm^2$), and the connecting shafts have torsional stiffness ($k_1 = 1200\dfrac{Nm}{rad}$, $k_2 = 2000\dfrac{Nm}{rad}$). The gear ratio is $n = 2$. To write a mathematical model of the system, first we build the equivalent system model (Figure 10-13) from the given physical model (Figure 10-12), and then we develop free-body diagrams of the system components (Figures 10-13 and 10-14). Using the free-body diagrams shown in Figures 10-13 and 10-14, we dervice the mathematical formulations (10-21) of the motion of the torsional model of the system.

$$\begin{cases} J_1\ddot{\theta}_1 = k_1(\theta_2 - \theta_1) \\ J_2'\ddot{\theta}_2 = k_1(\theta_1 - \theta_2) + k_2'(\theta_3 - \theta_2) \\ J_3'\ddot{\theta}_3 = k_2'(\theta_2 - \theta_3) \end{cases} \tag{10-21}$$

where $k_2' = n^2 k_2 = 4k_2 = 8000\dfrac{Nm}{rad}$, $J_2' = J_2 + J_3 n^2 = J_2 + J_3 4 = 4.5\, kgm^2$,

$J_3' = n^2 J_4 = 4J_4 = 2.0\, kgm^2$;and where θ_1, θ_2, and θ_3 are angular displacements of mass 1, mass 2, and mass 3, respectively.

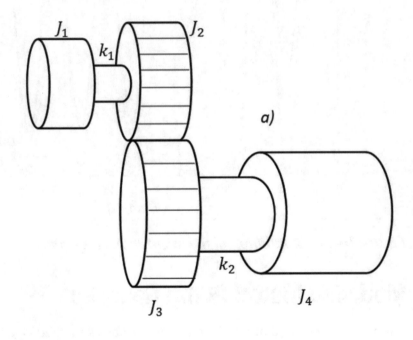

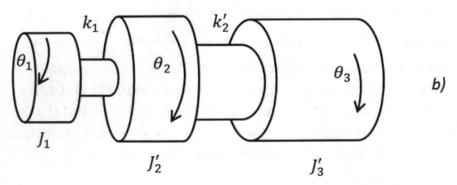

Figure 10-12. *a) The original system; b) equivalent system*

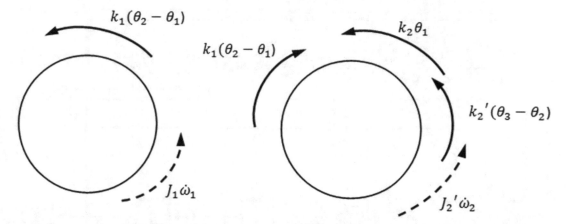

Figure 10-13. *Free-body diagram of mass 1 and mass 2*

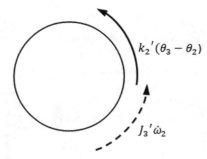

Figure 10-14. *Free-body diagram of mass 3*

Finally, the system shown in Equation (10-22) of differential equations is solved by simulation models.

$$
\begin{cases}
\ddot{\theta}_1 = \dfrac{k_1}{J_1}\left(\theta_2 - \theta_1\right) \\[2mm]
\ddot{\theta}_2 = \dfrac{1}{J_2'}\left[k_1\left(\theta_1 - \theta_2\right) + k_2'\left(\theta_3 - \theta_2\right)\right. \\[2mm]
\ddot{\theta}_3 = \dfrac{k_2'}{J_3'}\left(\theta_2 - \theta_3\right)
\end{cases}
\tag{10-22}
$$

A first simulation model. Using the system of equations in Equation (10-22), we build the Simulink model, `Motor_Pump_model.mdl`, shown in Figure 10-15. Note that in the Simulink model, k_2' is represented by k2p, J_2' by J2p, and J_3' by J3p.

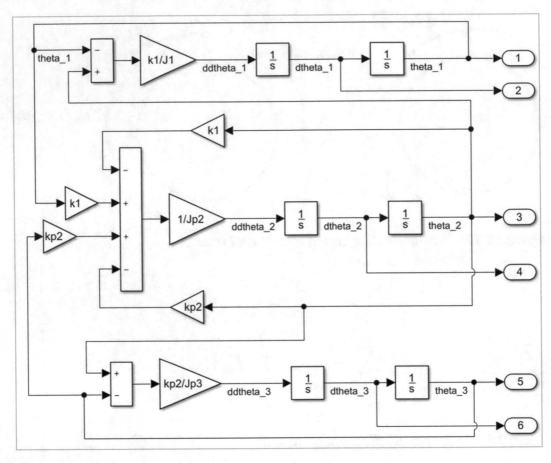

Figure 10-15. *Simulink model, Motor_Pump_model.slx, of the system motor pump*

After defining the numerical values of the parameters of stiffness and the moments of inertia, as well as the internal initial conditions of $\theta_1(0) = 0.005; \theta_2(0) = \theta_3(0) = 0; \dot{\theta}_1(0) = \dot{\theta}_2(0) = \dot{\theta}_3(0) = 0$ in integrator blocks of the model, we execute the Simulink model, and we obtain the results from Scope1 and Scope2. In the simulation, the results are shown in time versus $\theta_1(t)$, $\theta_2(t)$, $\theta_2(t)$, as shown in Figure 10-16, and in phase plots, as shown in Figures 10-17 and 10-18, such as $\theta_1(t)$ versus $\omega_1(t)$, $\theta_2(t)$ versus $\omega_2(t)$, $\theta_3(t)$ versus $\omega_3(t)$. Note to get smoother results from our simulations, we set tight limits, i.e., *RelTol* = 10^{-6} and *AbsTol* = 10^{-9}, on relative and absolute error tolerances, respectively.

The script Sim_Motor_Pump.m executes the Simulink model, Motor_Pump_model.slx, and plots the results.

```
clc; close all; clearvars
out=sim('Motor_Pump');
figure('name', 'Motor Pump')
yyaxis left
plot(out.yout{1}.Values, 'b--'), hold on
plot(out.yout{3}.Values, 'k-')
ylabel('\it Angular Displacement \theta_1, \theta_3, [rad]')
yyaxis right
plot(out.yout{5}.Values, 'r-.'), grid on
xlabel('\it Time, [s]')
ylabel('\it Angular Displacement \theta_2,  [rad]')
legend('\theta_1', '\theta_3', '\theta_1','location', 'southeast')
title('\it Motor Pump simulation')
xlim([0, 1.5])
%% Phase plots
theta1 = out.yout{1}.Values.Data;
dtheta1 = out.yout{2}.Values.Data;
theta2 = out.yout{3}.Values.Data;
dtheta2 = out.yout{4}.Values.Data;
theta3 = out.yout{5}.Values.Data;
dtheta3 = out.yout{6}.Values.Data;
figure('name', 'Motor Pump - Phase plot 1')
plot(theta1, dtheta1, 'b--'), grid on
ylabel('\it \omega_1')
title('Phase plot')
xlabel('\theta_1')
axis square
figure('name', 'Motor Pump - Phase plot 2')
plot(theta2, dtheta2, 'k-', theta3, dtheta3, 'r-.')
title('Phase plot')
xlabel(' \theta_2, \theta_3')
ylabel('\it \omega_2, \omega_3')
legend('\theta_2 vs. \omega_2', '\theta_3 vs. \omega_3', 'location',
'southeast')
axis square
grid on
```

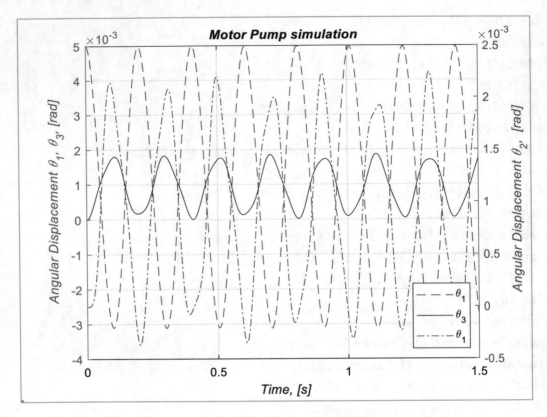

Figure 10-16. *Time versus angular displacements of three masses*

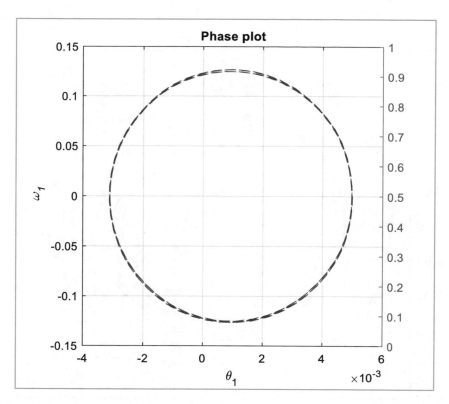

Figure 10-17. *Phase plot: $\theta_i(t)$ versus $\omega_i(t)$*

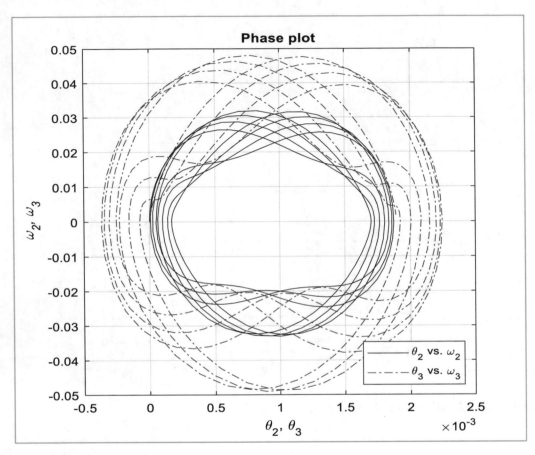

Figure 10-18. *Phase plot: $\theta_i(t)$ versus $\omega_i(t)$*

To simplify the process of MATLAB script writing, the system of differential equations, shown in Equation (10-22), is rewritten in a matrix form.

$$\begin{Bmatrix} \ddot{\theta}_1 \\ \ddot{\theta}_2 \\ \ddot{\theta}_2 \end{Bmatrix} = \begin{bmatrix} -\dfrac{k_1}{J_1} & \dfrac{k_1}{J_1} & 0 \\[2ex] \dfrac{k_1}{J_2'} & -\left(\dfrac{k_1}{J_2'}+\dfrac{k_2'}{J_2'}\right) & \dfrac{k_2'}{J_2'} \\[2ex] 0 & \dfrac{k_2'}{J_2'} & -\dfrac{k_2'}{J_2'} \end{bmatrix} \begin{Bmatrix} \theta_1 \\ \theta_2 \\ \theta_3 \end{Bmatrix} \qquad (10\text{-}23)$$

A second simulation model. The next model is a script called `Motor_Pump_Model_2.m` that embeds the system model in Equation (10-23) and computes numerical solutions of the problem with the solver `ode113`.

```
% Motor_Pump_Model_2.m
K1=1200;                  % [Nm/rad]
K2prime=8000;             % [Nm/rad]
J1=1.5;                   % [kgm^2]
J2prime=4.5;              % [kgm^2]
J3prime=2.0;              % [kgm^2]
t=[0, 1.5];               % Simulation time
theta0=[.005; 0; 0];      % Initial conditions
dtheta0=[0; 0; 0];        % Initial conditions
KJ=[-K1/J1, K1/J1, 0;
   K1/J2prime, -(K1/J2prime+K2prime/J2prime), K2prime/J2prime;
   0, K2prime/J3prime, -K2prime/J3prime;];
A=[zeros(3), eye(3); KJ, zeros(3)]; Fun=@(t, theta)(A*theta);
opts=odeset('reltol', 1e-6, 'abstol', 1e-9);
[time, THETA]=ode113(Fun, t, [theta0; dtheta0], opts);
figure; plot(time, THETA(:,1), 'b',time, THETA(:,2), 'm--', ...
    time, THETA(:,3), 'k:', 'linewidth', 1.5)
legend('\it \theta_1(t)','\theta_2(t)','\theta_3(t)', 'location',
'southeast'), grid on
title('Motor-pump model simulation MATLAB'), xlabel('t, time'),
ylabel('\it \theta_1(t), \theta_2(t), \theta_3(t) ')
figure; plot(THETA(:,1), THETA(:,4), 'b',...
    THETA(:,2), THETA(:,5), 'm--', THETA(:,3),...
    THETA(:,6), 'k:', 'linewidth', 1.5)
legend('\it \omega_1(t)','\omega_2(t)','\omega_3(t)', 'location',
'southeast')
title('Motor-pump model simulation MATLAB')
xlabel('\it \theta_1(t), \theta_2(t), \theta_3(t)')
ylabel('\it \omega_1(t), \omega_2(t), \omega_3(t)'), axis square
grid on
```

Note that the simulation results from the second simulation approach with ode113 are not displayed in this context.

A third simulation solution. The next example of this modeling problem uses the Runge-Kutta method and is implemented via a function file called Motor_Pump_Demo.m, shown here:

```matlab
function Motor_Pump_Demo
h=0.01;                  % Time step
t=0:h:1.5;               % Simulation time
theta0  = [0.005, 0, 0]; % Initial conditions
dtheta0 = [0, 0, 0];     % Initial conditions
u=[theta0, dtheta0];     % Whole Initial conditions
for m=1:length(t)-1
k1(m,:)=Motor_pump_shaft(u(m,:));
k2(m,:)=Motor_pump_shaft(u(m,:)+h*k1(m,:)./2);
k3(m,:)=Motor_pump_shaft(u(m,:)+h*k2(m,:)./2);
k4(m,:)=Motor_pump_shaft(u(m,:)+h*k3(m,:));
u(m+1,:)=u(m,:)+h*(k1(m,:)+2*k2(m,:)+2*k3(m,:)+k4(m,:))./6;
end
close all; figure(1)
plot(t,u(:,1),'b',t,u(:,3),'k:',t,u(:,5),'m-.','linewidth', 1.5)
title('Angular displacements of Inertia Masses')
xlabel('Time, t'),
ylabel('\theta_1(t), \theta_2(t), \theta_3(t)')
legend('\theta_1(t)','\theta_2(t)','\theta_3(t)','location', 'southeast'),
grid on
figure(2)
plot(t,u(:,2),'b',t,u(:,4),'k:',t,u(:,6),'m-.','linewidth', 1.5)
title('Angular velocities of Inertia Masses')
xlabel('Time, t'), ylabel('\omega_1(t), \omega_2(t), \omega_3(t)')
legend('\omega_1(t)','\omega_2(t)','\omega_3(t)','location', 'southeast'),
grid on
figure(3); plot(u(:,1), u(:,2), 'b', u(:,3), u(:,4), 'm--',...
    u(:,5), u(:,6), 'k:', 'linewidth', 1.5),
axis square, axis tight; title('Phase plots of Inertia Masses')
xlabel('\theta_1(t), \theta_2(t), \theta_3(t)')
ylabel('\it \omega_1(t), \omega_2(t), \omega_3(t)')
legend('\omega_1(t)','\omega_2(t)','\omega_3(t)','location', 'southeast')
function dudx=Motor_pump_shaft(u)
K1=1200;        % [Nm/rad]
K2prime=8000;   % [Nm/rad]
```

```
J1=1.5;         % [kgm^2]
J2prime=4.5;    % [kgm^2]
J3prime=2.0;    % [kgm^2]
dudx=zeros(1,6);
dudx(1)=u(2);
dudx(2)=(-K1/J1)*u(1)+(K1/J1)*u(3);
dudx(3)=u(4);
dudx(4)=(K1/J2prime)*u(1)-(K1/J2prime+K2prime/J2prime)*u(3)+...
    (K2prime/J2prime)*u(5);
dudx(5)=u(6);
dudx(6)=(K2prime/J3prime)*u(3)-(K2prime/J3prime)*u(5);
return
```

All three simulation models of the system produce identical results of angular displacement, velocity, and phase plots. As noted, the accuracy of numerical solutions depends on the step size of the employed solver.

Modeling Double Pendulum

Pendulums are used in vast units and mechanisms of machines, for instance, internal combustion engines, robot arms, and manipulators. Based on the double pendulum shown in Figure 10-19, we can formulate kinetic and potential energy equations of each mass by applying Lagrange's equation. Note that the system has two degrees of freedom defined by θ_1 and θ_2. The bars are assumed to be massless, and masses are concentrated in a part point. Joints have no friction.

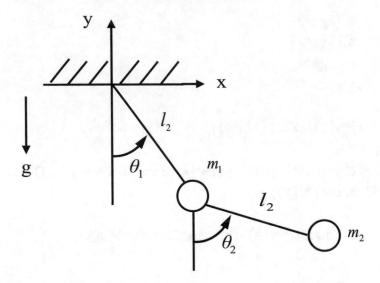

Figure 10-19. *Physical model of the double pendulum*

The equations of motion of the double pendulum for masses m_1 and m_2 are derived with the following formulations:

$$(m_1 + m_2)l_1^2\ddot{\theta}_1 - m_2l_1l_2\sin(\theta_2 - \theta_1)(\dot{\theta}_2 - \dot{\theta}_1)\dot{\theta}_2$$
$$+ m_2l_1l_2\cos(\theta_2 - \theta_1)\ddot{\theta}_2 + (m_1 + m_2)gl_1\sin(\theta_1) = 0 \tag{10-24}$$

$$-m_2l_1l_2\sin(\theta_2 - \theta_1)(\dot{\theta}_2 - \dot{\theta}_1)\dot{\theta}_1 + m_2l_1l_2\cos(\theta_2 - \theta_1)\ddot{\theta}_1 + m_2l_2^2\ddot{\theta}_2 + m_2gl_2\sin(\theta_2) = 0 \tag{10-25}$$

These two equations can be simplified by dividing the first equation by l_1 and the second one by l_2.

$$\begin{cases} (m_1 + m_2)l_1\ddot{\theta}_1 - m_2l_2\sin(\theta_2 - \theta_1)(\dot{\theta}_2 - \dot{\theta}_1)\dot{\theta}_2 + \\ m_2l_2\cos(\theta_2 - \theta_1)\ddot{\theta}_2 + (m_1 + m_2)g\sin(\theta_1) = 0 \\ -m_2l_1\sin(\theta_2 - \theta_1)(\dot{\theta}_2 - \dot{\theta}_1)\dot{\theta}_1 + m_2l_1\cos(\theta_2 - \theta_1)\ddot{\theta}_1 \\ + m_2l_2\ddot{\theta}_2 + m_2g\sin(\theta_2) = 0 \end{cases} \tag{10-26}$$

This system of equations, as shown in Equation (10-26), depicts the motion of the double pendulum.

For numerical simulations, the following values are taken:

$m_1 = 10[kg]; m_2 = m_1; l_1 = 5[m];$ and $l_2 = l_1$. The following are the initial conditions:
$\theta_1(0) = -0.5; \theta_2(0) = -1; \dot{\theta}_1(0) = 1; \dot{\theta}_2(0) = 2.$

Based on the system of equations of motion of the double pendulum, we can develop the Simulink model DOUBLE_pendulum.slx (Figure 10-20).

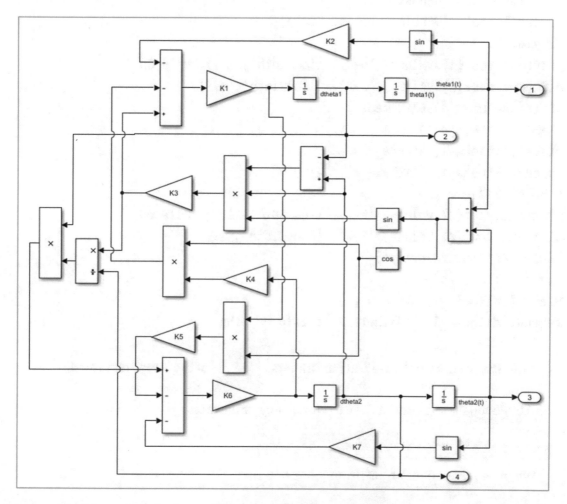

Figure 10-20. *A Simulink model of the double pendulum*

Note that in the model, shown in Figure 10-20, all of the initial conditions are set up via [Integrator] ([Integrator], [Integrator2], [Integrator3], [Integrator4]) blocks in the order of $\dot{\theta}_1, \theta_1, \dot{\theta}_2, \theta_2$. In addition, to simplify the model, [Gain] blocks are combined as

K_1, K_2, ..., K_7 whose values are defined in the Model Properties of the model DOUBLE_ pendulum.slx. Note the simulation results (Figure 10-21) we obtained via the channels of the [Out] block. The m-file that calls and executes the Simulink model is called Sim_ DOUBLE_Pendulum.m, shown here:

```
clc; close all; clearvars
out=sim('DOUBLE_pendulum');
figure('name', 'Double Pendulum Sim');
subplot(211)
plot(out.yout{1}.Values, 'b-', 'linewidth', 1.5), hold on
plot(out.yout{3}.Values, 'r-.', 'linewidth', 1.5)
title('Angular displacement')
xlabel('Time, [s]')
ylabel('\theta_1, \theta_2')
legend('\theta_1', '\theta_2'); grid on
subplot(212)
plot(out.yout{2}.Values, 'b-', 'linewidth', 1.5), hold on
plot(out.yout{4}.Values, 'r-.', 'linewidth', 1.5)
title('Angular velocity')
xlabel('Time, [s]')
ylabel('d\theta_1, d\theta_2')
legend('d\theta_1', 'd\theta_2'); grid on; shg
```

Note that in the Configuration parameters of the Simulink model,

DOUBLE_Pendulum.slx under Save to workspace or file tab:

Solver	Save to workspace or file
Data Import/Export	

.

We have taken off tick mark from [Save options]

▼ Additional parameters

Save options	
☐ Limit data points to last:	1000
Output options:	Refine output ▼

[Limit data points to last:] 1000.

Moreover, we have put a tick mark on ☑ Single simulation output: out

under variable name (default name) - out, and we have selected the format type to be

dataset: Format: Dataset ▼ .

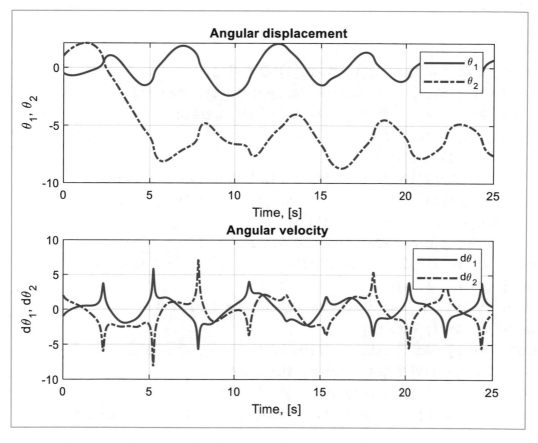

Figure 10-21. *Simulation results from the Simulink model*
DOUBLE_pendulum.slx

An alternative simulation model. Another approach to the problem can be developed via the solver ode15i, which is used for implicitly defined differential equations. Here is a script called DOUBLE_Pedulum_Model_2.m to simulate the motion of the double pendulum model. The simulation results of the script are shown in Figure 10-22 and Figure 10-23.

```matlab
%%  ODE15i IMPLICIT ODE Solver
% HELP:  This simulates a double pendulum model
% theta1(0)=-0.5; dtheta1(0)= -1
% theta2(0)= 1; dtheta2(0) = 2.
L1= 5; L2 = L1;         % meter
g = 9.81;               % meter/second^2
m1=10;                  % kg
m2=10;                  % kg
M=m1+m2;
test=@(t, theta, dtheta)([-dtheta(1)+theta(2);
    M*L1*dtheta(2)+m2*L2*cos(theta(3)-theta(1))*dtheta(4)-...
    m2*L2*(theta(4)*(theta(4)-theta(2)))*sin(theta(3)-...
    theta(1))+M*g*sin(theta(1));
    -dtheta(3)+theta(4);
    m2*L2*dtheta(4)+m2*L1*cos(theta(3)-theta(1))*dtheta(2)-...
    m2*L1*(theta(2)*(theta(4)-theta(2)))*sin(theta(3)- ...
    theta(1))+m2*g*sin(theta(3))]);
% theta10=-.5; dtheta10=-1; theta20=1; dtheta20=2;
tspan=[0:pi/30:5*pi];  % Simulation time with a fixed time step
the0=[-.5, -1, 1, 2];
dthe0=[0; 0; 0; 0];     % dtheta0 determined from theta0
theF0=[1 1 1 1]; dtheF0=[];
[the0, dthe0]=decic(test, 0, the0,theF0, dthe0, dtheF0);
[time, theta]=ode15i(test, tspan, the0, dthe0);
figure('name', 'Double pendulum 1')
plot(time, theta(:,1), 'b-', time, theta(:,2), 'k-.', 'linewidth', 1.5),
hold on
plot(time, theta(:,3),'r--',time,theta(:,4), 'm:', 'linewidth', 2)
legend('\theta_1(t)','d\theta_1(t)','\theta_2(t)','d\theta_2(t)',
'location', 'southeast')
title('A double pendulum problem solved with ODE15i')
xlabel('time'), ylabel('\theta(t)'), grid on

figure('name', 'Double pendulum 2'), yyaxis left
plot(theta(:,1), theta(:,2),'b-', 'linewidth', 1.5)
```

```
ylabel('d\theta_1'), yyaxis right
plot(theta(:,3), theta(:,4), 'm-.', 'linewidth', 1.5)
legend('\theta_1 vs. d\theta_1','\theta_2 vs. d\theta_2(t)', 'location',
'southeast')
title('Phase plot of a double pendulum problem')
xlabel('\theta_1, \theta_2'), ylabel('d\theta_2'),
axis tight; grid on; shg
```

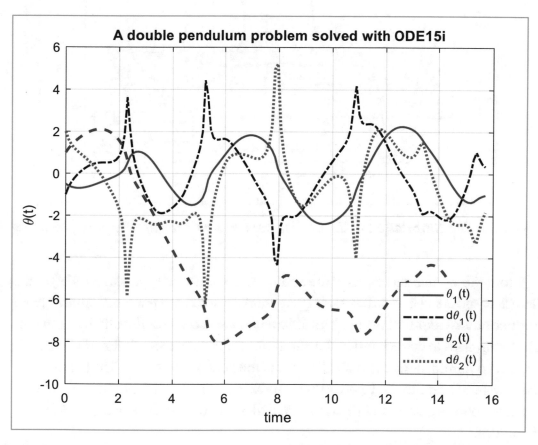

Figure 10-22. *Simulation of the double pendulum problem with a model using*
ode15i

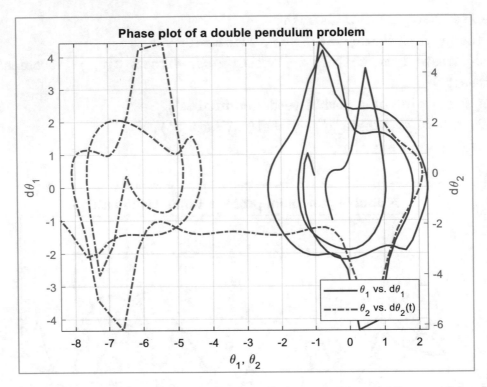

Figure 10-23. *Simulation of the double pendulum problem with a model using* `ode15i`

From the simulation results of the Simulink model (`DOUBLE_pendulum.slx`) and the `ode15i`-based `DOUBLE_Pedulum_model_2.m` model, we can see that both approaches produce almost identical results. The differences are because of the difference in our solver step sizes in `ode15i` and in the Simulink model. These simulation models are based on a nonlinear mathematical model of the double pendulum. The behavior of the system is nonlinear (Figures 10-21, 10-22, and 10-23) in terms of the angular displacements θ_1 and θ_2 and velocities $d\theta_1$ and $d\theta_2$ of the two bars of the double pendulum.

A linearized model of the double pendulum. For small oscillations of the two masses, we can approximate the system formulations in Equation (10-24), Equation (10-25), and Equation (10-26) and linearize them by taking $\sin\theta \approx \theta$ and $\cos\theta \approx 1$. Moreover, we ignore first-order derivative terms $\dot{\theta}_1, \dot{\theta}_2$ $((\dot{\theta}_2 - \dot{\theta}_1)\dot{\theta}_2)$ from the formulations.

$$\begin{cases} (m_1 + m_2)l_1\ddot{\theta}_1 + m_2l_2\ddot{\theta}_2 + (m_1 + m_2)g\theta_1 = 0 \\ m_2l_1\ddot{\theta}_1 + m_2l_2\ddot{\theta}_2 + m_2g\theta_2 = 0 \end{cases} \qquad (10\text{-}27)$$

Note that the linearized model in Equation (10-27) of the pendulum is a homogenous second-order ODE. That is implemented via a Simulink model called Double_Pendulum_Linear.slx, as shown in Figure 10-24, and its simulation results are shown in Figures 10-25 and 10-26.

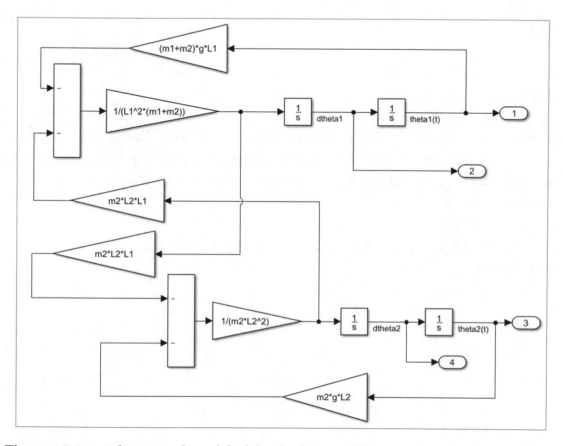

Figure 10-24. *A linearized model of the double pendulum to simulate in small oscillations*

Here is the m-file (Sim_Double_Pendulum_Linear.m) that runs the Simulink model and plots the results:

```
clc; close all; clearvars
out=sim('Double_Pendulum_Linear');
figure('name', 'Double Pendulum Sim');
subplot(211)
plot(out.yout{1}.Values, 'b-', 'linewidth', 1.5), hold on
```

```
plot(out.yout{3}.Values, 'r-.', 'linewidth', 1.5)
title('Angular displacement')
xlabel('Time, [s]')
ylabel('\theta_1, \theta_2')
legend('\theta_1', '\theta_2')
grid on
subplot(212)
plot(out.yout{2}.Values, 'b-', 'linewidth', 1.5), hold on
plot(out.yout{4}.Values, 'r-.', 'linewidth', 1.5)
figure('name', 'Phase plot')
plot(out.yout{1}.Values.Data, out.yout{2}.Values.Data, 'b-',...
    out.yout{3}.Values.Data,out.yout{4}.Values.Data, 'r-.', 'linewidth', 1.5)
legend('\theta_1 vs. d\theta_1', '\theta_2 vs. d\theta_2', 'location',
'southeast')
title('Phase plot')
xlabel('\theta_1, \theta_2'), ylabel('d\theta_1, d\theta_2')
axis square
```

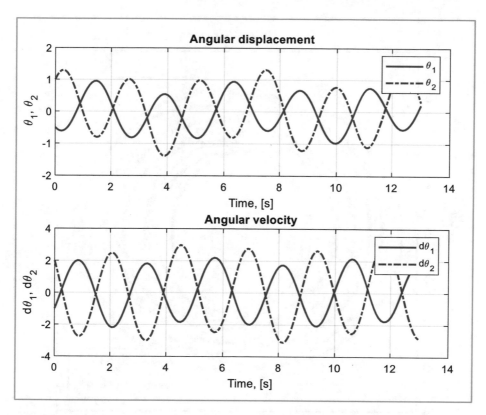

Figure 10-25. *Simulation results of the double pendulum with small oscillations*

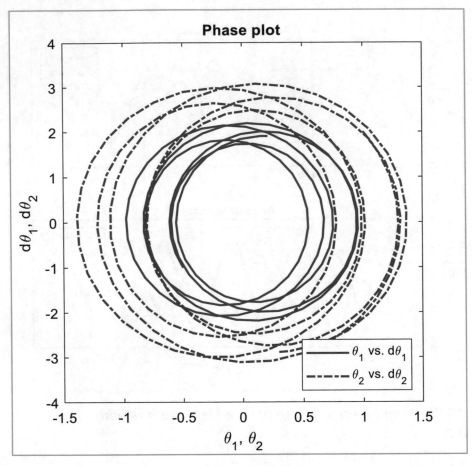

Figure 10-26. *Phase plot of the double pendulum with small oscillations*

Another approach. Another method to create a linearized model of the double pendulum is to employ ode15i (an implicit ODE solver) and to compute the analytical solution with the dsolve() function. Note that when the system's equation of motion is a linear ODE, then the dsolve() function can compute an analytical solution's formulation of the problem. Here is part II of the script DOUBLE_Pedulum_Linear_2.m that implements a linearized model of the double pendulum:

```
%% Part II.
%%  Linearized model of the double pendulum solved by ODE15i
%    DOUBLE_Pedulum_Linear_2.m
L1= 5; L2 = L1;        % m
g = 9.81;              % m/s^2
m1=10;                 % kg
```

```
m2=10;                    % kg
M=m1+m2;
test=@(t, theta, dtheta)([ dtheta(1)-theta(2);
      (m1+m2)*L1*dtheta(2)+m2*L2*dtheta(4)+(m1+m2)*g*theta(1);
dtheta(3)-theta(4); m2*L2*dtheta(4)+m2*L1*dtheta(2)+m2*g*theta(3)]);
tspan=[0, 2*pi];
% Initial Conditions:
theta10=-.5; dtheta10=-1; theta20=1; dtheta20=2;
the0=[theta10, dtheta10, theta20, dtheta20];
thep0=[0;0;0;0]; theF0=[1 1 1 1]; dtheF0=[];
[the0, dthep0]=decic(test,0, the0,theF0,thep0, dtheF0);
[time, theta]=ode15i(test, tspan, the0,dthep0);
plot(time, theta(:,1), 'k', time, theta(:,2), 'bs-',...
time, theta(:,3),'m--', time, theta(:,4), 'go-','linewidth', 1.5)
title('Linearized model simulation of double pendulum')
xlabel('time'), ylabel('\theta(t)'), axis tight;
grid on; hold on
%% Analytic Solution of the Linearized Model with DSOLVE
clearvars t
syms theta1(t)  theta2(t)
Dtheta1=diff(theta1, t, 1);
Dtheta2=diff(theta2, t, 1);
D2theta1=diff(theta1, t,2);
D2theta2=diff(theta2, t,2);
EQN = [D2theta1==(-1/(M*L1))*(m2*L1*D2theta2+M*g*theta1), ...
          D2theta2==(-1/(m2*L1))*(m2*L1*D2theta1+m2*g*theta2)];
ICs = [theta1(0)==-.5,Dtheta1(0)==-1, theta2(0)==1, Dtheta2(0)==2];
Theta=dsolve(EQN, ICs);
t=0:pi/20:2*pi;
Theta1=(vectorize(Theta.theta1)); Theta2=(vectorize(Theta.theta2));
% Compute values of Theta1 & Theta2:
theta1=eval(Theta1); theta2=eval(Theta2);
plot(t, theta1, 'mx', t, theta2, 'k*', 'linewidth', 1.5)
legend('\theta_1(t) via ODE15i','d\theta_1(t) via ODE15i',...
'\theta_2(t) via ODE15i','d\theta_2(t) via ODE15i',...
'\theta_1 dsolve','\theta_2 dsolve'), grid on; hold off
```

The simulation results shown in Figures 10-25, 10-26, and 10-27 of the Simulink model and the results of using the solver ode15i and dsolve() (analytical solver) are identical. The linearized model behavior is periodic and is valid only if the oscillation angles of the pendulum arms are very small, up to 5 degrees, or 0.0873 radians.

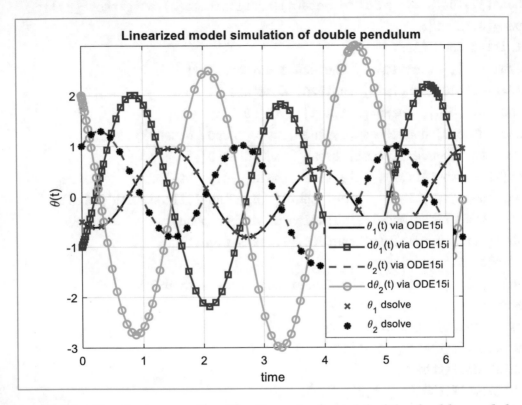

Figure 10-27. *Simulation results of the linearized model of the double pendulum from the m-file with* ode15i, dsolve()

Reference

[1] Franklin, Gene F., Powell, J. David, Emami-Naeni, Abbas, *Feedback Control of Dynamic Systems* (6th ed.), Prentice Hall (2009).

CHAPTER 11

Trajectory Problems

This chapter contains two examples: falling and thrown objects modeled via differential equations and falling and thrown objects implemented via simulation models in m-files and Simulink models.

Falling Object

Let's look at a simple physics exercise of a falling object. Consider a ball of mass m falling under the influence of Earth's gravity (g), as shown in Figure 11-1. Let h be the height that the ball falls, and let $v(t)$ be the velocity of the ball. The coordinate system will be positive downward.

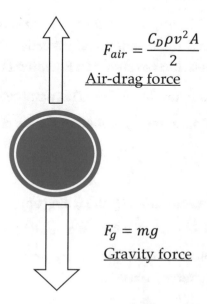

$$F_{air} = \frac{C_D \rho v^2 A}{2}$$
Air-drag force

$$F_g = mg$$
Gravity force

Figure 11-1. *Falling object and the forces acting upon it*

© Sulaymon L. Eshkabilov 2020
S. L. Eshkabilov, *Practical MATLAB Modeling with Simulink*, https://doi.org/10.1007/978-1-4842-5799-9_11

As shown in Figure 11-1, a falling object in the atmosphere is subject to two external forces: gravity/gravitational force and air-drag force. The net force of the falling object is equal to the acceleration multiplied by the mass of the falling object according to Newton's second law.

$$F_o = F_g - F_{air} \tag{11-1}$$

$$\frac{mdv}{dt} = mg - \frac{C_D \rho v^2 A}{2} \tag{11-2}$$

$$\frac{dv}{dt} = g - \frac{C_D \rho v^2 A}{2m} \tag{11-3}$$

where C_D is an air-drag coefficient of the falling object that is defined experimentally and is dependent of its shape, inclination, and flow conditions; ρ is the air density $\left[\dfrac{kg}{m^3}\right]$; A is the cross-sectional area of the falling object $[m^2]$; and m is the mass of the falling object $[kg]$.

The drag force depends on the square of the velocity of the falling object. Therefore, as the body accelerates its velocity, the drag force will increase. It quickly reaches a point (acceleration equals to 0) where the drag becomes equal to the weight (gravitational force). When drag force is equal to the weight (gravitational force) of the falling object, there will be no net external force $\left(\dfrac{mdv}{dt} = 0\right)$ on the object, and the acceleration of the object will be equal to 0. The object then falls at a constant velocity as described by Newton's first law of motion. The constant velocity is called the *terminal velocity*.

Equation (11-3) can be rewritten as shown in Equation (11-4) by substituting all air-drag constants with one parameter, $b = \dfrac{C_D \rho A}{2}$. This parameter is called Newton's drag. We obtain the next formulation, shown in Equation (11-4), of acceleration.

$$\frac{dv}{dt} = g - \frac{bv^2}{m} \tag{11-4}$$

Let's look at two case scenarios to study the falling object motion for two different values of b, i.e., $\dfrac{b_1}{m} = 0.1$ and $\dfrac{b_2}{m} = 0.001$ with a zero initial condition. We will find the terminal velocity for both cases. From the values of b_1 and b_2, in case #1, the drag force impacts 100 times of the drag force in case #2.

Here is case #1: $\dfrac{dv}{dt} = g - 0.1v^2$.

Here is case #2: $\dfrac{dv}{dt} = g - 0.001v^2$.

386

Here is the complete simulation model script, `Falling_Object.m`:

```matlab
% Falling_Object.m
%% Assignment. Falling object.
% Given: m*dv/dt=m*g-b*v^2; b/m=0.1 [1/sec]; Case # 1.
%        m*dv/dt=m*g-b*v^2; b/m=0.001 [1/sec]; Case # 2
%        For both cases: v(0)=0 and g=9.8 [m/sec^2];
%           Define a function. Case # 1. dv/dt=9.8-0.1*v^2
%           Define a function. Case # 2. dv/dt=9.8-0.001*v^2
% Step # 1.Function handle is used to define an equation
f_C1=@(t,v)(9.8-0.1*v^2');
f_C2=@(t,v)(9.8-.001*v^2);
% Step # 2. ODE23tb(f, timespan, IC) defined
[time1,v1]=ode23tb(f_C1,[0,30],0);
[time2,v2]=ode23tb(f_C2,[0,30],0);
% Step # 3. ODE15s(f, timespan, IC) defined
[time1a,v1a]=ode15s(f_C1,[0,30],0);
[time2a,v2a]=ode15s(f_C2,[0,30],0);
% Step # 4. Display simulation results
figure(1), plot(time1,v1,'k--', 'linewidth', 1.5); hold on
plot(time2,v2,'b-', 'linewidth', 1.5)
plot(time1a, v1a,'mo',time2a, v2a, 'rx')
legend('Case # 1 ode23tb','Case #2 ode23tb',...
'Case #1 ode15s', 'Case #2 ode15s', 'location', 'southeast')
xlabel('\it Time, [sec]'); ylabel('\it Velocity, v(t) [m/sec]');
title({'\it Case # 1: dv = 9.8-0.1*v^2', ...
  '\it Case # 2: dv = 9.8-0.001*v^2'})
grid on
% Step # 5. Locate terminal velocity values for both cases
v_WHICH=(v1==max(v1));
v_MAX=v1(v_WHICH);
% Step # 6. Add to the plot terminal velocity values
textmessage=['Case#1 - Terminal speed: '...
    num2str(v_MAX) '   [m/s]'];
gtext(textmessage);
v2_WHICH=(v2==max(v2));
```

```
v2_MAX=v2(v2_WHICH);
textmessage=['Case#2 - Terminal speed: '...
    num2str(v2_MAX) '    [m/s]'];
gtext(textmessage); hold off
```

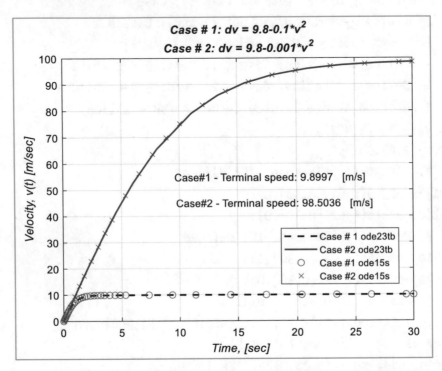

Figure 11-2. *Time versus velocity of the falling object in two environments with strong air drag (case #1) and small air-drag (case #2) forces*

The simulation results from the both solvers for both cases, as displayed in Figure 11-2, are well converged and almost identical. By increasing the drag force of the object by 100 times, its terminal speed will be decreased by about $90\frac{m}{s}$, which is about a 995 percent decrease in its speed. The influence of air-resistance (air-drag) force of a falling object is considerably great. The terminal speed is reached approximately after 2.5 and 30 seconds for cases #1 and #2, respectively.

Thrown Ball Trajectories

Several factors influence the flight trajectory of a hit or thrown ball. This exercise is a good example for the applications of differential equations. All good tennis players know how to hit the ball to make it fly according to a certain angle to get it over the net and hit the right spot of the court on the opponent's side. Good-quality hits are the result of a few years of hard practices and work. If a new player hits the ball about 1 percent faster or slightly stronger than a professional tennis player does, it will land considerably farther away (about 0.50 meter or more) beyond the baseline. The ball trajectory problem has been well formulated and explained by the theory of ball trajectory in a few sources, for instance [1].

Figure 11-3 shows three forces that are acting on a ball flying in air.

- There is gravitational force, as shown here:

$$\vec{F_g} = m\vec{g} \tag{11-5}$$

where $\vec{g} = (0, 0, -g)$ is a vector of the gravitational acceleration.

- The air-drag force, as shown here:

$$\vec{F_D} = \frac{-D_L(v)m\vec{v}}{v} \tag{11-6}$$

This is in the opposite direction of the velocity vector $\vec{v}$ of the ball.

- There is Magnus force, as shown here:

$$\vec{F_M} = M_L\vec{\omega} / (\omega \times \vec{v}) / v \tag{11-7}$$

The magnitudes of the drag force $D_L(v)$ and the Magnus force $M_L(v)$ usually have a form given by the theory of ideal fluids [2], as shown here:

$$D_L(v) = \frac{C_D}{2} \frac{\pi d^2}{4} \rho v^2 \tag{11-8}$$

$$M_L(v) = \frac{C_M}{2} \frac{\pi d^2}{4} \rho v^2 \tag{11-9}$$

where ρ is air density. The coefficients C_D and C_M are dependent on real fluids, for instance, air as well as the velocity v, the ball revolution, and the ball surface material. These coefficients are found experimentally and taken from the experimental results of the source [1]. For a tennis ball moving with a velocity of $v = 13.6\frac{m}{s}$ and with a ball revolution (spin) of $n = [800, 3250]$ rpm, the coefficients of C_D and C_M depend on the ratio of $\dfrac{v}{\omega}$ only, where $\omega = \dfrac{d}{2}\left|\vec{\omega} \times \dfrac{\vec{v}}{v}\right|$ in some cases the projection of the equatorial velocity

$\dfrac{\omega d}{2}$ of the spinning ball onto the velocity vector $\vec{v}$. The following empirical expressions for the coefficients are obtained by [1]:

$$C_D = 0.508 + \left(\cfrac{1}{22.053 + 4.196\left(\dfrac{v}{\omega}\right)^{\frac{5}{2}}}\right)^{\frac{2}{5}} \tag{11-10}$$

$$C_M = \cfrac{1}{2.022 + 0.981\left(\dfrac{v}{\omega}\right)} \tag{11-11}$$

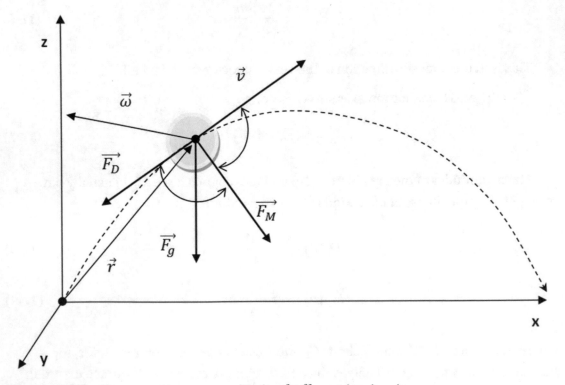

Figure 11-3. *Forces acting on a spinning ball moving in air*

We can neglect the deceleration of the ball revolution, and we take ω to be constant. The trajectory vector of the ball can be defined via Newton's second law equations for the position vector $\vec{r}(t)$ from [3].

$$\frac{md^2\vec{r}(t)}{dt^2} = -m\vec{g} - \frac{D_L v}{v} + \frac{M_L \vec{\omega}}{\omega} \times \frac{\vec{v}}{v} \tag{11-12}$$

with the initial conditions $\vec{r}(0) = \vec{r}_0$ and $\dfrac{d\vec{r}(0)}{dt} = \vec{v}_0$. The equation is a nonlinear system of three differential equations, and no analytical solution exists for it. Therefore, the only way to solve it is numerically. We choose the x-axis in the x-z plane, and then we obtain the final formulation [3] of Equations (11-13) and (11-14) from Equation (11-12) by plugging in D_L and M_L.

$$\ddot{x} = -C_D \alpha v \dot{x} + \eta C_M \alpha v \dot{z} \tag{11-13}$$

$$\ddot{z} = -g - C_D \alpha v \dot{z} - \eta C_M \alpha v \dot{x} \tag{11-14}$$

where $v = \sqrt{\dot{x}^2 + \dot{z}^2}$ and $\alpha = \dfrac{\rho \pi d^2}{8m}$. The parameter $\eta = \pm 1$ describes the direction of rotation, for instance, for the top spin, $\eta = 1$. As initial conditions for $t = 0$, the following initial values are used:

$$x(0) = 0, \ z(0) = h, \ \dot{x}(0) = v_0 \cos(\theta), \ \dot{z}(0) = v_0 \sin(\theta) \tag{11-15}$$

where v_0 is the magnitude of the initial velocity vector $\vec{v}_0$ and where θ is an angle between $\vec{v}_0$ and the x-axis.

For our simulations, we will use two conditions, which are the ball in a vacuum and the ball in the air.

For the case of a ball in a vacuum, one force acting upon the ball is gravitational force, and thus, Equations (11-13) and (11-14) will have a simple form: $\ddot{x} = 0$, $\ddot{z} = -g$.

For the case of a ball in the air, to find the numerical solutions of the problem, we rewrite Equation (11-13) and Equation (11-14) of second-order ODEs via a system of first-order ODEs by introducing new variable names, as shown here:

$$\begin{cases} \dot{x} = v_x \\ \dot{z} = v_z \\ \dot{v}_x = -C_D\, \alpha\, v\, v_x + \eta C_M\, \alpha\, v\, v_z \\ \dot{v}_z = -g - C_D\, \alpha\, v\, v_z - \eta C_M\, \alpha\, v\, v_x \end{cases} \qquad (11\text{-}16)$$

where $v = \sqrt{v_x^2 + v_z^2}$. The initial conditions are $x(0) = 0$, $z(0) = h$, $v_x(0) = v_0 \cos(\theta)$, and $v_z(0) = v_0 \sin(\theta)$. For our simulations, we use the following numerical values [1]:

$$g = 9.8 \left[\frac{m}{s^2} \right]$$

$$\text{Diameter of the tennis ball} = 0.063\,[m]$$

$$\text{Mass}: m = 0.05\,[kg]$$

$$\text{Air density: } \rho = 1.29 \left[\frac{kg}{m^3} \right]$$

Moreover, for the initial conditions, we will use $h = 1\,[m]$, $v_0 = 25 \left[\frac{m}{s} \right]$, and $\theta = 15^0$.

For the spin of the ball, we will use $\omega = 20 \left[\frac{1}{s} \right]$ and $\eta = 1$ for the top spin.

The simulation models of the tennis ball trajectory were developed in MATLAB and Simulink. The analytical solution for case #1 when a tennis ball is hit in a vacuum is in Thrown_Ball.m. The simulation results from the script are shown in Figure 11-4.

```
% Thrown_Object.m. Case 1. Analytical solution - Ball traj. in %vacuum
x_z_vac=dsolve('D2x=0', 'D2z=-g', 'x(0)=0',...
'Dx(0)=v0*cos(theta*pi/180)', 'z(0)=h', ...
'Dz(0)=v0*sin(theta*pi/180)', 't');
xt=x_z_vac.x   %#ok
zt=x_z_vac.z   %#ok
g=9.8;         % gravitational acceleration
h=1;           % ball hit 1 m above ground
theta=15;      % ball hit under 15 degrees
v0=25;         % initial velocity of ball
```

```
xt=vectorize(xt); zt=vectorize(zt); t=linspace(0,1.5, 200);
xt=eval(xt); zt=eval(zt); z_i=find(zt==min(abs(zt)));
touch_gr=xt(z_i); t_touch = t(z_i);
plot(xt, zt,'ko', 'markersize', 4,'MarkerFaceColor','y'), grid on
xlim([0, touch_gr]);
text0=('Ball is hit h=1[m] above ground & under \theta=15^0');
gtext(text0, 'fontsize', 13);
text1=['Ball hits the ground in distance (of x) ' ... num2str(touch_gr)
'[m]']; gtext(text1);
text2=['Ball hits the ground after : 'num2str(t_touch)' [sec]'];
gtext(text2);
title('Trajectory of TENNIS ball hit in VACUUM, v_0=25 [m/s] ')
xlabel('x, (horizontal) [m]'), ylabel('z, [m]')
```

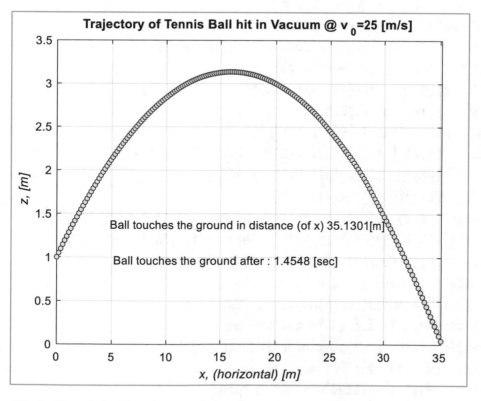

Figure 11-4. *Tennis ball trajectory in a vacuum*

The next simulation model, Thrown_Ball_All.m, computes the trajectory of the ball in a vacuum, in the air with a spin, and in the air without spin. The simulation results obtained from the script are shown in Figure 11-5.

```
% Thrown_Ball_All.m
h =1;        % [m];
v0=25;       % [m*s^-1];
theta=15;    % [deg];
ro=1.29;     % [kg*m^-3];
g=9.81;      % [m/s^2]
d=.063;      % [m];
m=.05;       % [kg];
w=20;        % [m*s^-1]; angular velocity of a spinning ball
eta=1;       % describes direction(+-) and presence of rotation;
                  % eta=1 is for top-spin.
alfa=ro*pi*d^2/(8*m); time=0:.01:1.5;
ICs=[0, 1, v0*cos(theta*pi/180), v0*sin(theta*pi/180)];
u   = @(t,x)sqrt(x(3).^2+x(4).^2);
CD = @(t,x)(0.508+(1./(22.053+4.196*(u(t,x)./w).^(5/2))).^(2/5));
CM = @(t,x)(1/(2.022+.981*(u(t,x)./w))));
vacuum = @(t,x)([x(3); x(4); 0; -g]);
nospin = @(t,x)([x(3); x(4); (-1)*CD(t,x)*alfa*u(t,x)*x(3);
            (-1)*g-CD(t,x)*alfa*u(t,x)*x(4)]);
topspin= @(t,x)([x(3); x(4);
    (-1)*CD(t,x)*alfa*u(t,x)*x(3)+eta*CM(t,x)*alfa*u(t,x)*x(4);
    (-1)*g-CD(t,x)*alfa*u(t,x)*x(4)-eta*CM(t,x)*alfa*u(t,x)*x(3)]);
[tvac, XZvac]= ode45(vacuum, time, ICs, []);
[tns, XZns]  = ode45(nospin, time, ICs, []);
[tts, XZts]  = ode23(topspin, time, ICs, []);
% Case #1. Ball is hitted in the Vacuum
z_i=find(abs(XZvac(:,2))<=min(abs(XZvac(:,2)))));
t_gr=XZvac(z_i,1); t_t = time(z_i);
% Case #2. Ball is hitted without a spin
z1_i =find(abs(XZns(:,2))<=min(abs(XZns(:,2)))));
t_gr1=XZns(z1_i,1); t_t1 = time(z1_i);
% Case #3. Ball is hitted with a Top-spin
```

```
z2_i =find(abs(XZts(:,2))<=min(abs(XZts(:,2)))));
t_gr2=XZts(z2_i,1); t_t2 = time(z2_i);
plot(XZvac(:,1), XZvac(:,2), 'bo', 'linewidth', 1), grid; hold on
plot(XZns(:,1), XZns(:,2), 'm', 'linewidth', 2)
plot(XZts(:,1), XZts(:,2), 'ko-', 'linewidth', 1.0)
legend('In Vacuum','No-Spin in Air','Top-Spin in Air'),
ylim([0, 3.35])
title 'Trajectory of a Tennis Ball hit under \theta=15^0, h=1 m'
xlabel 'Length, [m]'; ylabel 'Height, [m]'
tt1=['Vacuum: Ball hits the ground (x-axis):' num2str(t_gr) '[m]'];
gtext(tt1, 'backgroundcolor', 'w');
tt1a=['Vacuum: Ball hits the ground after: ' num2str(t_t) '[s]'];
gtext(tt1a, 'backgroundcolor', 'w');
tt2=['No-spin: Ball hits the ground: ' num2str(t_gr1) ' [m]'];
gtext(tt2, 'backgroundcolor', 'w');
tt2a=['No-spin: Ball hits the ground: ' num2str(t_t1) ' [s]'];
gtext(tt2a, 'backgroundcolor', 'w');
tt3=['Top-spin: Ball hits the ground: ' num2str(t_gr2) ' [m]'];
gtext(tt3, 'backgroundcolor', 'w');
tt3a=['Top-spin: Ball hits the ground: ' num2str(t_t2) ' [s]'];
gtext(tt3a, 'backgroundcolor', 'w'); hold off
```

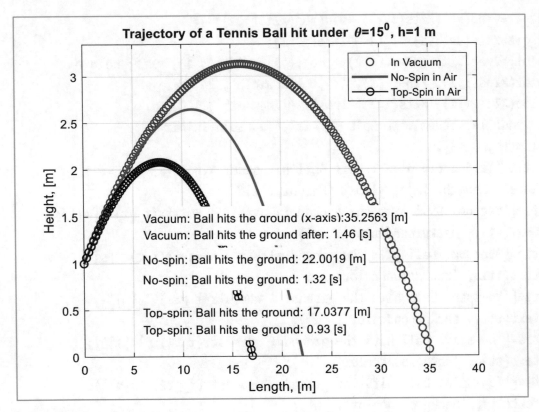

Figure 11-5. *Tennis ball trajectories in a vacuum and in the air with a top spin and no spin*

Another simulation approach to modeling a thrown ball trajectory is to build a Simulink model, TENNIS_Ball.slx, as shown in Figure 11-6 (and two subsystems, as shown in Figure 11-7 and Figure 11-8), to simulate three cases: in a vacuum, with a top spin in the air ($\eta = 1$), and without a spin in the air ($\eta = 0$).

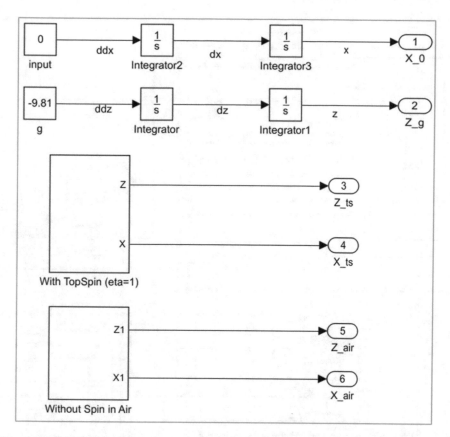

Figure 11-6. *The main Simulink model of the tennis ball trajectory*

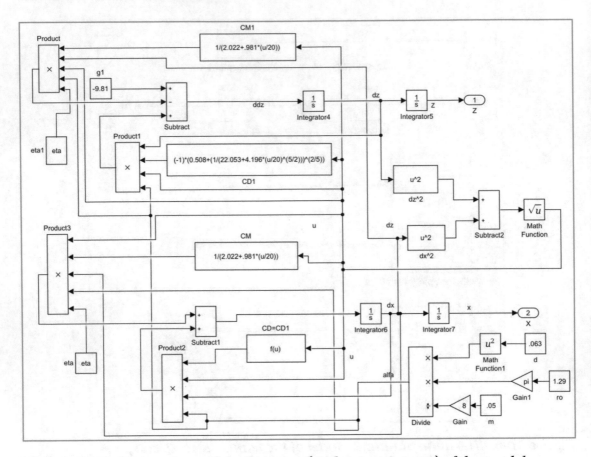

Figure 11-7. *A top-spin model subsystem (with top-spin: η=1) of the model TENNIS_Ball.slx*

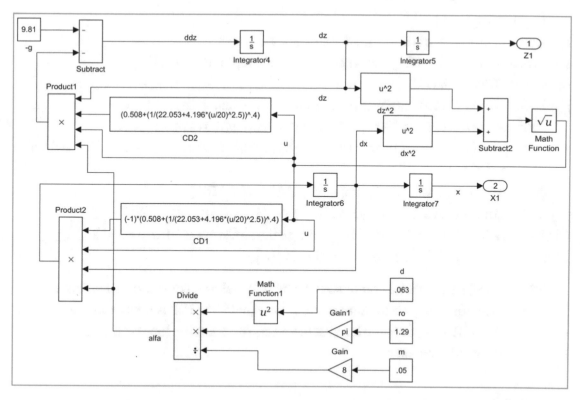

Figure 11-8. *No-spin (without spin: $\eta = 0$) subsystem of the model TENNIS_Ball.slx*

Note that when building the Simulink model `TENNIS_Ball.slx` shown in Figure 11-6, the [Function] block is employed for algebraic expressions of C_D and C_M to simplify the model construction. In addition, all initial conditions including $\dot{z}(0) = v_0 \cos(\theta)$ and $\dot{x}(0) = v_0 \sin(\theta)$ are inserted into the model via [Integrator] blocks by internal initial condition sources. The vacuum part of the main Simulink model of the problem is composed of [Constant] input, [Integrator], and [Scope] blocks. The air part with a top spin and no spin are modeled in subclasses and put under two subsystems, as shown in Figure 11-7 and Figure 11-8.

When the Simulink model `TENNIS_Ball.slx` with the fixed step-size solver `ode8`, as shown in Figure 11-6, Figure 11-7, and Figure 11-8, is executed by using the following script, `Sim_TENNIS_Ball.m`, we obtain identical simulation results (the plot figure is not displayed in this context) as the previous results displayed in Figure 11-5.

```
% Sim_TENNIS_Ball.m
% SIMULINK models with a fixed step solver ode8
set_param('TENNIS_ball','Solver','ode8','StopTime','1.5')
out=sim('TENNIS_ball');
plot(out.yout{2}.Values.Data,out.yout{1}.Values.Data, 'g',...
    out.yout{4}.Values.Data,out.yout{3}.Values.Data, 'c:',...
    out.yout{6}.Values.Data,out.yout{5}.Values.Data,'r-.', ... 'linewidth',
    3), grid on
legend('In Vacuum via SIM','NoSpin in Air via SIM',...
'TopSpin in Air via SIM'), ylim([0, 3.35])
title('Trajectory of Tennis Ball hit under \theta=15^0 & h=1 m')
xlabel('Length, [m]'), ylabel('Height, [m]')
```

We can see, again, that a higher spin of the ball makes it cover a shorter distance. Moreover, the numerical simulation results show that if the ball is hit in a vacuum (or air drag is neglected), it reaches the longest distance and takes the longest flying time, as known from physics courses.

References

[1] Stepanek, A., "The Aerodynamics of Tennis Balls: The Topspin Lob," *American Journal of Physics*, 56, 1988, pp. 138–142.

[2] Richardson, E. G., *Dynamics or Real Fluids*, Edward Arnold (1961).

[3] Gander, W., Hrebicek, J., *Solving Problems in Scientific Computing Using Maple and MATLAB* (4th ed.), Springer (2004).

CHAPTER 12

Simulation Problems

This chapter contains mathematical and simulation models of two examples: the Lorenz system and the Lotka-Voltera problem.

Lorenz System

The Lorenz system is a system of ODEs first studied and observed by Edward Lorenz while performing weather forecasting analyses in 1963. He then developed a simplified mathematical formulation for atmospheric convection [1]. The system currently has wide applications in forecasting chaotic processes observed in nature and in modeling atmospheric convection phenomena, model dynamos [2], brushless DC motors [3], electric circuits [4], lasers [5], and chemical reactions [6] in simplified forms.

The Lorenz system is composed of three Lorenz equations, as shown here:

$$\begin{cases} \dfrac{dx}{dt} = \sigma\left(y - x\right) \\[2mm] \dfrac{dy}{dt} = x\left(\rho - z\right) - y \\[2mm] \dfrac{dz}{dt} = xy - \beta z \end{cases} \tag{12-1}$$

where x, y, and z are the system state variables. t is time, and σ, β, and ρ are the system parameters.

The Lorenz system is nonlinear and three-dimensional. Depending on the values of system parameters σ, β, and ρ, the Lorenz system behavior differs. These parameter values are shown for demonstration purposes in this section: $\sigma = 10$, $\beta = \dfrac{8}{3}$, $\rho = 28$. As noted, very small changes in the Lorenz system parameter values will lead to dramatic changes in the overall system behavior but not always chaotic behaviors in solutions over longer evolutions.

© Sulaymon L. Eshkabilov 2020
S. L. Eshkabilov, *Practical MATLAB Modeling with Simulink*, https://doi.org/10.1007/978-1-4842-5799-9_12

We have developed a simulation script called the `LORENZ_functions.m`
model of the Lorenz equations (with the sample initial conditions of
$x(0) = -8, y(0) = 8, z(0) = 27$ and $= 10, \beta = \dfrac{8}{3}, \rho = 28,$ and t=0:.005:25). It takes two input
arguments, specifically, time ranges and initial conditions.

```
function LORENZ_functions(t,ICs)
% LORENZ Functions
if nargin<1
ICs=[-8, 8, 27]; t=0:.005:25;
end
Sigma=10; Beta=8/3; Ro=28;
% Dx/Dt=-Sigma*X+Sigma*Y;
% Dy/Dt=- Y-X*Z;
% Dz/Dt=-Beta*Z+X*Y-Beta*Ro;
Lorenz_sys = @(t, X)([-Sigma*X(1)+Sigma*X(2);-X(2)-X(1).*X(3); ...
-Beta*X(3)+X(1).*X(2)-Beta*Ro]);
[time, fOUT]=ode45(Lorenz_sys, t, ICs, [ ]);
close all; figure
plot3(fOUT(:,1), fOUT(:,2), fOUT(:,3)), grid
xlabel('x(t)'), ylabel('y(t)'), zlabel('z(t)')
title('LORENZ functions x(t) vs. y(t) vs. z(t)'), axis tight
figure; comet3(fOUT(:,1), fOUT(:,2), fOUT(:,3))% WATCH Animations
figure; subplot(311); plot(time, fOUT(:,1), 'b','linewidth', 3), grid
minor; title('LORENZ functions x(t), y(t), z(t)'), xlabel('time'),
ylabel('x(t)')
subplot(312); plot( time', fOUT(:,2), 'r', 'linewidth', 2 ),
grid minor; xlabel 'time', ylabel 'y(t)'
subplot(313); plot(time, fOUT(:,3),'k', 'linewidth', 2),
grid minor
xlabel('time'), ylabel('z(t)')
figure; plot(fOUT(:,1), fOUT(:,2), 'b', 'linewidth', 1.5)
grid minor, title('LORENZ functions'), xlabel('x(t)')
ylabel('y(t)'), axis square
figure; plot(fOUT(:,1), fOUT(:,3), 'k', 'linewidth', 1.5)
grid minor, title('LORENZ functions'), xlabel('x(t)')
ylabel('z(t)'),axis square
```

```
figure; plot(fOUT(:,2), fOUT(:,3), 'm', 'linewidth', 1.5)
grid minor, title('LORENZ functions'), xlabel('y(t)')
ylabel 'z(t)',axis square
```

After executing this script, we obtain the plots shown in Figures 12-1, 12-2, 12-3, 12-4, and 12-5.

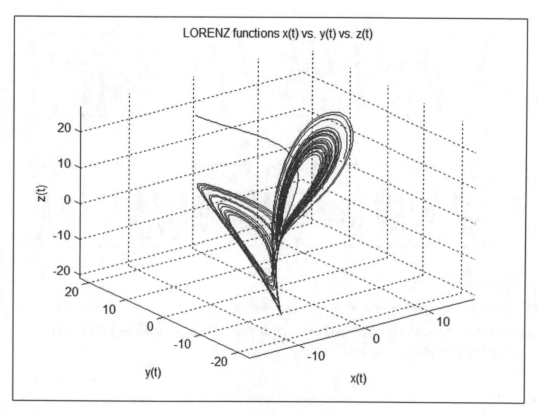

Figure 12-1. *Lorenz functions x(t) versus y(t) versus z(t)*

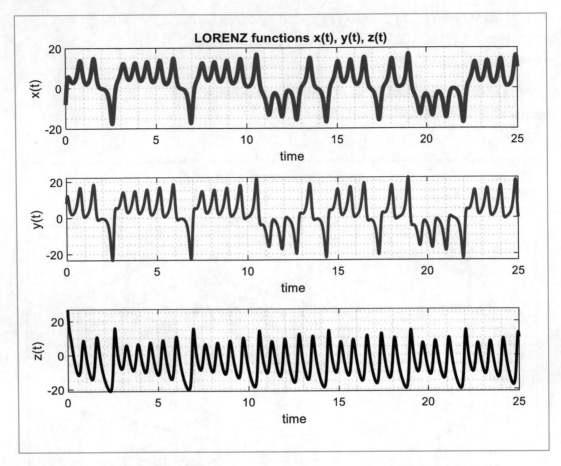

Figure 12-2. *Simulation resulsts of Lorenz functions (equations) x(t), y(t), z(t) vs. time shown in subplots*

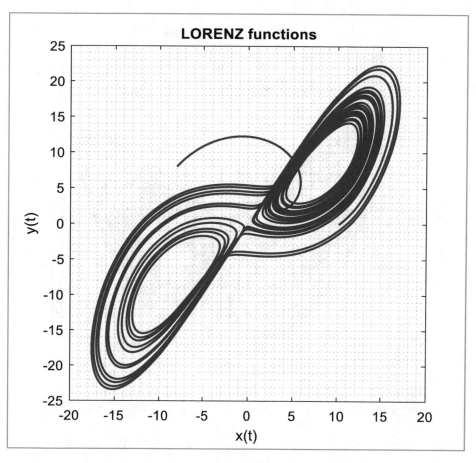

Figure 12-3. *Lorenz functions x(t) versus y(t)*

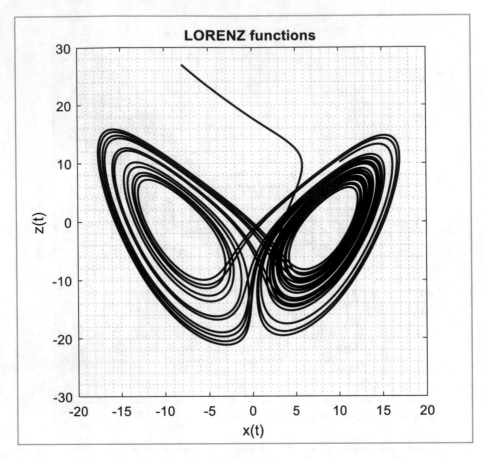

Figure 12-4. *Lorenz functions x(t) versus z(t)*

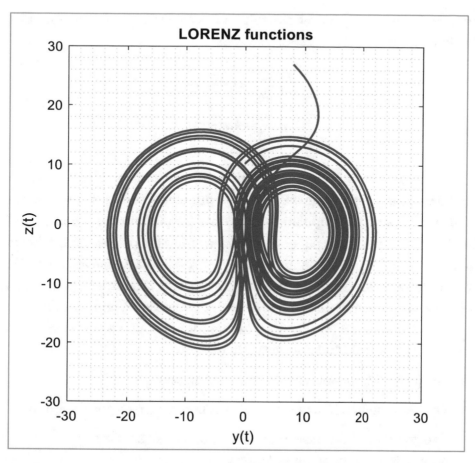

Figure 12-5. *Lorenz functions y(t) versus z(t)*

The accuracy of the numerical solutions of the developed simulation models, based on the ode45 solver, is very much dependent on the step size. Similarly, the numerical solutions of this exercise can be obtained by using other ODE solvers and Simulink modeling or by writing MATLAB scripts with the Euler, Runge-Kutta, Adams, or Adams-Moulton method.

Lotka-Voltera Problem

The Lotka-Voltera model was initially proposed by Alfred Lotka in 1910 [7]. The Lotka-Voltera model of differential equations is one of the specific examples of the Kolmogorov model [8]. The Lotka-Voltera problem equations, known as the *predator-prey equations*, depict predator-prey relationships observed in nature and are formulated in the following way with a pair of first-order ODEs:

$$\begin{cases} \dfrac{dx}{dt} = \alpha x - \beta xy \\ \dfrac{dy}{dt} = \delta xy - \gamma y \end{cases}$$ (12-2)

where the x variable can be considered as the density or number of preys (e.g., rabbits); the y variable represents the density or number of predators (e.g., wolves); α is the intrinsic rate of prey population increase; β is a predator rate coefficient; δ is reproduction rate of predators per one prey eaten; and γ is a predator mortality rate.

The nonlinear differential equations in the Lotka-Voltera model are frequently employed to describe the dynamics of biological systems of nature [8, 9] where two species interact to survive. One of the two species is a predator, and the other is a prey.

The Lotka-Voltera model makes a few assumptions about the environment and evolution of the preys and predators, listed here:

- The preys find an abundance of food at all times.

- The food supply of the predators depends on the prey's population size.

- The growth and decrease rates of the population are directly proportional to the population size.

- During the whole process, the environment does not change to favor any one species, and none of the species adapts to the environment in the model. The solutions of the Lotka-Voltera equations are deterministic and continuous, which implies that the number of the preys and the number of the predators overlap each other continually.

To simulate the Lotka-Voltera model, we'll use the following numerical values and initial conditions (number of preys and predators): $\alpha = 1$, $\beta = 0.05$, $\delta = 0.02$, $\gamma = 0.5$ and $x(0) = 10$, $y(0) = 10$. Here is the model:

```
% Lotka_Voltera.m
% LOTKA-VOLTERA ~ Predator-Prey ~ two body simulation %%%%
% dx/dt = alfa*x-beta*x*y;
% dy/dt = delta*x*y-gamma*y;
Alfa = 1; Beta = 0.05; Delta = 0.02; Gamma = 0.5;
```

```
% NOTE: x(1)=x and x(2)=y
dxy=@(t,x)([Alfa*x(1)-Beta*x(1).*x(2);...
    Delta*x(1).*x(2)-Gamma*x(2)]);
time=[0, 20];
ICs=[10 10];        % Initial number of preys and predators
%   Since this has two populations, we pass the array [1 2].
%   Options=odeset('reltol', 1e-5),'NonNegative', [1 2]);
opts=odeset('reltol', 1e-5, 'NonNegative', [1 2]);
[t, xy]=ode45(dxy, time, ICs, opts);
plot(t, xy, 'linewidth', 2),
grid minor, legend('Prey','Predator')
title('Lotka-Voltera function: predator/prey two-body sim.')
figure
plot(xy(:,1),xy(:,2)), grid on; hold on
ICs=[40 20];
opts=odeset('reltol', 1e-5, 'NonNegative', [1 2]);
[t, xy]=ode45(dxy, time, ICs, opts);
plot(xy(:,1),xy(:,2), 'r--', 'linewidth', 1.5), grid on
ICs=[60 20];
opts=odeset('reltol', 1e-5, 'NonNegative', [1 2]);
[t, xy]=ode45(dxy, time, ICs, opts);
plot(xy(:,1),xy(:,2), 'mo-', 'linewidth', 1.5), grid on
ICs=[80 20];
opts=odeset('reltol', 1e-5, 'NonNegative', [1 2]);
[t, xy]=ode45(dxy, time, ICs, opts);
plot(xy(:,1),xy(:,2), 'kx:', 'linewidth', 1.5), grid on
ICs=[Alfa/Beta Gamma/Delta];
opts=odeset('reltol', 1e-5, 'NonNegative', [1 2]);
[t, xy]=ode45(dxy, time, ICs, opts);
plot(xy(:,1),xy(:,2), 'gx-', 'linewidth', 1.5), grid on
title('Lotka-Voltera model: 20 predators vs.~ preys')
xlabel('x(t), prey rate'), ylabel('y(t), predator rate')
legend('x_0/y_0=1','x_0/y_0=2','x_0/y_0=3','x_0/y_0=4',...
'ICs=[\alpha/\beta \gamma/\delta]'), hold off
```

Figure 12-6 and Figure 12-7 show the simulation results from the script.

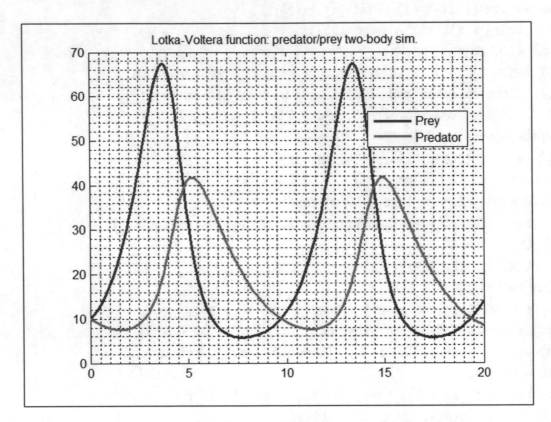

Figure 12-6. *Lotka-Voltera simulation with 20 preys and 20 predators*

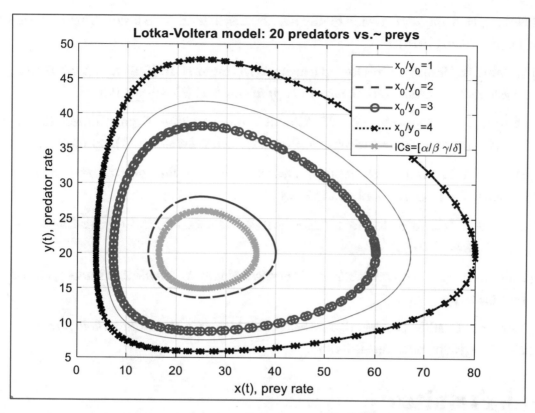

Figure 12-7. *Lotka-Voltera model simulation with 20 predators versus 20, 60,*
and 80 preys, as well as a special case, which is that preys are $\frac{\alpha}{\beta} = 20$ and predators
are $\frac{\gamma}{\delta} = 25$

The model approach of the Lotka-Voltera problem can be applied not only in studies of biological processes in nature but also in the modeling dynamics of development trends in various industries with disturbing factors.

References

[1] Berge, P., Pomeau, Y., Vidal, Ch., *Order within Chaos: Towards a Deterministic Approach to Turbulence.* New York: John Wiley & Sons (1984).

[2] Gorman, M., Widmann, P.J., Robbins, K.A., "Nonlinear dynamics of convection loop: A quantitative comparison of experiment with theory." Physica D 19 (2): 255–267 (1968).

[3] Haken, H. "Analogy between higher instabilities in fluids and lasers." *Physics Letters* A 53 (1): pp. 77–78 (1975).

[4] Hemati, N. "Strange attractors in brushless DC motors." *IEEE Transactions on Circuits and Systems I: Fundamental Theory and Applications* 41 (1): 40–45 (1994).

[5] Cuomo, K. M., Oppenheim, A. V., "Circuit implementation of synchronized chaos with applications to communications." Physical Review *Letters* 71 (1): pp. 65–68 (1993).

[6] Hilborn R. C., *Chaos and Nonlinear Dynamics: An Introduction for Scientists and Engineers* (2nd ed.) Oxford University Press (2000).

[7] Lotka, A.J., "Contribution to the Theory of Periodic Reaction," *J. Physics-Chemistry*, 14 (3) pp. 271–274 (1910).

[8] Freedman, H. I., *Deterministic mathematical Models in Population Ecology*, Marcel Dekker (1980).

[9] Brauer F., Castillo-Chavez, C., *Mathematical Models in Population Biology and Epidemiology*. Springer-Verlag (2000).

Drill Exercises

The following are exercises to try on your own.

Exercise 1

Perform the following:

- Build two simulation models of the Lorenz system in Simulink with the solver ode2 and in MATLAB using Heun's method by writing a script (not using ODEx solvers).

- Simulate both models and compare their results to the simulation results from a MATLAB script model using Heun's method (not using ODEx solvers).

Exercise 2

Perform the following:

- Build a Simulink model of the Lotka-Voltera problem with ode8 and develop a four-step Adams-Moulton method based on a MATLAB script model (not using ODEx solvers, e.g. ode23, ode45, ode113, etc.) of the Lotka-Voltera problem.

- Simulate both models and compare their results.

Exercise 3

Control is an important part of engineering, and there are many systems that are modeled with differential equations. Here is one such example for a complex control problem that is formulated with the following equation: $\ddot{y} + \left(3 + \dot{y}^2\right)\dot{y} + \left(1 + y + y^2\right)u(t) = 0$, where $u(t) = \sin(t)$. The given initial conditions are $y(0) = 1$ and $\dot{y}(0) = 0$.

a. Build a simulation model of the problem in Simulink and employ a function block to compute algebraic operations. Use ode23.

b. Write a simulation script of the problem using ode45.

c. Build a simulation model of the problem using Euler's improved method.

d. Simulate the models from (a), (b), and (c). What is the adequate step size for Euler's improved method to converge with solutions from the other two methods?

Exercise 4

Perform the following:

a) Add one additional mass ($m_0 = 1[kg]$) in a three-DOF mass-spring-damper system (shown in Figure 10-16 in Chapter 10) before mass 1. Add two dampers ($C_0 = 1\left[\dfrac{Nm}{s}\right]$) on both sides of a new mass (m_0), add one spring ($K_0 = 25\left[\dfrac{N}{m}\right]$) in between two masses, do not add applied force ($F = 0$ [N]), and remove a damper (C_1) and spring (K_1). The rest of the system will remain the same as given. Build the physical model's schematic view, build a free-body diagram, and develop a mathematical formulation of the four-DOF system's equation of motion.

b) Build a simulation model in Simulink using ode23 and develop a simulation script with the ode45 solver of a new four-DOF spring-mass-damper system from (a).

c) Simulate both models from (b) and compare their results.

Exercise 5

Perform the following:

– Build a MATLAB ode23 solver–based simulation model of the microphone model from the formulations of (9-3) and (9-4) from Chapter 9.

– Simulate your model and compare with the simulation results with the Simulink model of the microphone given in Chapter 9.

– Write a MATLAB script to simulate the system response in terms of membrane displacement and the current flow of the microphone model from formulations of (9-3) and (9-4) from Chapter 9 using MATLAB's built-in ODE solvers ode23tb, ode45, and ode113, and then compare the solutions.

– Simplify the Simulink model of the microphone using an embedded MATLAB function block to compute the current flow in the system.

Exercise 6

Perform the following:

– Build a MATLAB ode23 solver–based simulation model and three-step Adams-Moulton method–based MATLAB script model of a DC motor with a flexible load.

– Simulate the models and compare them with the simulation results of the Simulink model of the DC motor with flexible load.

Exercise 7

Perform the following:

   a) By writing a MATLAB script and employing ode23, develop a simulation model of a single-DOF spring-mass-damper system excited by a step function (Heaviside function $\Phi(t - 1)$) with a magnitude of 10 N, a mass of M=10 kg, a damping of C=50 $\left[\dfrac{Nm}{s}\right]$, and a stiffness of K=100 $\left[\dfrac{N}{m}\right]$,. Hint: Use the built-in function stepfun().

   b) Develop a simulation model in Simulink for the case when the system in part (a) is excited by an impulse force (Dirac delta function $\delta(t)$), e.g., with a magnitude of 1000 N at $t = 3$ sec.

   c) Simulate the models in parts (a) and (b) for a period of 0 to 90 seconds.

Exercise 8

Perform the following:

   a) Write a mathematical formulation of a two-DOF spring-mass-damper system by considering dry friction (c_1, c_3, and $c_1 = c_3$) under two masses to be nonlinear (Coulomb friction) damping. Hints: Use a signum function.

   b) Build a simulation model of the system from part (a) in Simulink.

   c) Compute the system responses using the Simulink model from part (b) and using the following parameter values: $F = 10\cos(25t)$, $C_2 = 1\left[\dfrac{Nm}{s}\right]$, $K_1 = 85\left[\dfrac{N}{m}\right]$, $K_2 = 75\left[\dfrac{N}{m}\right]$, $m_1 = 5[kg]$, $m_2 = 3.5[kg]$, and friction coefficient value $\mu = 1.05$.

Exercise 9

Perform the following:

a) Redesign the Simulink model 3-DOF spring-mass-damper system given in Chapter 9 and simplify it for the forced vibration case.

b) Write a MATLAB script using the 4/5-order Runge-Kutta (not using ode45) to compute the system response of the three-DOF spring-mass-damper system for the free vibration case.

c) Write a MATLAB script to compute the system response of three masses using the backward Euler method for the forced vibration case.

d) Make the necessary changes to animate the system responses using the simulation results from parts (a) and (b).

Exercise 10

Define a system response of the two-DOF spring-mass-damper system shown in Figure 9-25 using modal analysis for the following three cases:

Case #1. $F(t) = 10 \sin (20t)[N]$; initial conditions:
$x_1(0) = x_2(0) = 0; \dot{x}_1(0) = \dot{x}_2(0) = 0$

Case #2. $F(t) = \delta(t)\ [N]$ (Impulse force : *Dirac Delta* function); initial conditions: $x_1(0) = x_2(0) = 0; \dot{x}_1(0) = \dot{x}_2(0) = 0$

Case #3. $F_1(t) = 0\ [N]$; initial conditions:
$x_1(0) = 0, x_2(0) = 0.5; \dot{x}_1(0) = \dot{x}_2(0) = 0$

Exercise 11

Define a system response of a 3-DOF spring-mass-damper system (shown in Figure 9-18) defined with the following:

$$[M] = \begin{bmatrix} 2.5 & 0 & 0 \\ 0 & 2 & 0 \\ 0 & 0 & 3 \end{bmatrix}, [C] = \begin{bmatrix} 3 & -0.5 & 0 \\ -0.5 & 2.5 & -0.5 \\ 0 & -0.5 & 3 \end{bmatrix}, [K] = \begin{bmatrix} 30 & -5 & 0 \\ -5 & 25 & -5 \\ 0 & -5 & 30 \end{bmatrix},$$

Is it feasible to use modal analysis to compute the system responses analytically for this system if the given initial conditions are as follows?

Case #1. $F_1(t) = 0[N]$; $F_2(t) = 0\ [N]$; $F_3(t) = \delta(t)[N]$ (Impulse force: Dirac delta function); with initial conditions:

$$x_1(0) = x_2(0) = x_3(0) = 0;\ \dot{x}_1(0) = \dot{x}_2(0) = \dot{x}_3(0) = 0$$

Case #2. $F_1(t) = F_2(t) = F_3(t) = 0\ [N]$; Initial conditions:

$$x_1(0) = x_3(0) = 0,\ x_2(0) = 0.5;\ \dot{x}_1(0) = \dot{x}_2(0) = \dot{x}_3(0) = 0$$

Compute the numerical solutions of the system by employing the ode113 solver for both cases.

Exercise 12

Perform the following:

- Write a MATLAB script that computes the system response of the DC motor model using ode23 and ode45.

- Make the necessary changes in the Simulink model of the DC motor to compute the load by employing an embedded MATLAB function block.

Exercise 13

Perform the following:

- Write a MATLAB script to simulate the displacement and velocity of three different falling objects (with three different air-drag coefficient values: 0.2, 0.35, and 0.52) with a mass (M=5 kg) in animation plots.

- Write a MATLAB script to simulate the trajectory of a thrown ball with top spin and with no spin in animation plots.

Exercise 14

Multiple-story buildings subject to earthquakes can be modeled as a system of mass-spring coupled systems with displacements only in the horizontal direction. Let's consider a five-story building excited by the external force $F_i(t) = [F_0 \cos(\omega t); 1.12F_0 \cos(\omega t); 0.88F_0 \cos(\omega t); 0.5F_0 \cos(\omega t); 0.5F_0 \cos(1.15\omega t); 0.45F_0 \cos(1.25\omega t)]$. The mathematical formulation of the system can be written with the following system of differential equations:

$$\begin{cases} m_1\ddot{x}_1 = -k_1 x_1 + k_2(x_2 - x_1) + F_1(t) \\ m_2\ddot{x}_2 = -k_2(x_2 - x_1) + k_3(x_3 - x_2) + F_2(t) \\ m_3\ddot{x}_3 = -k_3(x_3 - x_2) + k_4(x_3 - x_2) + F_3(t) \\ m_4\ddot{x}_4 = -k_4(x_4 - x_3) + k_5(x_4 - x_3) + F_4(t) \\ m_5\ddot{x}_5 = -k_5(x_5 - x_4) + F_5(t) \end{cases}$$

where $x_1, x_2, ..., x_5$ are the locations of masses at the first, second, ..., fifth floor, respectively. $m_1, m_2, ..., m_5$ are the point masses of the first, second, ... fifth floor, respectively. Each floor is treated to have the point mass located at its center of mass. $k_1, k_2, ..., k_5$ are the stiffness of the first, second, ..., fifth floor, respectively, and the damping effects of each floor are ignored. Use the following numerical values of the building's physical parameters: $k_1 = k_2 = ... = k_5 = 10^6$ [N/m] (identical stiffness in each floor), $m_1 = 5500$ [kg], $m_2 = 5000$ [kg], $m_3 = 5500$ [kg], $m_4 = 6000$ [kg], $m_5 = 4750$ [kg], and external force $F_0 = 5000$ N applied with $\omega = 13$ [Hz].

Build simulation models of the system and simulate it for a period of $t = 0...15$ [sec] using the following methods:

1) By writing a script based on Euler's method

2) By creating a Simulink model (using the fixed-step solver ode5)

3) By writing a script with the 4/5-order Runge-Kutta method

Then compare the found numerical solutions from 1 (Euler's method), 2 (Simulink model), 3 (4/5-order Runge-Kutta method) and check the efficiency of each approach in terms of computation time.

Define the natural frequencies and mode shapes of the building.

Exercise 15

Let brine tanks A, B, and C be given volumes of 80, 60, and 30, respectively, as shown in Figure 12-8 (the problem context[1] and the problem conditions have been modified).

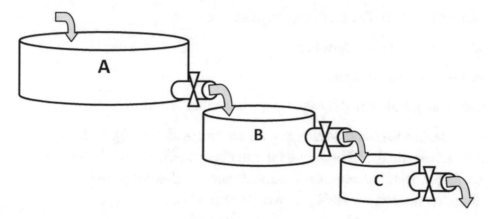

Figure 12-8. *Three brine tanks in cascade*

It is assumed that water enters tank A at the constant rate of R *(R=10)*, drains from A to B at the rate of R, drains from B to C at the rate of R, and then drains from tank C at the rate of R. Hence, the volumes of the tanks remain unchanged. Uniform stirring of each tank is assumed, which implies a uniform salt concentration throughout each tank. Let $y_1(t)$, $y_2(t)$, $y_3(t)$ denote the amount of salt at time t in each tank. Water is added to tank A containing 0 percent of salt. Therefore, the salt in all the tanks is eventually lost from the drains. The cascade is modeled by the chemical balance law: rate of change = input rate − output rate.

By applying the previous written balance law, we can write the system's model as the triangular deferential system.

$$\begin{cases} \dot{y}_1 = -\dfrac{1}{8}y_1 \\[2mm] \dot{y}_2 = \dfrac{1}{8}y_1 - \dfrac{1}{6}y_2 \\[2mm] \dot{y}_3 = \dfrac{1}{6}y_2 - \dfrac{1}{2}y_3 \end{cases}$$

---

[1]Taken from www.math.utah.edu/~gustafso/2250systems-de.pdf.

Let's assume that the following initial conditions (the amount of salt in tanks) are given: $y_1(0) = 1.5$, $y_2(0) = 1.5$, $y_3(0) = 1.5$.

Compute numerical solutions of the system depicted by a system of ODEs for $t = 0...15$.

- Write a function file called, e.g., BrineODE.m.

- Write an anonymous function.

- Write an inline function.

- Create a Simulink model called, e.g., BrineODEsim.mdl.

- Use ode23, ode45, and ode113 and a Simulink model (using the fixed-step solver ode3), and compare the solutions, checking the efficiency of each approach. Take smaller time steps and simulate your created Simulink model (BrineODEsim.mdl) from an m-file.

- Write a script file based on the 4/5-order Runge-Kutta method and compare the solutions found from the Runge-Kutta method script with the ones found with ode23, ode45, and ode113 and the Simulink model.

- Compute the analytical solutions of the system using dsolve() and Laplace transforms if it is feasible.

Exercise 16

Given a two-DOF rotational system (see Motor-Pump Gear Box), there is applied torque only on disk 2 that is $\tau_a(t) = 10$ Nm; mass moments of inertia: $J_1 = 1.5$ kgm^2, $J_2 = 0.5$ kgm^2, torsional stiffness: $k_1 = 1200 \dfrac{Nm}{rad}$, $k_2 = 2000 \dfrac{Nm}{rad}$, frictional damping: $c_1 = 12$ $Nm/$ sec, $c_2 = 22$ $Nm/$ sec. All initial conditions are zero.

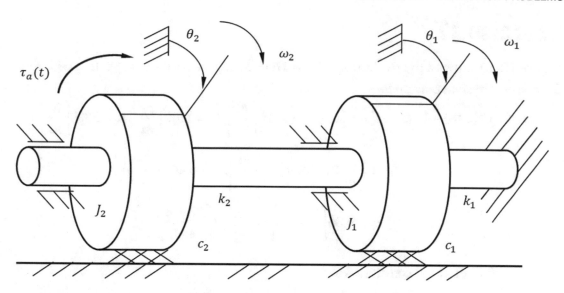

Figure 12-9. *Rotational system example*

Perform these exercises:

1) Derive the equations of motion of the system.

2) Derive the transfer function formulation $\dfrac{\Theta_2(s)}{T_a(s)}$ by applying the Laplace transforms and simulate it in Simulink and MATLAB. (Hint: use tf() and step().)

3) Develop a simulation model of the system by using the three-step Adams-Moulton method (by writing a script) and simulate it for a period of $t = 0\ldots2\ [sec]$ with a step size of $\Delta t = 0.01$.

4) Develop a Simulink model (use a variable step solver such as ode23s) with adjusted absolute and relative error tolerances.

5) Write a function file for the system's equations of motion by using the matrix approach and simulate the system by employing ode113 with a time-fixed step of $\Delta t = 0.01$ for the period of $t = 0\ldots2\ [\text{sec}]$.

6) Compare the computed numerical solutions from the above exercise 2, 3, 4, and 5, and assess the accuracy of each approach.

7) Define the natural frequencies and mode shapes of the building system.

Exercise 17

Figure 12-10 shows a lightly damped three-DOF spring-mass-damper system with the following physical parameters:

$$m_1 = 1\,[kg],\ m_2 = 1\,[kg],\ m_3 = 3\,[kg], k_1 = 1110\left[\frac{N}{m}\right], k_2 = 4440\left[\frac{N}{m}\right], k_3 = 3330\left[\frac{N}{m}\right];$$

$$k_4 = 5550\left[\frac{N}{m}\right]; c_1 = 0.4\left[\frac{N-m}{\sec}\right]; c_2 = 0.44\left[\frac{N-m}{\sec}\right];$$

$$c_3 = 0.66\left[\frac{N-m}{\sec}\right]; c_4 = 0.88\left[\frac{N-m}{\sec}\right].$$

Note that all initial conditions are zero.

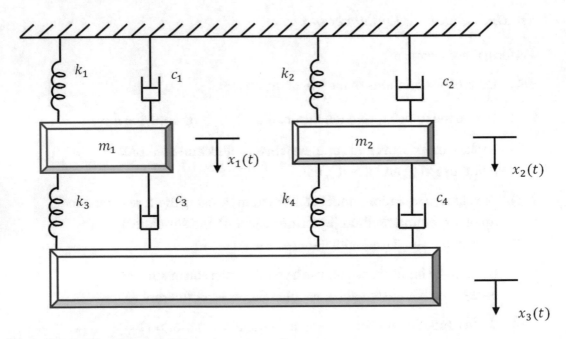

Figure 12-10. *Three-DOF spring-mass-damper system*

Perform these exercises:

- Derive the equations of motion of the system.

- Derive the transfer function formulation by applying the Laplace transforms: $\dfrac{X_3(s)}{F(s)}$.

– By applying the modal analysis, define the natural frequencies and modal damping ratios.

– Compute the frequency response of the system for 0... 100 $\left[\dfrac{rad}{sec}\right]$ and the phase plot by using the bode() and freqresp() functions of the control system toolbox or by substituting $s = j\omega$ in the system's transfer function formulation.

– Simulate the system's free responses (displacements) of $x_1(t)$, $x_2(t)$, and $x_3(t)$ in the time domain by using ode45 if the system has zero initial conditions, e.g., $x_2(0) = 0$ $[m]$; $x_3(0) = 0$ $[m]$; $x_1(0) = 0$; $\dot{x}_1(0) = 0[m/\sec]$; $\dot{x}_2(0) = 0[m/\sec]$; $\dot{x}_3(0) = 0[m/\sec]$. Why are the system's free responses zero in this simulation?

– Simulate the system's free responses for displacements of $x_1(t)$, $x_2(t)$, and $x_3(t)$ in the time domain by using ode23tb if the system has nonzero initial conditions, e.g., $x_2(0) = 0.05$ $[m]$; $x_3(0) = -0.025$ $[m]$; $x_1(0) = 0$; $\dot{x}_1(0) = 0[m/\sec]$; $\dot{x}_2(0) = 0[m/\sec]$; $\dot{x}_3(0) = 0[m/\sec]$.

Exercise 18

This example is a dynamics of love story. This is an interesting and fun example to study the dynamics of nonlinear systems based on a historical story set in Italy in the middle ages. As the story goes, there was a scholar and poet, Petrarch, who at the age of 23 fell in love with a beautiful, married, French woman named Laura. Over 21 years, Petrarch would write a collection of 366 poems to honor his love for Laura. These poems express a wide range of emotions, from extreme joy to deep depression. Rinaldi[2] developed a relatively simple model with differential equations expressing the emotions of Petrarch and Laura, re-creating a chronology for the various poems.

The model depicting the dynamics of love is composed of three coupled first-order differential equations developed by Rinaldi. In the model, Laura is described by a single variable $L(t)$ representing her attitude toward the poet Petrarch. Positive values indicate sympathy or affection, while negative values mean coldness and antagonism. The poet Petrarch is described by two variables: his love $P(t)$ for Laura, and his poetic

---

[2]Rinaldi, S. (1998). "Laura and Petrarch: An intriguing case of cyclical love dynamics." *SIAM Journal on Applied Mathematics*, 58, 1205–1221.

inspiration $I_P(t)$. High values of P correspond to delighted love, while low (negative) values indicate depression. The full Petrarch-Laura model developed by Rinaldi is as follows:

$$
\begin{cases}
\dfrac{dL}{dt} = -\alpha_L L(t) + \beta_L \left[P(t) \left(1 - \dfrac{P(t)^2}{\gamma_L^2} \right) + A_P \right] \\[4mm]
\dfrac{dP}{dt} = -\alpha_P P(t) + \beta_P \left[L(t) + \dfrac{A_L}{1 + \delta_P I_P(t)} \right] \\[4mm]
\dfrac{dI_P}{dt} = -\alpha_{IP} I_P(t) + \beta_{IP} P(t)
\end{cases}
\tag{a}
$$

where α_L, α_P, α_{IP} and β_L, β_P, β_{IP} are the feedback parameters that are positive numbers and represent emotions of Laura and Petrarch and Petrarch's inspiration to write poems. The constants A_P and A_L describe the physical appearance of Petrarch and Laura. The other two parameters, γ_L and δ_P, are Laura's and Petrarch's feedback emotions toward affection or inspiration.

Note that the system of differential equations in (a) is composed of three coupled first-order differential equations. For simulation models, take the following values:

$$
\alpha_L = 3\alpha_P, \ \alpha_{IP} = 0.1\alpha_P, \alpha_P = 1.2, \beta_L = \alpha_P, \beta_P = 5\alpha_P, \beta_{IP} = 10\alpha_P
$$

$$
A_L > 0, \ A_P < 0 \text{ and so}, \ A_L = 2, \ A_P = -1, \gamma_L = \delta_P = 1.
$$

Moreover, consider the initial conditions $L(0) = 3$, $P(0) = 1$, and $I_P(0) = 0$, and for the initial simulation use $t=[0, 10^4]$ sec.

- Develop a computer simulation model to study the dynamics of love story using Euler's improved and 4/5-order Runge-Kutta methods by writing scripts.

- Develop a simulation model using Ralston's and Milne's methods.

- Create a Simulink model with a variable step size solver (ode8).

– Compare the numerical solutions from (1) the improved Euler and Runge-Kutta methods, (2) Ralston's and Milne's methods, and (3) the Simulink model by plotting in reasonable time periods in 3D line plots (hint: use `plot3`) of $P(t)$ versus $L(t)$ versus $I_P(t)$ and simulate $P(t)$ versus $L(t)$ versus $I_P(t)$ (hint: use `comet3`).

– What if Petrarch's love for Laura is greater than Laura's love for Petrarch? How will the dynamics of love change?

– Is that feasible to compute analytical solutions of the problem using Symbolic MATH toolbox or MuPAD notes?

PART IV

Partial Differential Equations

CHAPTER 13

Solving Partial Differential Equations

A partial differential equation is "...a differential equation that contains unknown multivariable functions and their partial derivatives," according to Wikipedia [1]. There are many PDE applications in physics, engineering, and computer science. For example, PDEs are employed to formulate various phenomena, such as heat transfer, machine dynamics, wave propagation, elasticity, fluid dynamics, and many more.

The following is a generalized formulation of a PDE for a given function. It depends on more than one independent variable and on the partial derivative of the function with respect to more than one independent variable.

$$F\left(x, y, z, t, u(x, y, z, t), \frac{\partial u}{\partial x}, \frac{\partial u}{\partial y}, \frac{\partial u}{\partial z}, \frac{\partial u}{\partial t}, \frac{\partial^2 u}{\partial x^2}, \frac{\partial^2 u}{\partial y^2}, \frac{\partial^2 u}{\partial z^2}, \frac{\partial^2 u}{\partial t^2}\right) = 0 \qquad (13\text{-}1)$$

This is an example of a second-order or degree PDE. The solution of Equation (13-1) is the function $u(x, y, z, t)$, whose partial derivatives will satisfy the formulation given in Equation (13-1).

In Part 3, we demonstrated how to solve PDEs numerically using MATLAB's built-in PDE solver `pdepe()` and the Laplacian operator `del2()` and how to write scripts based on Gauss-Seidel and finite difference methods in the examples of one-dimensional and two-dimensional heat transfer problems and one-dimensional and two-dimensional wave propagation problems.

© Sulaymon L. Eshkabilov 2020
S. L. Eshkabilov, *Practical MATLAB Modeling with Simulink*, https://doi.org/10.1007/978-1-4842-5799-9_13

pdepe()

pdepe() can solve partial differential equations of the following form:

$$c\left(x,t,u,\frac{\partial u}{\partial x}\right)\frac{\partial u}{\partial t}=x^{-m}\frac{\partial y}{\partial x}\left(x^{m}f\left(x,t,u,\frac{\partial u}{\partial x}\right)\right)+s\left(x,t,u,\frac{\partial u}{\partial x}\right) \qquad (13\text{-}2)$$

where m represents the geometry of the system. If it is slab, $m = 0$ (Cartesian coordinate system), $m = 1$ is for a cylindrical geometry (a cylindrical coordinate system), and $m = 2$ is for a spherical shape (a spherical coordinate system).

There should be two boundary conditions; one is the initial condition, and the other is the boundary condition at each boundary.

$$u(x,t_0)=u_0(x) \qquad (13\text{-}3)$$

$$p(x,t,u)+q(x,t)f\left(x,t,u,\frac{\partial u}{\partial x}\right)=0 \qquad (13\text{-}4)$$

where $p(x, t, u)$ and $q(x, t)$ are some arbitrary functions. $f\left(x,t,u,\frac{\partial u}{\partial x}\right)$ is the function given in the main partial differential equation.

The following is the general syntax of pdepe():

```
Solution = pdepe(m, PDEfun, ICfun, BCfun, Xmesh, Tspan)
```

where m is the power in Equation (13-2) and indicates the geometry/coordinate system of the problem. PDEfun is a PDE expression of the given problem. ICfun is a given/known initial condition (s) of the given problem. BCfun is a given/known boundary condition (s) of the given problem. Xmesh is the space (distance) for which the solutions are to be computed, and Tspan is the time span over that period for which the given problem is to be simulated.

One-Dimensional Heat Transfer Problem

Let's consider the system shown in Figure 13-1 describing the heat transfer from one edge to the other edge (boundary) of the wall with a thickness of L. The system has two boundary conditions, $T(x, 0) = 0$ and $T(L, t) = 0$, which represent temperature values on the left and right edges, respectively. There is one heat source on the left side providing energy.

$$-k\frac{\partial T}{\partial x}\bigg|_{x=0} = q'' \qquad (13\text{-}4)$$

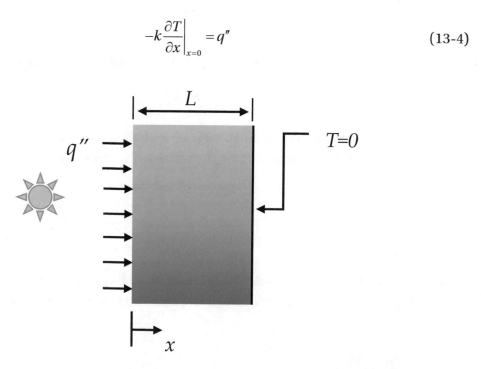

Figure 13-1. *Heat transfer from the left edge to the right edge of the wall*

Here is the heat transfer model of the system expressed with the next second-order partial differential equation:

$$k\frac{\partial^2 T}{\partial x^2} - \rho c_p \frac{\partial T}{\partial t} = 0 \qquad (13\text{-}5)$$

where T is the temperature, ρ is a density, c_p is the key heat conductivity component, and k is the thermal conductivity.

In the heat transfer problem shown in Equation (13-5),

$$c = \rho c_p \tag{13-6}$$

the parameter c given in Equation (13-2) is a constant, and thus, we can write the following as shown here:

$$f = k\frac{\partial T}{\partial x} \tag{13-7}$$

The source term is not present in our problem, and thus, it is equal to zero.

$$s = 0 \tag{13-8}$$

There are two boundary conditions: $T(L, t) = 0$ and $T(x, 0) = 0$.

Now we can rewrite Equation (13-4) by considering Equation (13-5) and the formulation $p + qf = 0$.

$$q'' + k\frac{\partial T}{\partial x}\bigg|_{x=0} = 0 \tag{13-9}$$

where $x = 0$, $p = q''$, and $q = 1$. Moreover, from the boundary condition $T(L, t) = 0$, we know that $x = L$, $p = T(L, t) = u_r$, and $q = 0$. Note the u_r boundary condition value at the right end.

Example 1

Let's consider that the temperature at the right end of the wall is $+20°C$, which means $p = T(L, t) = u_r - 20$.

Here is the solution script (Heat_Transfer.m) of the problem:

```
% Given Data:
global ro cp k q
k  = 666;        % [W/m-K]
L  = 0.25;       % [m]
ro = 1000;       % [kg/m^3]
cp = 500;        % [J/kg-K]
q  = 5e6;        % [W/m^2]
```

```matlab
tend = 5;          % [s]
m=0;               % Cartesian coordinate system
x = linspace(0, L, 200);
t = linspace(0, tend, 50);
% Solution Computed:
SOL=pdepe(m, @PDEfun, @ICfun, @BCfun, x, t);
TEMP=SOL(:,:,1);
figure(1)
plot(x, TEMP(end,:), 'b-', 'linewidth', 2), grid on
title(sprintf('Computed Solution @ t = %s', num2str(tend)))
xlabel('\it Distance, [m]')
ylabel('\it Temperature (\circC)')
figure(2)
plot(t, TEMP(:,1),  'k-', 'linewidth', 2), grid on
title(sprintf('Computed Solution @ t = %s', num2str(tend)))
xlabel('\it Time, [s]')
ylabel('\it Temperature (\circC)')
figure(3)
mesh(x, t, SOL)
xlabel('\it Distance, [m]')
ylabel('\it Time, [s]')
zlabel('\it Temperature (\circC)')
title('\it Solution of: $$ k*\frac{\partial^2{T}}{\partial{x^2}}-\rho*c_p\
frac{\partial{T}}{\partial{t}}=0 $$', 'Interpreter' ,'latex')
axis tight
function [c, f, s] = PDEfun(x, t, u, DuDx)
% PDE1
global ro cp k
c=ro*cp;
f=k*DuDx;
s=0;
end
function u0=ICfun(x)
% Initial Conditions:
u0=0;
end
```

```
function [pl, ql, pr, qr] = BCfun(xl, ul, xr, ur, t)
% Boundary Conditions
global q
pl=q;       % At the left edge where the heat source is: q+k*dT/dx=0
ql=1;
pr=ur;
qr=0;       % At the right edge: ur=0
end
```

Figures 13-2, 13-3, and 13-4 show the simulation results of the script over distance, time, and space and time, respectively.

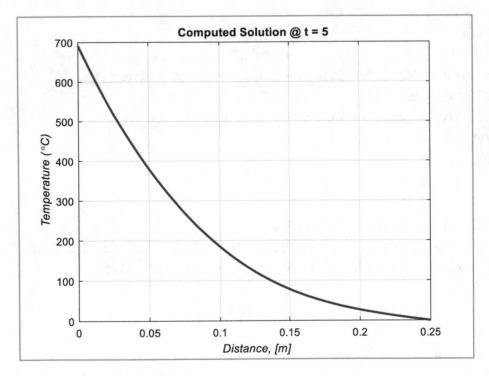

Figure 13-2. *Solution over distance*

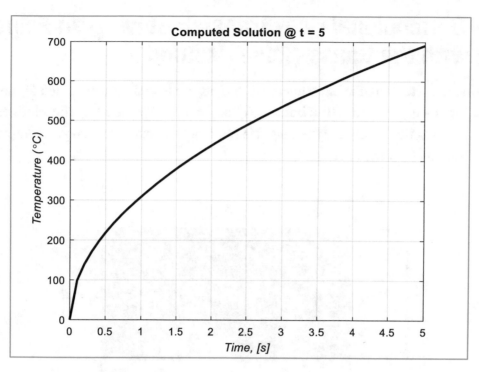

Figure 13-3. *Solution over time*

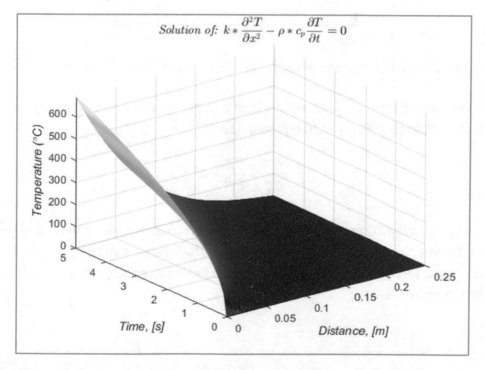

Figure 13-4. *Solution over space and time*

Two-Dimensional Heat Transfer: Solving an Elliptic PDE with the Gauss-Seidel Method

Let's consider a rectangular plate heated differently on four sides (on the left T_L, on the right T_R, on the top T_t, and on the bottom T_b), as shown in Figure 13-5. The width of the plate is W along the x-axis, and the height is H along the y-axis. How do you compute the temperature inside the plate?

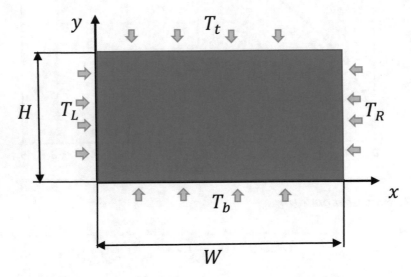

Figure 13-5. *Physical system: plate heated from all sides differently*

The mathematical formulation of the heat transfer in this given problem is the second-order elliptic PDE given in Equation (13-10).

$$\frac{\partial^2 T}{\partial x^2} + \frac{\partial^2 T}{\partial y^2} = 0 \tag{13-10}$$

Note that this is a 2D problem. The temperature is changing along the x- and y-axes. To solve this problem numerically, we can employ the Gauss-Seidel method [2]. To discretize the given system, we split the plate into finite rectangular elements and nodes, as shown in Figure 13-6.

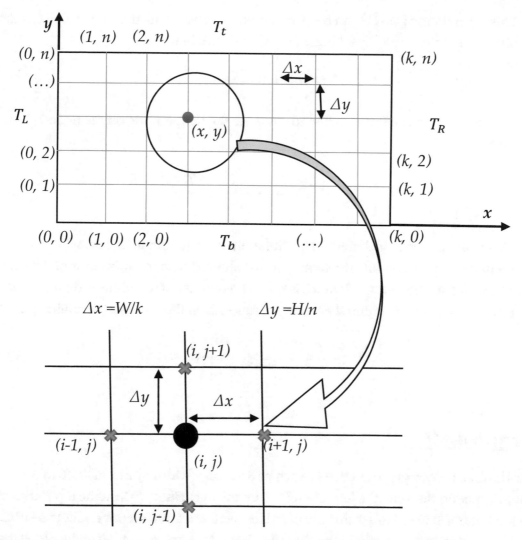

Figure 13-6. *Discretization scheme of the elliptic PDE*

Using the discretization scheme of the elliptic PDE of the heat transfer illustrated in Figure 13-7, we write the following discretized equation to compute the temperature on the node (i, j):

$$\left[\frac{T^k_{i-1,j} - 2T^k_{i,j} + T^k_{i+1,j}}{\Delta x^2} + \frac{T^k_{i,j-1} - 2T^k_{i,j} + T^k_{i,j+1}}{\Delta y^2} \right] = 0 \qquad (13\text{-}11)$$

This formulation (13-11) can be also written by considering the same values for Δx and Δy, or in other words, by taking $\Delta x = \Delta y$, as shown here:

$$T_{i+1,j} + T_{i-1,j} + T_{i,j+1} + T_{i,j-1} - 4T_{i,j} = 0 \qquad (13\text{-}12)$$

Now we can rewrite the equation to compute the temperature on the node $T_{i,j}$, as shown here:

$$T_{i,j} = \frac{\left(T_{i+1,j} + T_{i-1,j} + T_{i,j+1} + T_{i,j-1}\right)}{4} \qquad (13\text{-}13)$$

This is the Gauss-Seidel method formulation of the problem. Equation (13-13) is solved iteratively to compute the temperature values at each of the interior nodes (m by n) until a prespecified error tolerance is met. The number of iterations is defined with the error value that is computed for every iteration from the following formulation:

$$\varepsilon = \left| \frac{T_{i,j}^{\,current} - T_{i,j}^{\,previous}}{T_{i,j}^{\,current}} \right| * 100\% \qquad (13\text{-}14)$$

Example 2

Let's look at a physical system that is a 3 m by 5 m plate (width by height) with these boundary conditions: on the left side 100° C, on the right side 75° C, on the top 250° C, and on the bottom 300° C. We assume the initial temperature at all interior nodes to be 0° C.

We discretize the system for k=30 nodes along the x-axis and n=50 nodes along the y-axis, which gives the following:

Δx = W/k =3/30= 0.1 [m]; Δy = H/n =5/50= 0.1 [m];

Note that the temperature values along the edges of the plate are known and equal to the applied temperature values on each side.

Along the left edge:

T(0,0) = 100° C; T(0, 1) = 100° C; T(0, 2)= 100° C, .. T(0, n)= 100° C;

Along the right edge:

T(k, 0) = 75° C; T(k, 1) = 75° C; T(k, 2)= 75° C, .. T(k, n-1)= 75° C;

Along the top edge:

T(1, n) = 250° C; T(2, n) = 250° C; T(3, n)= 250° C, .. T(k, n)= 250° C;

Along the bottom edge:

T(1, 0) = 300° C; T(2, 0) = 300° C; T(3, 0)= 300° C, .. T(k-1, n-1)= 300° C;

We set the error tolerance to be $0.001°$ C and the initial error value to be $1°$ C.

Now, based on all these formulations, as shown in Equation (13-13) and Equation (13-14), and based on these boundary conditions (temperature values along the edges of the plate), we write the next script, called Heat_Transfer_GS_method.m:

```
% Heat flow simulation using the Iterative Method based on
% Gauss-Seidel iterative method.
% Two dimensional heat transfer on a rectangular plate (3m-by-5m)
% with the temperature/heat flow from different sides
clc; clearvars; close all
TL = 100;     % [C] Left side
TR = 75;      % [C] Right side
Tt = 250;     % [C] Top
Tb = 300;     % [C] Bottom
k  = 30;      % number of nodes along x axis
n  = 50;      % number of nodes along y axis

W = 3;        % [m] width
H = 5;        % [m] height
x = linspace(0, W, k);  % Mesh size along x axis
y = linspace(0, H, n);  % Mesh size along y axis
[xx, yy] = meshgrid(x, y);
% Note: the mesh size differs along x axis from the one along y axis
% The mesh size along x and y axes can be set uniform
% by setting k = n.
T=zeros(k, n); % Memory allocation
T(1,1:n) = TL; % Left Side of the Plate
T(k,1:n) = TR; % Right Side of the Plate
T(1:k,n)=Tt;   % Top Side of the Plate
T(1:k,1)=Tb;   % Bottom Side of the Plate
```

```matlab
tol = 1e-3;     % Error tolerance
Error = 1;      % Initial Error Value
m=0;
while Error>tol
     Told=T;
    for ii=2:k-1
        for jj=2:n-1
    T(ii,jj)=(1/4)*(T(ii,jj-1)+T(ii-1, jj)+T(ii+1, jj)+T(ii, jj+1));
        end
    end
    Error = max(max(abs((T-Told)./T)))*100;
end
figure(1)
surf(x, y, T'); colorbar;
title('\it 2D Temperature distribution on a plate')
xlabel('\it Width, [m]')
ylabel('\it Height, [m]')
zlabel('\it Temperature, [\circC]')
grid on
axis tight
figure(2)
ribbon(yy, T')
title('\it 2D Temperature distribution on a plate')
xlabel('\it Width in # of grids')
ylabel('\it Height, [m]')
zlabel('\it Temperature, [\circC]')
grid on
axis tight
figure(3)
contour(x, y, T'); colormap;
title('\it 2D Temperature distribution on on a plate')
xlabel('\it Width, [m]')
ylabel('\it Height, [m]')
zlabel('\it Temperature, [\circC]')
grid on
```

```
figure(4)
pcolor(x, y, T'), shading interp;
title('\it 2D Temperature distribution on on a plate')
xlabel('\it Width, [m]')
ylabel('\it Height, [m]')
zlabel('\it Temperature, [\circC]')
grid on; shg
```

Figures 13-7, 13-8, 13-9, and 13-10 illustrate the script's simulation results.

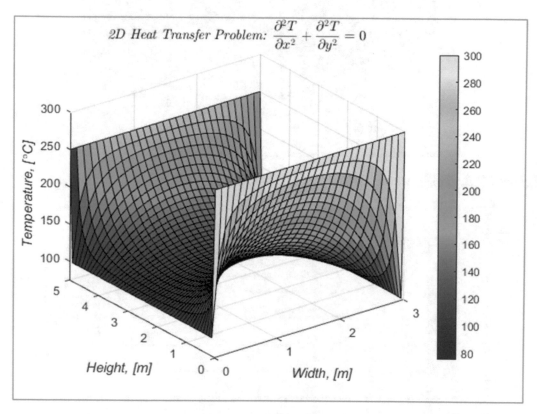

Figure 13-7. *Temperature distribution: surf() plot*

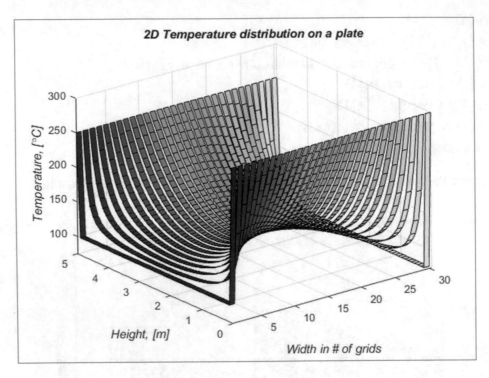

Figure 13-8.   *Temperature distribution: ribbon() plot*

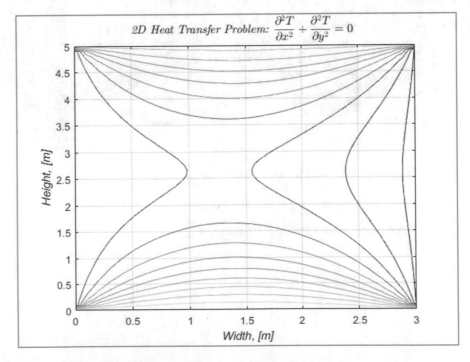

Figure 13-9.   *Temperature distribution: contour() plot*

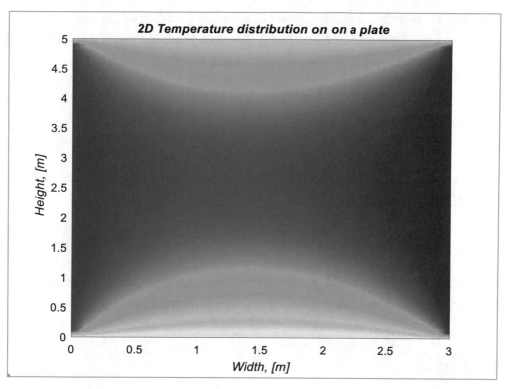

Figure 13-10. *Temperature distribution: pcolor() plot*

Note that the accuracy of the computed solutions with the Gauss-Seidel method highly depends on our set error tolerance value, which is set to be 10^{-3} in this example. Moreover, the accuracy of the solutions improves with the increase of the number of grids or, in other words, by decreasing the mesh size Δx and Δy.

del2(): Laplace Operator

We can also use the discrete Laplace operator to solve heat transfer equations with respect to given boundary conditions and mesh grids. In the previously shown example, the matrix $T(x, y)$ was computed at the points of a square grid, i.e., $\Delta x = \Delta y$. In this case, a finite difference approximation of the Laplace differential operator (Δ to compute second-order partial derivatives with regard to ∂x^2, ∂y^2) can be used to compute the Laplacian (L) function, as shown here:

$$L = \frac{\Delta T}{4} = \frac{\partial^2 T}{\partial x^2} + \frac{\partial^2 T}{\partial y^2} \tag{13-15}$$

In MATLAB, there is a built-in Laplacian operator function del2() that computes the interior values of L on a node point of (i, j) with respect to the difference between the value of the point (i, j) and its four neighbor points, which are $(i+1, j)$, $(i-1, j)$, $(i, j+1)$, and $(i, j-1)$.

$$L_{i,j} = \frac{T_{i+1,j} + T_{i-1,j} + T_{i,j+1} + T_{i,j-1} - 4T_{i,j}}{4} \qquad (13\text{-}16)$$

The final value of T is computed from the computed interior values of L on each node and the respective values of T.

$$T_{i,j}{}^{final} = T_{i,j}{}^{previous} + L_{i,j} \qquad (13\text{-}17)$$

Using Equation (13-16) and Equation (13-17) and using the Laplacian operator del2() is straightforward. Let's consider the following example.

Example 3

The example is a 3 m by 5 m plate (width by height) with these boundary conditions: on the left side 100^0 C, on the right side 75^0 C, on the top 250^0 C, and on the bottom 300^0 C. We assume the initial temperature at all interior nodes to be 0^0 C. (This is the same exercise given above in Example 13-2.)

We write the following script (Heat_Transfer_Laplacian_Operator.m) using the Laplacian operator del2():

```
% Heat flow simulation using the Laplacian operator
% Two dimensional heat transfer on a rectangular plate (3m-by-5m)
% with the temperature/heat flow from different sides
%%
clc; clearvars; close all
TL = 100;      % [C] Left side
TR = 75;       % [C] Right side
Tt = 250;      % [C] Top
Tb = 300;      % [C] Bottom
k  = 30;       % number of nodes along x axis
n  = 50;       % number of nodes along y axis
```

```
W = 3;          % [m] width
H =  5;         % [m] height
x = linspace(0, W, k);    % Mesh size along x axis
y = linspace(0, H, n);    % Mesh size along y axis
[xx, yy] = meshgrid(x, y);
% Here the mesh size differs along x axis from the one along y axis
% Note: the mesh size along x and y axes can be set uniform
% by setting N accordingly
T=zeros(k, n);       % Memory allocation
T(1,1:n) = TL;       % Left Side of the Plate
T(k,1:n) = TR;       % Right Side of the Plate
T(1:k,n) = Tt;       % Top Side of the Plate
T(1:k,1) = Tb;       % Bottom Side of the Plate
N_iter=0;            % Initial iteration value
while N_iter<1000
    N_iter=N_iter+1;
    if mod(N_iter, 2)==0
    contourf(x, y, T'); colorbar;
    drawnow
    end
    L= del2(T);
    T(2:k-1, 2:n-1) = T(2:k-1, 2:n-1)+ L(2:k-1, 2:n-1);
end
title('\it 2D Heat Transfer Problem Solution:  $$ \frac{\partial^2{T}}{\partial
{x^2}}+\frac{\partial^2{T}}{\partial{y^2}} =0 $$ ', 'interpreter', 'latex')

xlabel('\it Width')
ylabel('\it Height')
zlabel('\it Temperature, [\circC]')
grid on
figure(2)
surf(x, y, T'); colorbar;
title('\it 2D Heat Transfer Problem Solution:  $$ \frac{\partial^2{T}}{\
partial{x^2}}+\frac{\partial^2{T}}{\partial{y^2}} =0 $$ ', 'interpreter',
'latex')
```

```
xlabel('\it Width, [m]')
ylabel('\it Height, [m]')
zlabel('\it Temperature, [\circC]')
grid on
axis tight
```

The preceding script produces an animated plot of the simulation results demonstrating how heat moves within the interior points (nodal points/grids) of the plate, as shown in Figure 13-11.

Note that the accuracy of the computed numerical solutions with the Laplacian operator (del2) increases with the increase of the number of iterations. This is similar to the finite difference method that gives a higher accuracy with a decrease of the error tolerance. Moreover, the simulated iteration process with the Laplacian operator shows hyperbolic contour lines (Figure 13-11), which are similar to the contour lines shown in Figure 13-9. The simulation results illustrated in Figure 13-12 match perfectly with the ones in Figure 13-7.

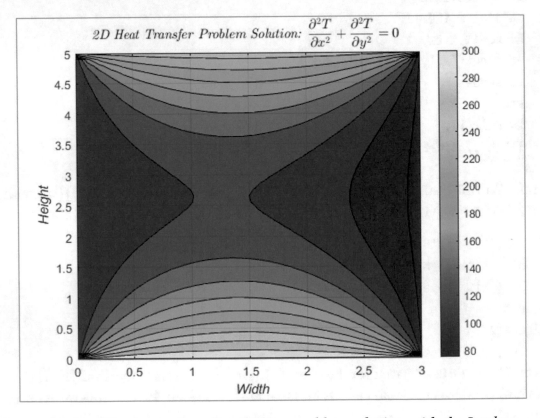

Figure 13-11. *2D temperature distribution problem solution with the Laplace operator*

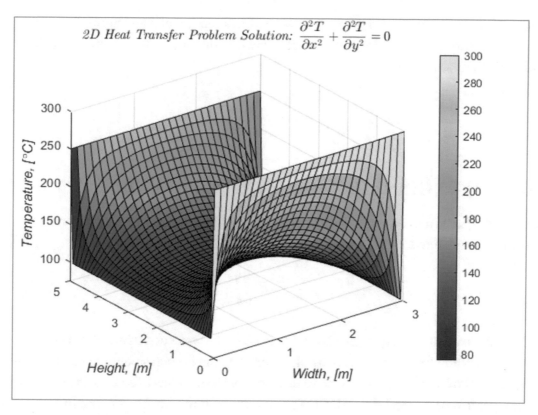

Figure 13-12. *2D temperature distribution problem solution with the Laplace operator, surf() plot*

Wave Equation

The wave equation is a second-order PDE that is used to formulate waves, such as sound, mechanical, and light waves. It is used widely in mechanics, acoustics, electromagnetics, and fluid dynamics computations. The wave equation, unlike the heat equation, may contain many spatial variables ($x_1, x_2, x_3, ..., x_n$) along with the time variable t.

A general wave equation is formulated with the following:

$$\frac{\partial^2 u}{\partial t^2} = c^2 \left(\frac{\partial^2 u}{\partial x_1^2} + \frac{\partial^2 u}{\partial x_2^2} + \frac{\partial^2 u}{\partial x_3^2} + ... + \frac{\partial^2 u}{\partial x_n^2} \right) \tag{13-18}$$

447

where c is a fixed non-negative (wave propagation speed) real coefficient. There is an alternative, more compact formulation of the wave equation that uses the Laplacian operator, ∇, as shown here:

$$\frac{\partial^2 u}{\partial t^2} = c^2 \nabla^2 u \qquad (13\text{-}19)$$

Solving a One-Dimensional Wave Equation

The wave equation in one space (only with respect to the x space variable) can be formulated as follows:

$$\frac{\partial^2 u}{\partial t^2} = c^2 \frac{\partial^2 u}{\partial x^2} \qquad (13\text{-}20)$$

The formulation in Equation (13-20) describes the wave occurring in only one dimension along the x-y plane; it was discovered by the French scientist Jean-Baptiste le Rond d'Alembert [3]. Let's look at an example of Hooke's law of a vibrating spring to demonstrate a numerical simulation of a one-dimensional wave equation. In fact, from Newton's and Hooke's laws [4], we can derive the one-dimensional PDE wave equation. Let's assume that there is an array of finite masses, m, that are interconnected with massless springs of length Δx, and let's assume the spring elements have a uniform spring constant of k.

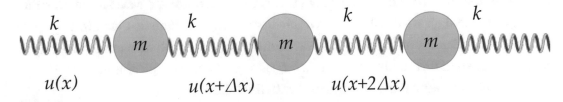

Figure 13-13. *Deriving the wave equation from Newton's and Hooke's laws*

Figure 13-13 shows the unit masses interconnected with a uniform stiff spring/string. Displacements of mass elements are *u(x)* that measure the magnitude of strains (waves) traveling along the spring elements. Using a free-body diagram of the system shown in Figure 13-13, we can compute the forces generated/disturbed on the mass element at $x + \Delta x$ from Newton's and Hooke's law (at x and $x + 2\Delta x$) [4], as shown here:

$$F_{x+\Delta x}^{Newton} = m\ddot{x} = m\frac{\partial^2 u(x+\Delta x, t)}{\partial t^2} \tag{13-21}$$

$$F_x^{Hooke} = ku(x+\Delta x, t) - ku(x, t) \tag{13-22}$$

$$F_{x+2\Delta x}^{Hooke} = ku(x+2\Delta x, t) - ku(x+\Delta x, t) \tag{13-23}$$

Now we compute the force at $x + \Delta x$ based on the finite difference of forces computed at x, $x + 2\Delta x$, from Hooke's law, as shown in Equations(13-22) and (13-23).

$$F_{x+\Delta x}^{Hooke} = F_{x+2\Delta x}^{Hooke} - F_x^{Hooke} = k\big[u(x+2\Delta x, t) - u(x+\Delta x, t) - u(x+\Delta x, t) - u(x, t)\big] \tag{13-24}$$

Moreover, by equating the forces at $x + \Delta x$ computed with Newton's law in Equation (13-21) and Hooke's law in Equation (13-24), we can write the following:

$$\frac{\partial^2 u(x+\Delta x, t)}{\partial t^2} = \frac{k}{m}\big[u(x+2\Delta x, t) - 2u(x+\Delta x, t) + u(x, t)\big] \tag{13-25}$$

The formulation in Equation (13-25) can be rewritten if *n* number of masses are located uniformly, $L = n\Delta x$; if total mass (*M*) is evenly distributed (*m*) along *n* nodes (*M = nm*); and if the total spring stiffness (*k*) is uniformly distributed (*K=k/n*).

$$\frac{\partial^2 u(x+\Delta x, t)}{\partial t^2} = \frac{KL^2}{M\Delta x^2}\big[u(x+2\Delta x, t) - 2u(x+\Delta x, t) + u(x, t)\big] \tag{13-26}$$

By considering the case of Δx being infinitely small ($\Delta x \to 0$) and *n* being infinitely many ($n \to \infty$), we can re-write Equation (13-26) in a more compact and continuous form, as shown here:

$$\frac{\partial^2 u(x, t)}{\partial t^2} = \frac{KL^2}{M}\frac{\partial^2 u(x, t)}{\partial x^2} \tag{13-27}$$

where the constant part of the equation $\dfrac{KL^2}{M} = c$ is the wave propagation speed dependent on the stiffness and length of the spring and on the mass values of the string/spring. Note that by substituting $\dfrac{KL^2}{M} = c$ in Equation (13-27), we get Equation (13-20), which is the 2D wave equation.

By discretizing Equation (13-27) and replacing the partial derivatives with their finite (approximate) differences on both sides (left and right) of Equation (13-27), we can write the following equations:

$$\frac{\partial^2 u(x,t)}{\partial t^2} \approx \frac{u_i^{n+1} - 2u_i^n + u_i^{n-1}}{\Delta t^2} \tag{13-28}$$

$$\frac{KL^2}{M} \frac{\partial^2 u(x,t)}{\partial x^2} \approx \frac{KL^2}{M} \frac{u_{i+1}^n - 2u_i^n + u_{i-1}^n}{\Delta x^2} \tag{13-29}$$

By equating Equation (13-28) and Equation (13-29), we can write an algebraic or discretized formulation of Equation (13-27).

$$\frac{u_i^{n+1} - 2u_i^n + u_i^{n-1}}{\Delta t^2} = \frac{KL^2}{M} \frac{u_{i+1}^n - 2u_i^n + u_{i-1}^n}{\Delta x^2} \tag{13-30}$$

By solving Equation (13-30) for u_i^{n+1}, we can formulate the recursive algorithm for numerical computations.

$$u_i^{n+1} = \frac{KL^2}{M} \frac{\Delta t^2}{\Delta x^2} \left(u_{i+1}^n - 2u_i^n + u_{i-1}^n \right) + 2u_i^n - u_i^{n-1} \tag{13-31}$$

By substituting the constant terms $\dfrac{KL^2}{M}$ and $\dfrac{\Delta t^2}{\Delta x^2}$ with C^2 (C is called the Courant number [5]), we can rewrite Equation (13-31), as shown here:

$$u_i^{n+1} = C^2 \left(u_{i+1}^n - 2u_i^n + u_{i-1}^n \right) + 2u_i^n - u_i^{n-1} \tag{13-32}$$

Example 4

Let's look at a numerical example of an elastic string with a length of 20 [m], a uniform stiffness 100 [Nm], and a uniform distributed mass of 2 kg. There are no displacements on both ends of the string. The simulation period for all three cases is T=30 [s]. Consider these three disturbance cases:

- A small displacement disturbance present on the left side on the (fifth, sixth, and seventh) node points $u_3 = 0.05$, $u_4 = 0.5$, and $u_5 = 0.05$

- Sinusoidal excitation force applied from the center of the string:
  $u_{100} = 0.5sin\left(\dfrac{10\pi t}{T}\right)$ with T = 30 [s]

- Combined disturbances: $u_3 = 0.05$, $u_4 = 0.1$, and $u_5 = 0.05$ and
  $u_{100} = 0.5sin\left(\dfrac{10\pi t}{T}\right)$ with T = 30 [s]

By using Equation (13-31) and assigning zero boundary conditions and disturbances for each case, we can write the next script, called Wave_Propagation_1D.m:

```
%% Wave propagation in a string - Spring-Mass system simulation (Hooke's law):
clearvars; clc;
% Given parameters
k = 100;                        % [Nm] stiffness
mass=2;                         % [kg] stiffness
Lx = 20;                        % [m] String length
dx=.1;                          % [m] Distance between nodes
Nodes=Lx/dx;                    % Number of nodes
x = linspace(0, Lx, Nodes);
K = k/Nodes;
c = K*Lx^2*dx^2/mass;
dt=0.05;                        % Simulation time step
u = zeros(Nodes,1);             % Boundary Conditions
um1= u;
up1 = u;
C = c*dt/dx;                    % Wave propagation speed
t=0;                            % Simulation start time
T=30;                           % Simulation period
```

451

```
Case = input('Enter case # (1, 2, 3):    ');
if Case==1
    % Small disturbance:
    u(5:7)=[0.05 0.5 0.05];    % Case 1
    up1(:)=u(:);
    while t<T
        % Time evolutions:
        t=t+dt;
        % Saving the iteration results:
        un1=u; u=up1;
        for ii=2:Nodes-1
        up1(ii)=(C^2)*(u(ii+1)-2*u(ii)+u(ii-1)) - un1(ii)+2*u(ii);
        end
        clf;
        plot(x, u, 'g-o', 'linewidth', 1), shg
    end
elseif Case == 2
    u(5:7)=0;                              % Case 2
    up1(:)=u(:);
    while t<T
        t=t+dt;
        % Saving the simulation results:
        un1=u; u=up1;
        % Disturbance Source:
        u(100)=0.5*sin(10*pi*t/T);
        for ii=2:Nodes-1
        up1(ii)=(C^2)*(u(ii+1)-2*u(ii)+u(ii-1)) - un1(ii)+2*u(ii);
        end
        clf;
        plot(x, u, 'k-o', 'linewidth', 1), shg
    end
else   % Case 3
    u(5:7)=[0.05 0.5 0.05];
    up1(:)=u(:);
    while t<T
        t=t+dt;
```

```
        % Saving the simulation results:
        un1=u; u=up1;
        % Disturbance Source:
        u(100)=0.5*sin(10*pi*t/T);
        for ii=2:Nodes-1
        up1(ii)=(C^2)*(u(ii+1)-2*u(ii)+u(ii-1)) - un1(ii)+2*u(ii);
        end
        clf;
        plot(x, u, 'r-o', 'linewidth', 1), shg
    end
end
xlabel('\it x space, [m]')
ylabel('\it u(t, x) Displacement, [m]')
title(['\it Wave propagation after t = ' num2str(T) ' [s]. ', ...
    'Case # ' num2str(Case)])
grid on
```

This script gives an animated plot of the simulation results with respect to the three cases. Figures 13-14, 13-15, and 13-16 show the simulation results of the script.

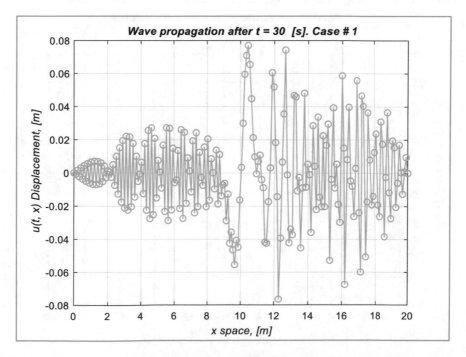

Figure 13-14. *One-dimensional wave propagation problem, case 1*

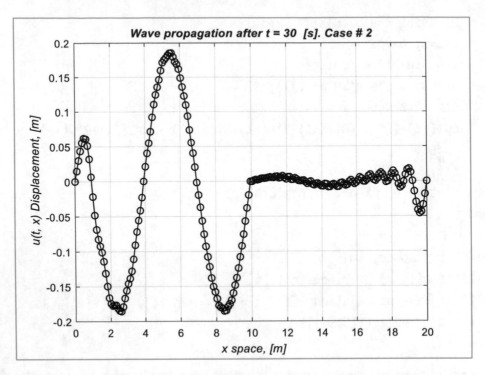

Figure 13-15. *One-dimensional wave propagation problem, case 2*

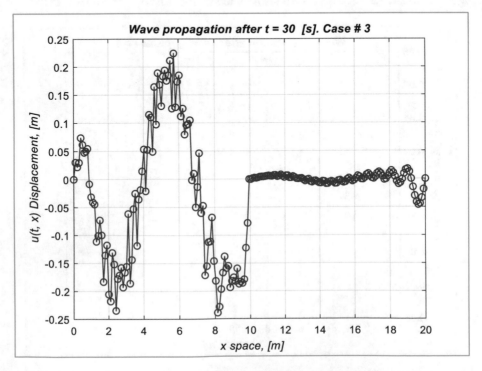

Figure 13-16. *One-dimensional wave propagation problem, case 3*

Solving a Two-Dimensional Wave Equation

Let's consider a 2D wave equation with a disturbance source–free region with the following formulation:

$$\frac{\partial^2 u(x,y,t)}{\partial t^2} = c^2\left(\frac{\partial^2 u(x,y,t)}{\partial x^2} + \frac{\partial^2 u(x,y,t)}{\partial y^2}\right) + f \tag{13-33}$$

By discretizing Equation (13-33), we can write the next one:

$$\frac{u_{i,j}^{n+1} - 2u_{i,j}^{n} + u_{i,j}^{n-1}}{\Delta t^2} \approx c^2\left(\frac{u_{i+1,j}^{n} - 2u_{i,j}^{n} + u_{i-1,j}^{n}}{\Delta x^2} + \frac{u_{i,j+1}^{n} - 2u_{i,j}^{n} + u_{i,j-1}^{n}}{\Delta y^2}\right) + f_i^n \tag{13-34}$$

For the sake of simplicity, we set $\Delta x = \Delta y$ and rewrite Equation (13-34).

$$u_{i,j}^{n+1} = 2u_{i,j}^{n} - u_{i,j}^{n-1} + \left[c\frac{\Delta t}{\Delta x}\right]^2\left(u_{i+1,j}^{n} - 4u_{i,j}^{n} + u_{i-1,j}^{n} u_{i,j+1}^{n} + u_{i,j+1}^{n} + u_{i,j-1}^{n}\right) + \Delta t^2\, f_i^n \tag{13-35}$$

Equation (13-35) is a recursive algorithm to compute the wave propagation in 2D. Here are the Dirichlet boundary conditions:

$$u(0,y,t)=0, u(Lx,y,t)=0 \text{ and } u(x,0,t)=0, u(x,Ly,t)=0 \tag{13-36}$$

Furthermore, for absorbing/stopping the propagated waves along four boundaries, let's consider the following zero boundary conditions:

$$\left.\frac{\partial u}{\partial x}\right|_{x=0} = c\left.\frac{\partial u}{\partial t}\right|_{x=0}, \left.\frac{\partial u}{\partial x}\right|_{x=Lx} = -c\left.\frac{\partial u}{\partial t}\right|_{x=Lx} \text{ and } \left.\frac{\partial u}{\partial y}\right|_{y=0} = c\left.\frac{\partial u}{\partial t}\right|_{y=0}, \left.\frac{\partial u}{\partial y}\right|_{y=Ly} = -c\left.\frac{\partial u}{\partial t}\right|_{y=Ly} \tag{13-37}$$

By considering the previously written boundary conditions, introducing $C^2 = \left[c\frac{\Delta t}{\Delta x}\right]^2$, and using the discretized Equation (13-34), we can write the following equations:

$$u_{1,j}^{n+1} = 2u_{2,j}^{n} + \frac{C^2-1}{C^2+1}\left[u_{2,j}^{n+1} - u_{1,j}^{n}\right] \tag{13-38}$$

$$u_{i,1}^{n+1} = 2u_{i,2}^{n} + \frac{C^2-1}{C^2+1}\left[u_{i,2}^{n+1} - u_{i,1}^{n}\right] \tag{13-39}$$

455

$$u_{n_x,j}^{n+1} = 2u_{n_x-1,j}^{n} + \frac{C^2-1}{C^2+1}\left[u_{n_x-1,j}^{n+1} - u_{n_x,j}^{n}\right] \tag{13-40}$$

$$u_{i,n_y}^{n+1} = 2u_{i,n_y-1}^{n} + \frac{C^2-1}{C^2+1}\left[u_{i,n_y-1}^{n+1} - u_{i,n_y}^{n}\right] \tag{13-41}$$

where C is the wave propagation speed, also called the Courant number.

Example 5

Let's take a 5 m by 5 m elastic plate with Dirichlet boundary conditions. The plate is excited in the center with a sinusoidal force, $f = 0.5\ sin\ (1.5\pi t)$. The wave propagation speed is $c = 0.25$.

Here is the solution script of this exercise, called Wave_Propagation_2D.m:

```
% 2D Wave propagation problem simulation
clearvars; clc
% Given parameters:
Lx = 5;   % Length along x axis
Ly=Lx;    % Length along y axis
dx=.05;   % Mesh size along x axis
dy=dx;    % Mesh size along y axis
Nx = Lx/dx;   % Number of node points along x axis
Ny=Nx;            % Number of node points along y axis
x=linspace(0, Lx, Nx);
y=x;
% Simulation period:
T=50;
% Displacement:
un = zeros(Nx, Ny);     % Memory allocation
unm1=un;
unp1 =un;

C  =0.25;      % Courant number
c  = 1;        % Wave propagation speed
dt=C*dx/c;     % Time increment
```

```matlab
t=0;              % Initial simulation time value
% Simulation Loop:
while t<T
    % Boundary conditions:
    un(:, [1,end])  = 0;
    un([1, end], :) = 0;
    % Time evolution:
    t=t+dt;
    % Save the simulation results:
    unm1=un; un=unp1;
    %  Excitation source:
    un(50,50)=0.5*sin(1.5*pi*t);
    for ii=2:Nx-1
        for jj=2:Ny-1
           unp1(ii, jj)=2*un(ii, jj)-unm1(ii, jj)+...
           (C^2)*(un(ii+1,jj)+un(ii,jj+1)-4*un(ii, jj)+ ...
           un(ii-1, jj)+un(ii,jj-1));
        end
    end
    surfc(x, y, un'), colorbar; shg
    caxis([-0.15 0.25])   % Pseudocolor axis scaling.
    title([' Simulation results after t = '  num2str(t) '  [s]'])
    pause(dt)
    axis tight
end
xlabel('Width, [m] ')
ylabel('Length, [m] ')
zlabel('Displacement, [m]')
%% Contour plot
figure(2)
contourf(x, y, un'), colorbar; shg
title([' Simulation results after t = '  num2str(t) '  [s]'])
axis tight
xlabel('Width, [m] ')
ylabel('Length, [m] ')
```

The solution script solves the given 2D wave propagation problem and plots an animated plot of the simulation results, as shown in Figure 13-17 and Figure 13-18. Note that this simulation is a live demonstration (Figure 13-17) of computed data; thus, it is slow.

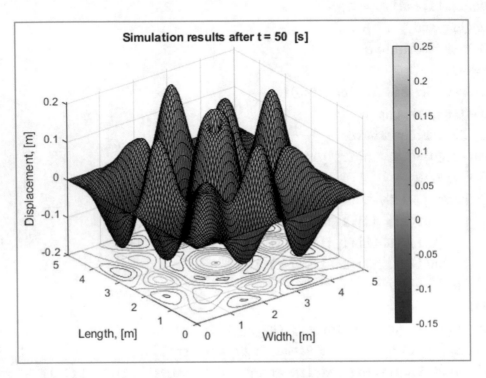

Figure 13-17. *Two-dimensional wave propagation problem simulation*

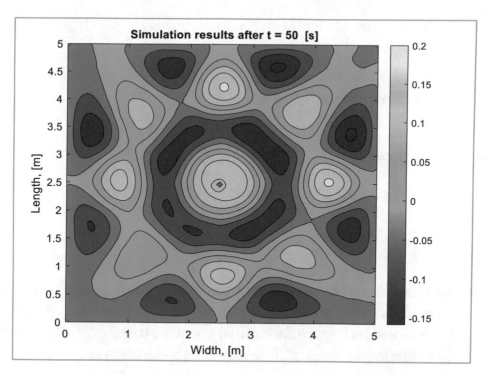

Figure 13-18. *Two-dimensional wave propagation problem simulation*

This ends the PDE example. Note that there is a MATLAB toolbox called the PDE toolbox that has many GUI entry tools and built-in functions, such geometry building (rectangle, circle, polygon), boundary condition setting, PDE equations setting, meshing, solving, plotting, and many more user-friendly options.

References

[1] Wikipedia: `https://en.wikipedia.org/wiki/Partial_differential_equation`, viewed on December 11, 2019.

[2] Barrett, R.; Berry, M.; Chan, T. F.; Demmel, J.; Donato, J.; Dongarra, J.; Eijkhout, V.; Pozo, R.; Romine, C.; and van der Vorst, H. *Templates for the Solution of Linear Systems: Building Blocks for Iterative Methods* (2nd ed). Philadelphia, PA: SIAM (1994). `http://www.netlib.org/linalg/html_templates/Templates.html`.

[3] Speiser, D. *Discovering the Principles of Mechanics 1600–1800*, p. 191, Basel: Birkhäuser (2008).

[4] Wikipedia: `https://en.wikipedia.org/wiki/Wave_equation#cite_note-`
Speiser-5, viewed on Dec 13, 2019.

[5] Wikipedia: `https://en.wikipedia.org/wiki/Courant-Friedrichs-Lewy_`
condition, viewed on December 13, 2019.

Drill Exercises

The following are exercises to try on your own.

Exercise 1

Solve the one-dimensional heat problem. Say you have a wire with a length of $L = 2$
$[m]$ $(0 < x \le L)$ and $k=0.001$. Use this initial temperature distribution: $T(x, 0) = 65x$ $[^0C]$.
Consider that both ends of the wire are inserted into an ice basket $(T(0)=T(L) = 0[^0C])$.
Compute the numerical solutions of the problem as a series of solutions.

Exercise 2

Solve this one-dimensional wave equation: $\dfrac{\partial^2 y}{\partial t^2} = 9 \dfrac{\partial^2 y}{\partial x^2}$.

Consider that $y(0, t) = y(1, t) = 0$, $y(x, 0) = \cos(2\pi x) + 0.25 \cos(5\pi x)$ for $0 < x < 1$ and
$\dfrac{\partial y}{\partial t}(x, 0) = 0$ for $0 < x < 1$.

Exercise 3

Use PDEPE(). Solve the given one-dimensional heat equation $\dfrac{\partial u}{\partial t} - k \dfrac{\partial^2 u}{\partial x^2} = 0$ for $0 < x < L$
and $0 < t$. Consider $L = 10$ and $k=0.55$. The boundary conditions are $u(0, t) = 75$ and $u(L, t)=125$ for all $t > 0$. The initial condition of the form is $u(x, 0)=u_0(x)$ for $0 < x < L$, where $u_0(x)$
is a known/given function depicting the initial temperature distribution throughout the
length of the rod. Consider the following: $u_0(x) = 5x-1.20$.

Exercise 4

Use the Laplacian operator del2(). Solve the Laplace equation $\dfrac{\partial^2 u}{\partial x^2} + \dfrac{\partial^2 u}{\partial y^2} = 0$ for u(x, y)

defined on $x \in [0, 2]$, $y \in [0, 2]$ with these boundary conditions: *u(x, 0)= 5, u(x, 2)=15; u(0, y)=5, u(2, y)=15.*

Exercise 5

Use the Gauss-Seidel method. Solve the Laplace equation $\dfrac{\partial^2 u}{\partial x^2} + \dfrac{\partial^2 u}{\partial y^2} = 0$ for u(x, y)

defined on $x \in [0, 2]$, $y \in [0, 2]$ with these boundary conditions: *u(x, 0)= 5, u(x, 2)=15; u(0, y)=5, u(2, y)=15.*

Exercise 6

Use the recursive algorithm. Solve the one-dimensional wave equation $\dfrac{\partial u}{\partial t} + a\dfrac{\partial u}{\partial x} = 0$ for

u(t, x) defined on $x \in [0, 10]$ with these boundary conditions: *u(t, 0)= 0* and *u(t, 10)=0*. The string is excited in the middle with $u(t, 5) = 0.25\, sin\, (2\pi t)$. Take =0.75.

Exercise 7

Use PDEPE(). Solve the given one-dimensional heat equation $\pi^2 \dfrac{\partial u}{\partial t} - \dfrac{\partial^2 u}{\partial x^2} = 0$ for $0 < x < L$

and $0 < t$. Consider *L=5*. The boundary conditions are *u(0, t) = 0* and $\pi e^{-t} + \dfrac{\partial u(5, t)}{\partial t} = 0$

for all $t > 0$. The initial conditions are $u(x, 0) = sin\,(2\pi x)$ for *0<x<L*.

Exercise 8

Use the Gauss-Seidel method. Solve the Laplace equation $\dfrac{\partial^2 u}{\partial x^2} + \dfrac{\partial^2 u}{\partial y^2} = 0$ for *u(x, y)*

defined on $x \in [0, 2]$, $y \in [0, 2]$ with these boundary conditions: *u(x, 0)= 5, u(x, 2)=15; u(0, y)=5, u(2, y)=15.*

Exercise 9

Use the finite difference method. Solve the Laplace equation $\dfrac{\partial^2 u}{\partial x^2} + \dfrac{\partial^2 u}{\partial y^2} = 0$ for $u(x, y)$

defined on $x \in [0, 2]$, $y \in [0, 2]$ with these boundary conditions: $u(x, 0)= 5$, $u(x, 2)=15$; $u(0, y)=5$, $u(2, y)=15$.

Exercise 10

Solve the diffusion equation (in the example, of a 5 m long thin metallic rod) using the numerical methods. The diffusion equation for a temperature scalar field (T) is expressed with the following:

$$\rho C_p \frac{\partial T}{\partial t} = K \frac{\partial^2 T}{\partial x^2} + \dot{q}$$

where ρ is the density of the material, C_p is the specific heat capacity (at the constant pressure, in case of a gas), and $\dot{q}$ is a heat source.

Consider the one-dimensional mixed Dirichlet and Neumann boundary conditions. A thin, 5 m long, metal rod is at a temperature of 100^0C. Compute how the temperature of the rod will change over time with the following boundary condition values: at $t = 0$, both ends of the rod are put on ice (0^0C). There is no heat source present, i.e., $\dot{q} = 0$. Moreover, consider that heat flows only along the rod. Here,

$$\rho = 7800 \left[\frac{kg}{m^3}\right], C_p = 502 \left[\frac{J}{kg * K}\right], \text{and } K = 30 \left[\frac{W}{m * K}\right].$$

Exercise 11

Consider the problem (diffusion equation and thin rod) given in Exercise 10. The rod is made of copper, and the initial temperature of the rod is 100^0C. At $t = 0$, the left end of the rod is put on ice (0^0C). Compute the heat flow across the rod.

$$\rho = 8885 \left[\frac{kg}{m^3}\right], C_p = 376.812 \left[\frac{J}{kg * K}\right]; K = 384.1 \left[\frac{W}{m * K}\right].$$

Exercise 12

Solve the Laplace equation using the finite difference method $\dfrac{\partial^2 u}{\partial x^2} + \dfrac{\partial^2 u}{\partial y^2} = 0$ for $u(x, y)$ defined on $x \in [0, 1]$, $y \in [0, 1]$ with these boundary conditions: $u(x, 0) = 2$, $u(x, 2) = 5$; $u(0, y) = 2$, $u(2, y) = 5$.

Exercise 13

Consider the following two-dimensional wave propagation problem:

$$\frac{\partial^2 u(x,y,t)}{\partial t^2} = c^2 \left(\frac{\partial^2 u(x,y,t)}{\partial x^2} + \frac{\partial^2 u(x,y,t)}{\partial y^2} \right) + f$$

The size of the elastic plate is 10 m by 10 m, and it is excited with a periodic force of $f = sin\,(\pi t) + 2 * \, cos\,(2\pi t)$ at $x = 3$ m and $y = 3$ m. The wave propagation speed is $c = 0.125$.

Index

© Sulaymon L. Eshkabilov 2020
S. L. Eshkabilov, *Practical MATLAB Modeling with Simulink*, https://doi.org/10.1007/978-1-4842-5799-9

Printed in the United States
by Baker & Taylor Publisher Services